Algorithms for Intelligent Systems

This book series publishes research on the analysis and development of algorithms for intelligent systems with their applications to various real world problems. It covers research related to autonomous agents, multi-agent systems, behavioral modeling, reinforcement learning, game theory, mechanism design, machine learning, meta-heuristic search, optimization, planning and scheduling, artificial neural networks, evolutionary computation, swarm intelligence and other algorithms for intelligent systems.

The book series includes recent advancements, modification and applications of the artificial neural networks, evolutionary computation, swarm intelligence, artificial immune systems, fuzzy system, autonomous and multi agent systems, machine learning and other intelligent systems related areas. The material will be beneficial for the graduate students, post-graduate students as well as the researchers who want a broader view of advances in algorithms for intelligent systems. The contents will also be useful to the researchers from other fields who have no knowledge of the power of intelligent systems, e.g. the researchers in the field of bioinformatics, biochemists, mechanical and chemical engineers, economists, musicians and medical practitioners.

The series publishes monographs, edited volumes, advanced textbooks and selected proceedings.

Indexed by zbMATH.

All books published in the series are submitted for consideration in Web of Science.

Uma N. Dulhare · Essam Halim Houssein
Editors

Machine Learning and Metaheuristics: Methods and Analysis

Springer

Editors
Uma N. Dulhare
Muffakham Jah College of Engineering
and Technology
Hyderabad, Telangana, India

Essam Halim Houssein
Faculty of Computers and Information
Minia University
Minia, Egypt

ISSN 2524-7565 ISSN 2524-7573 (electronic)
Algorithms for Intelligent Systems
ISBN 978-981-99-6644-8 ISBN 978-981-99-6645-5 (eBook)
https://doi.org/10.1007/978-981-99-6645-5

This Springer imprint is published by the registered company Springer Nature Singapore Pte Ltd.
The registered company address is: 152 Beach Road, #21-01/04 Gateway East, Singapore 189721, Singapore

Preface

In recent years, there has been a growing research interest in meta-heuristics (MHs) for solving combinatorial optimization problems in industry, biomedical, engineering, image processing. Many engineering optimization problems are more complex and difficult to solve by conventional methods. Nowadays, more focus on innovative methods derived from nature inspired by social or the natural system, which provides the good result in solving complex optimization problem. Several meta-heuristics methods are existing and new methods also under development. Meta-heuristic algorithms are a type of random algorithm, approximate algorithm that can better escape from the local optimum points to solve a huge range of engineering problems. They are classified into bio-inspired, physics/chemistry-based, socio-inspired algorithms and swarm-based algorithms. The bio-inspired algorithms are motivated from genetic evolution of animals and plants. The evolution-based algorithms are artificial neural network, genetic algorithms and so forth. The Swarm intelligence algorithms are particle swarm optimization, ant colony optimization, whale search algorithm, cuckoo search algorithm, artificial bee colony algorithm, whale search algorithm and so forth and simulated annealing as physics/chemistry-based and election algorithm as socio-inspired algorithm.

Machine learning is a data analytics technique to use for computational methods. Therefore, machine learning is integrated for fostering the meta-heuristics towards enhancing the search performance of optimization and make them "smarter" and "well informed" by improving their performance in terms of solution quality, convergence speed and robustness.

This book will cover the original topics in both science and engineering, such as machine learning, soft computing like Fuzzy system, Genetic algorithms, Artificial Neural Networks, which is the backbone of smart systems, meta-heuristics, multi-criteria decision-making, water pumping systems, renewable energy optimization system for sustainable society development applications, etc. The main contribution of this book is to emphasize the advantage and importance of machine learning fusion with meta-heuristics in various sustainable societal applications. This book will be a valuable reference for researchers in industry and academia, as well as for all graduate, undergraduate and Ph.D. students who are working in the metaheuristics and machine learning communities to solve real-world problems in different engineering fields, including mechanical, electrical and civil engineering and earth/environmental science and covers topics such as water management, robotics, process optimization, etc.

Hyderabad, India Uma N. Dulhare
Minia, Egypt Essam Halim Houssein

Introduction

This book "MLMH 2022" collects machine learning (ML) with various metaheuristics (MHs) for several real-world applications. Hence, the MHs have become an important tool for solving hard optimization problems in industry, biomedical, engineering, image processing. Besides, machine learning (ML) is a data analytics technique to use for computational methods. Therefore, machine learning is integrated for fostering the metaheuristics towards enhancing the search performance of optimization to deal with different global and engineering optimization problems. However, this book addresses the issues of two strategies as ML and MHs.

This book is a proficient guide on the relationship between machine learning and metaheuristics algorithms beside this, various examples of how machine learning can be used and leverage the performance of optimization method, which helps to assess the development of machine learning in metaheuristics applications of societal development. Moreover, each chapter provides applications that include comparisons. Broader audiences of this book will widely vary from individuals, researchers, scientists, academics, graduate and postgraduate students, libraries and development practitioners. Additionally, professionals who want to learn and explore the advances in concepts and implementation by fusion of ML and MH algorithms applied to societal applications. This book will generate tremendous impetus in terms of solutions for recent trend applications, thus will have highly acceptable scholarly value and at the same time potentially contribute to this different sector of research.

This book has 14 chapters that consist of an overview of machine learning and metaheuristics methods applied to solve applications. Chapter 1 explores various soft computing techniques for Biomedical Imaging Segmentation and Classification. Chapter 2 explains how to optimize the sleep parameters used in sleep clinics and daily sleep monitoring system by using parameter optimization techniques based on correlation analysis of primary and dependent parameters. Chapter 3 presents various optimization techniques of renewable energy system (RES) for electric necessity of farm house and irrigation load stipulation situated at remote zone. Chapter 4 examines the integration of multiple and single WT and PV in RDS under different scenarios, using various recent metaheuristics algorithms such as AOA, SMA and BWO. Chapter 5 presents the comprehensive review by analysing the working and performances of the different metaheuristics algorithms for the Classification and Prediction of Skin Lesions.

Chapter 6 presents automatic prediction of non-alcoholic fatty liver disease (NAFLD) using CNN models. Chapter 7 explores the development and evaluation of an off-grid PV generator pumping system that utilizes a submersible motor pump system, with a focus on analysing the efficiency of system components under varying solar irradiation conditions. Chapter 8 assembles renewable energy systems along with their benefits and drawbacks, hybrid wind and solar energy systems with various hybrid energy system components to identify the best possible combination of energy components for a typical rural community in order to minimize the total net present cost of the system over its lifetime with case study.

The use of artificial intelligence (AI) approaches to improve the security of Internet of Things (IoT) systems is explored in Chap. 9. Chapter 10 demonstrates the findings of this study of developmental trajectory of manufacturing memristor devices using the Catalytic Free Angle Toppling Technique (ATT) process. The various machine learning and deep learning techniques have been discussed in Chap. 11. Chapter 12 focuses on the analysis of feature selection techniques applied to high-dimensional binary classification tasks in the field of education. Chapter 13 explains the meta-heuristic algorithms' role in medical care and diagnostics. Chapter 14 deals with the chiral nanoplasmonics separation,synthesis and its biomedical application as theranostic in treatement and management of diseases.

This book is an important reference to emphasize the advantage and importance of machine learning fusion with meta-heuristic in various sustainable societal applications. I congratulate the authors for their effort and dedication to assembling the topics for addressing the book.

Uma N. Dulhare
Essam Halim Houssein

Contents

Contributors

Hussein Abdel-Mawgoud Electrical Engineering Department, Faculty of Engineering, Aswan University, Aswan, Egypt

Doaa A. Abdelkareem Faculty of Computers and Information, Luxor University, Luxor, Egypt

Hasina Adil Department of Basic Science, Oriental University, Indore, Madhya Pradesh, India

Mustafa Adil System Engineer, Tata Consultancy Services (TCS), Indore, Madhya Pradesh, India

AlaaDdin Al-Shidaifat Department of Nanoscience and Engineering, Inje University, Gimhae-si, Korea

Jagdish R. Baheti Kamla Nehru College of Pharmacy, Butibori, Nagpur, MS, India

Bhavana Balimidi Chaitanya Bharathi Institute of Technology, Hyderabad, Telangana, India

Parul Bhanarkar Department of Computer Science and Engineering, SGBAU Amravati University, Amravati, India

Ashima Bhatnagar Bhatia Vivekananda Institute of Professional Studies—TC, Delhi, India

Bhupesh Bhatia NSUT, Delhi, India

Shubhro Chakrabartty School of Computer Science Engineering and Applications, DY Patil International University, Pune, India

Disha M. Dhabarde Kamla Nehru College of Pharmacy, Butibori, Nagpur, MS, India

Uma N. Dulhare Department of Computer Science & Artificial Intelligence, Muffakham Jah College of Engineering & Technology, Hyderabad, Telangana, India

Sunil Gautam Department of Computer Science and Engineering, Institute of Technology, Nirma University, Ahmedabad, India

Ramesh Chandra Goswami Kadi Sarva Vishwavidyalaya, Gandhinagar, India

Alok Gupta Research and Development, 3VTechnix, Indore, Madhya Pradesh, India

Mohamed Abdel Hameed Faculty of Computers and Information, Luxor University, Luxor, Egypt

Essam H. Houssein Faculty of Computers and Information, Minia University, Minia, Egypt

Hiren Joshi Gujarat University, Ahmedabad, India

D. Naga Jyothi Chaitanya Bharathi Institute of Technology, Hyderabad, Telangana, India

Salah Kamel Electrical Engineering Department, Faculty of Engineering, Aswan University, Aswan, Egypt

Rama Rithika Kesaraju Chaitanya Bharathi Institute of Technology, Hyderabad, Telangana, India

Animesh Masih Department of Electrical Engineering, Medi-Caps University, Indore, India

Animesh Kumar Masih Department of Electrical Engineering, Medi-Caps University, Indore, Madhya Pradesh, India

Monika P. Maske Kamla Nehru College of Pharmacy, Butibori, Nagpur, MS, India

Azmath Mubeen Department of Computer Science and Engineering, University College of Engineering, Osmania University, Hyderabad, Telangana, India

M. Murali Department of Electrical Engineering, Medi-Caps University, Indore, Madhya Pradesh, India

Prashant Panse Department of Information Technology, Medi-Caps University, Indore, India

Santosh S. Raghuwanshi Department of Electrical Engineering, Medi-Caps University, Indore, Madhya Pradesh, India

Kiran Sharma School of Dentistry and Medical Sciences, Charles Sturt University, Orange, NSW, Australia

Pavika Sharma Department of ECE, BPIT, Delhi, India

Ayush Kumar Singh Department of Electrical and Electronics Engineering, National Institute of Technology Goa, Ponda, India

Hanjung Song Department of Nanoscience and Engineering, Inje University, Gimhae-si, Korea

Reena Thakur Department of Computer Science and Engineering, Jhulelal Institute of Technology, Nagpur, India

Pawan Whig Vivekananda Institute of Professional Studies—TC, Delhi, India

Mina Younan Faculty of Computers and Information, Minia University, Minya, Egypt

Chapter 1
Biomedical Imaging Segmentation and Classification Framework Based on Soft Computing Techniques

Essam H. Houssein, Doaa A. Abdelkareem, Mohamed Abdel Hameed, and Mina Younan

Abstract Biomedical imaging is an essential tool for the detection and diagnosis of various diseases in modern clinics, especially with the diversity and multiplicity of forms of medical imaging. Soft computing techniques have become more prevalent for solving complicated real-life problems, especially biomedical imaging issues. Segmentation and classification are two key stages in any biomedical imaging framework. Therefore, the purpose of this study is to give a review of the soft computing techniques used for the segmentation and classification of medical imaging as well as an overview of the development of this field. It first gives an overview of the various soft computing methodologies, followed by an overview of the various image segmentation and classification methods. Moreover, we concentrate this review on some of the methods, such as multilevel thresholding image segmentation and classification using convolutional neural networks (CNNs), and we carefully examine their architectural designs to learn how they were able to accomplish their reported performances. Finally, this study provides a summary of the future challenges and trends in the segmentation and classification of medical images.

Keywords Biomedical images · Computer-aided diagnosis system (CAD) image · Classification · Image segmentation convolutional neural network deep learning · Soft computing techniques

---

E. H. Houssein · M. Younan
Faculty of Computers and Information, Minia University, Minia, Egypt
e-mail: essam.halim@mu.edu.eg

M. Younan
e-mail: mina.younan@mu.edu.eg

D. A. Abdelkareem (✉) · M. A. Hameed
Faculty of Computers and Information, Luxor University, Luxor, Egypt
e-mail: Doaa.Abdelkareem@fci.luxor.edu.eg

M. A. Hameed
e-mail: m.abdelhamid@fci.luxor.edu.eg

U. N. Dulhare and E. H. Houssein (eds.), *Machine Learning and Metaheuristics: Methods and Analysis*, Algorithms for Intelligent Systems,
https://doi.org/10.1007/978-981-99-6645-5_1

1 Introduction

Soft computing is a set of imperfect computing techniques that can be used to evaluate and solve extremely complicated real-life problems, it was first introduced by Zadeh [1]. More traditional approaches have not provided efficient, optimal, or comprehensive solutions for these complex problems. Therefore, soft computing is the best approach that can be utilized to find tractability, robustness, and a low-cost solution with acceptance criteria for imperfection, uncertainty, partial truth, and approximation. So, over the past three decades, soft computing has been intensively researched and used in engineering computers and scientific research [2]. Medical imaging is a crucial method for detecting and diagnosing diseases in modern clinics, especially with the diversity and multiplicity of forms of medical imaging such as Magnetic Resonance Imaging (MRI), Ultrasound, Computed Tomography (CT), Mammography, Positron Emission Tomography (PET), Single Photon Emission Computed Tomography (SPECT), Optical Coherence Tomography (OCT), and Microwave Imaging. Therefore, the most recent developments in medical imaging are largely based on soft computing [3].

Image segmentation is a crucial preprocessing phase in pattern recognition, computer vision, and image processing for a variety of applications, including the analysis of images for medical purposes [4]. Furthermore, it is an essential stage in many primary scientific fields, such as face recognition, medical diagnosis, optical character recognition, satellite image segmentation, etc. [5]. From its basic form, it is a method of separating an image into a number of discrete, non-overlapping segments or structures of interest based on the image's grayscale, color, texture, shape, size, or position [6]. There are numerous ways to address the image segmentation problem. The most critical approaches for addressing this problem are threshold-based techniques, clustering-based techniques, edge-based techniques, region-based techniques, watershed-based techniques, partial differential equation-based techniques, and artificial neural network-based techniques [7].

Contrarily, the classification of medical images is an important area of study that is receiving increasing attention from both the scientific and medical communities. It covers the issues of medical diagnosis, analysis, and educational reasons. Due to promising applications in data mining, financial forecasting, organizing and retrieving multimedia, and bio-informatics, classification has been resurrected as a prominent and important machine learning field. Additionally, the public's access to an enormous volume of medical image data makes the development of new tools and classification methods necessary for the accurate diagnosis, evaluation, and classification of medical data [8]. Three main categories can be used to group classification techniques: texture classification, neural networks, and data mining methods. Texture classification is an image processing approach that aids in distinguishing between various image regions using the characteristics of the texture. A promising alternative to traditional classification techniques is the neural network. Furthermore, data mining is a discipline that makes use of statistical, machine learning, visualization, and other data manipulation and extraction methods in order to reduce the complexity of data and find hidden patterns in it [9].

The rest of this chapter is organized as follows: Sect. 2 focuses on soft computing techniques, Sect. 3 briefs about image segmentation methods, Sect. 4 discusses biomedical image classification framework, finally, Sect. 5 provides a concluding summary of the chapter.

2 Soft Computing Techniques

Soft computing is used in medical imaging because it provides a practical method for addressing the uncertainties included in the obtained image data [10]. Fuzzy computing, artificial neural networks, swarm intelligence, and evolutionary computing are the four primary categories of soft computing that are covered in this section, as illustrated in Fig. 1.

2.1 Fuzzy Computing

Soft computing, which can deal with information that is unclear, imprecise, vague, or only partially true, is built on fuzzy logic. Moreover, it can be used in a strict sense, referring to a logical framework that expands on traditional two-Boolean logic to support reasoning under uncertainty. Fuzzy sets, which are classes with rounded edges, are used by all theories and technologies and are referred to as fuzzy logic from a broader perspective [11]. It can be used to construct intelligent systems on the basis of information expressed in human language [12].

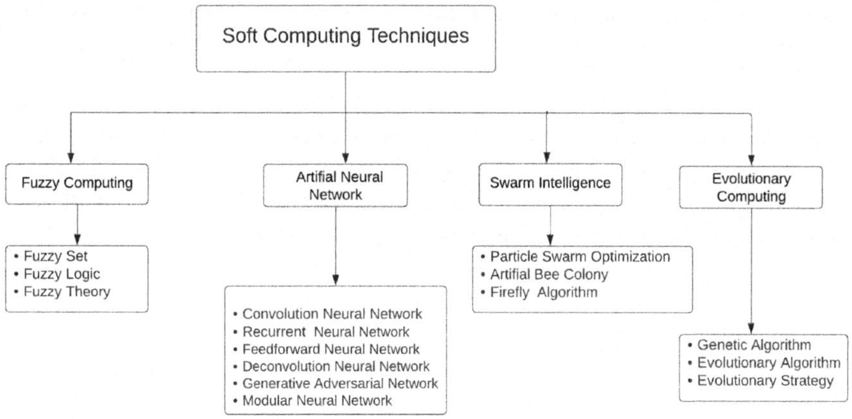

Fig. 1 Soft computing techniques

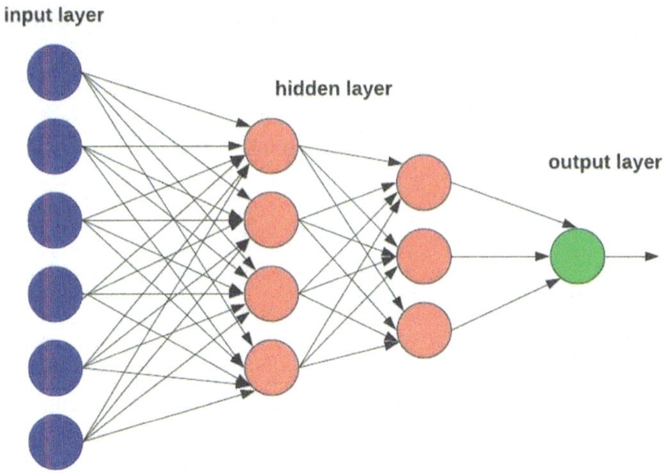

Fig. 2 Artificial neural network

2.2 *Artificial Neural Networks*

Artificial neural networks (ANNs) are an artificial intelligence technique that instructs computers to process data in a manner that is inspired by how the biological nervous system and the brain work. Clearly, it is a type of machine learning process called deep learning, which uses interconnected nodes or neurons in a layered framework to mimic the human brain. The structure of ANNs is shown in Fig. 2. It consists of a node layer with an input layer, one or more hidden layers, and an output layer. Consequently, each layer carries pertinent data and transmits it to the following layer. ANNs learn from training data and gradually increase their accuracy. However, if these learning algorithms are improved for accuracy, they can be useful computer science and AI tools that enable efficient data classification and clustering [13]. It needs to be trained so that it can provide the expected results when given a known set of inputs. Training typically involves feeding the network teaching and instructional patterns and allowing the network to modify its weighting function in accordance with some previously established learning criteria. There are three learning possibilities; supervised, semi-supervised, and unsupervised [14]. Researchers frequently use ANN in various fields such as robotics, speech recognition, facial identification, medical applications, manufacturing, and economics [15].

2.3 *Swarm Intelligence*

Swarm intelligence (SI) is a sub-field of artificial intelligence that simulates the behavioral patterns of social swarms that occur naturally, such as fish schools, ant colonies, bacterial hordes, herds of animals, and flocks of birds [16]. Furthermore, SI-based algorithms are a collection of population-based, nature-inspired algorithms that are used to generate optimal, time-saving, and cost-effective solutions to solve hard problems of NP that cannot be effectively solved using conventional methods [17]. Although each agent in a swarm has limited intelligence and abilities, they interact with each other and their environment to create social patterns that help them achieve common objectives, such as finding a food source or determining the best path to their destination [18].

Wang and Beni [19] first introduced the term "swarm intelligence" in 1989 to address the problems with global optimization in the field of robotics. Since then, academics have given SI-based algorithms a lot of attention, and they are now regarded as one of the most promising sub-fields of AI. This is demonstrated by the rising number of SI-based studies, which prompted the establishment of specialized publications and conferences on this crucial topic [20]. Popular and effective swarm intelligence algorithms include the following: Particle Swarm Optimization (PSO) [21], which simulates the behavior of schooling fish or flocks of birds, Artificial Bee Colony (ABC) [22], which is inspired by honey bees' foraging habits, Ant Colony Optimization (ACO) [23], that is motivated by ant foraging behavior, Bacterial Foraging Optimization (BFO) [24] is motivated by the bacterial group foraging behavior, Firefly algorithm (FFA) [25] is inspired by the way that fireflies flash, and the Spider Monkey Optimization (SMO) [26] technique is influenced by how spider monkeys forage.

2.4 Evolutionary Computing

Evolutionary computation (EC) algorithms are a state-of-the-art and effective method for determining near-optimal solutions. The computational models that employ evolutionary algorithms use evolutionary processes to address complicated issues. The theory of biological evolution serves as inspiration for these evolutionary processes. To increase the quality of the solution, EC algorithms encode the problem in terms of potential solutions. A fundamental aspect of EC approaches is the ability to construct strong termination criteria to control computation time while finding a nearly optimal solution [27]. There are many families of algorithms that fall under the umbrella of EC, such as genetic programming (GP) [28], genetic algorithms (GAs) [29], evolutionary programming (EP) [30], as well as evolutionary strategy (ES) [31].

3 Image Segmentation Techniques

Image segmentation is a necessary step in any computer-aided diagnosis (CAD) system to identify regions of interest from medical images. It is undeniable that accurate segmentation improves the system's overall performance and accuracy. Therefore, the results of segmentation are the main focus of all processes. There are six types of segmentation techniques: thresholding-based, edge-based, partial differential equation (PDE)-based, cluster-based, region-based, and neural network-based [32].

3.1 Threshold-Based Methods

The thresholding technique is the most common and extensively used segmentation methodology in the specified category, mainly due to its reliability and simplicity of use [33]. It has been observed that threshold-based segmentation techniques are quite successful in segmenting a variety of images, including color images, medical images, and satellite images [34]. Segmentation techniques can be divided into parametric and nonparametric techniques [35]. Parametric techniques use the probability density function (PDF) to identify each class, but this technique is computationally expensive and time-consuming. However, nonparametric techniques use a number of criteria, including between-class variance, entropy, and error rate, to determine the quality of a threshold value [36]. The two main categories of thresholding models are bi-level and multi-level. Bi-level thresholding (BT) divides the image into two classes by using a threshold (th) value. Pixels with intensities greater than (th) are labeled as belonging to the first class, while the remaining pixels are labeled as belonging to the second class. Only one (th) value is needed for bilevel thresholding. On the contrary, multilevel thresholding (MT), which requires more than one (th) value, is the task that is required when pixels have been divided into more than two classes [36]. There are a number of methods that have been used in MT to find the optimal thresholds, some of which are the entropies Otsu (variance between classes) [37], Kapur [38], and Tsallis [39], some of the most popular thresholding criteria. Otsu's method chooses the best thresholds for an image based on maximizing the between-class variance of the image to as clearly separate the segmented classes as feasible under the assumption that the number of pixels in each class is comparable, while the Kapur technique determines the optimal threshold by maximizing the entropy of the object and background pixels. In order to define the Tsallis entropy concept, Tsallis generalized the Boltzmann–Gibbs standard entropy and devised the Tsallis approach for the thresholding problem. Therefore, MT is an optimization problem that is solved using optimization algorithms [40]. According to Scopus databases, Fig. 3a presents the statistics for a thresholding-based method for biomedical image studies from 2013 to 2022, and Fig. 3b shows the distribution of the thresholding technique for the biomedical image research area.

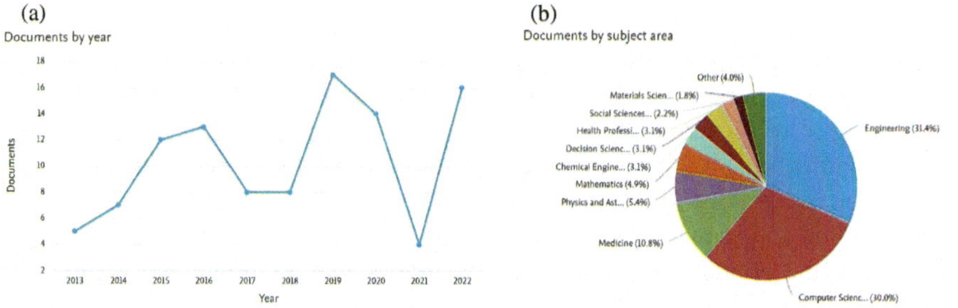

Fig. 3 Research Impact using Scopus from 2013 to 2022 on thresholding-based image segmentation: **a** Histogram, **b** Distribution for biomedical image research areas

A number of academics also focus on segmenting medical images using optimization algorithms; for instance, in [4], the authors suggested skin cancer image segmentation using an improved golden jackal optimization (IGJO) algorithm using opposition-based learning technique to improve the performance of the original golden jackal optimization. The performance of the proposed method was compared with other seven metaheuristic algorithms, including WOA, SOA, SSA, HHO, GTO, MPA, and the original GJO. The proposed algorithm outperformed other algorithms in terms of PSNR, SSIM, FSIM, and MSE segmentation metrics, effectively solving the segmentation issue. In [41], an effective skin cancer segmentation method was developed using Mayfly Optimization (MFO) with Kapur multilevel thresholding for dermoscopic images. In 2022, Uplaonkar et al. [42] proposed a liver tumor ultrasound image segmentation approach using the Otsu method. In [43], a hybrid coronavirus optimization algorithm (COVIDOA) with the Harris Hawks optimization algorithm (HHOA) has been proposed to find the best threshold values using the entropy methods of Otsu and Kapur as fitness functions. An effective technique for CT image segmentation with multiple thresholds was provided [44] using a modified multiverse optimizer (MVO) for processing COVID-19 chest radiography in order to aid doctors in the COVID-19 diagnosis. In [45], the authors suggested a technique for segmenting MRI images using the LASHED optimization algorithm. The results, which have been statistically validated, show that the suggested strategy improves segmentation quality and consistency. An evolutionary grey gradient algorithm (EGGA)-based multilevel thresholding technique for brain MRI image segmentation was presented by the authors in [46]. They applied an adaptive swallow swarm optimization (ASSO) method to enhance the fitness function. The ASSO was evaluated using 25 MRI scans, and the results showed that it outperformed the original SSO. In contrast, the authors of [47] provided an improved approach to the FPA algorithm to address the problem of medical image segmentation. They used eight CT images to test the suggested method and Otsu's approach as an objective function. This approach outperforms the original FPA, PSO, GA, and DE as well as other meta-heuristic algorithms. Furthermore, a hybrid of SMA and WOA known as HSMA-WOA is suggested [48] to improve the behavior of the search agent and maintain the balance between the exploration and exploitation phases of the original SMA. The suggested method is used to solve multi-level segmentation issues by maximizing Kapur's entropy, which improves the regions of interest within a set of COVID-19 chest X-ray images. The HSMA-WOA can successfully complete the segmentation task, according to results comparing it to state-of-the-art algorithms in terms of fitness values and two quality measures. Dáz-Cortés et al. [49] suggested a segmentation technique that uses the spatial information of pixels included in the image to improve thresholding and segment the thermal image. This study used the DA approach to determine the optimal threshold values for energy curves in thermal imaging for the detection of breast cancer. A proposed [50] automatic segmentation system for liver tumors using compressed tomography and magnetic resonance imaging (MRI) uses electromagnetic optimization (EMO) with Otsu and Kapur functions. Table 1 summarizes some of the best state-of-the-art research works in biomedical image segmentation using thresholding-based techniques.

3.2 Cluster-Based Methods

In the clustering approach, several pixels are grouped into a number of clusters, each of which shares the same feature characteristics and has a high similarity within its own cluster and a low similarity with other clusters [59]. There are two basic types of clustering techniques: hard clustering and soft clustering. In hard clustering, a pixel can only belong to one cluster at a time. In light of this, membership between clusters and elements is binary (i.e., 0 or 1). This method computes all cluster centers before allocating the elements to the closest cluster. K-means is one of the most notable instances of hard clustering [60]. As opposed to hard clustering, soft clustering uses fractional membership, which makes it feasible for usage in practical applications. With varying degrees of belongingness, one pixel might be a part of many clusters. This degree of belonging is characterized by fractional membership. One of the most crucial instances of soft clustering is fuzzy C-means [61].

3.3 Region-Based Methods

In region-based approaches, segmentation is often accomplished using clusters of adjacent pixels with comparable brightness. The most popular methods in this category are region-growing and split-and-merge [62]. In regional growing methods, images are divided into various parts based on the seed. These seeds are then either manually chosen or automatically chosen based on experience or for a specific application. On the other hand, in split-and-merge methods, an image is iteratively split into several regions based on common characteristics. After that, related regions are merged together to remove any more unneeded segments [63].

Table 1 Summary of recent related work on biomedical image segmentation using the thresholding techniques

References	Methodologies	Year	Problems solved	Type of images
[41]	Mayfly Optimization (MFO) with Kapur's thresholding method	2022	Skin Cancer	Dermoscopy Images
[42]	Modified Otsu thresholding segmentation	2022	Liver Cancer	Ultrasound images
[43]	Coronavirus Optimization Algorithm (COVIDOA) and Harris Hawks Optimization Algorithm (HHOA) based on Otsu and Kapur methods	2022	Different Medical diseases	CT, MRI, and X-ray
[4]	Improved golden jackal optimization algorithm based on otsu method	2022	Skin Cancer	Dermoscopy Images
[44]	Enhanced multiverse optimizer (CCMVO) with multiple thresholds using nonlocal means, 2D histogram, and 2D Kapur's entropy	2022	COVID-19	Computed Tomography (CT)
[51]	Adaptive opposition slime mold algorithm (AOSMA) using a threshold score (TS)	2022	Brain Tumor	Magnetic Resonance Imaging (MRI)
[52]	Opposition African vulture optimization algorithm with minimum generalized cross entropy	2022	Brain and Skin Cancer	Magnetic Resonance Imaging (MRI) and Dermoscopic Images
[4]	Improved equilibrium optimizer (EO) using dimension learning hunting (DLH) and fuzzy entropy	2022	COVID-19	Computed Tomography (CT)
[53]	Image segmentation using otsu thresholding method	2021	Skin Cancer	Dermoscopy Images
[54]	Sine cosine optimization Algorithm with otsu method	2021	Brain Tumor	Computed Tomography (CT)
[55]	Improved chimp optimization algorithm (ChOA) using opposition and Lévy Flight with orsu and Kapur thresholding methods	2021	Breast cancer	Thermography Images
[56]	A modified differential evolution (MDE) algorithm using slime mould algorithm with Kapur's entropy	2021	Breast cancer	Thermography Images
[57]	Modified black widow optimization algorithm and Otsu method	2021	Brain tumors	MRI and visual images
[58]	Improved manta ray foraging optimization and Otsu as objective dunction	2021	COVID-19	Computed Tomography (CT)

(continued)

Table 1 (continued)

References	Methodologies	Year	Problems solved	Type of images
[46]	Linear population size reduction (LSHADE) algorithm and minimum cross entropy function	2021	Brain Tumor	MIR
[48]	Hybrid novel Slime mould algorithm with whale optimization algorithm and Kapur's function	2020	COVID-19	X-ray images
[49]	Dragonfly algorithm and two different thresholding methods Otsu and Kapur	2018	Breast Cancer	Thermography Images
[50]	Electromagnetism optimization (EMO) with otsu and Kapur functions	2020	Liver Cancer	MRI and CT images
[45]	Evolutionary gray gradient algorithm (EGGA) using Otsu as objective function	2017	Brain Tumor	MIR
[47]	Flower pollination algorithm with a randomized location modification and Otsu method	2015	Different Medical diseases	MRI and CT images

3.4 Partial Differential Equation (PDE)-Based Methods

For image segmentation tasks, PDE-based approaches are thought to be the fastest, making them appropriate for use in real-time applications [64]. The most prevalent and versatile methods for PDF are the non-linear isotropic diffusion filter and convex non-quadratic variation. The non-linear isotropic diffusion filter improves the edges of an image, while convex non-quadratic variation reduces noise in images.

3.5 Edge-Based Methods

Edge-based technique is a popular image-processing method that recognizes the edges of numerous objects in an image. Using data gathered from the edges, it aids in locating characteristics of related objects in the image. In such cases, edge-based segmentation techniques are regarded as efficient methods that have been used for a variety of image analysis and processing tasks, such as the identification of blood vessels in ophthalmoscope images [7].

3.6 *Artificial Neural Network (ANN)-Based Methods*

Another group of image segmentation approaches is artificial neural networks (ANN), in which each pixel in the image can function as a network neuron. The result is that the image creates a neural network. Using training samples, the training process is completed, and after that, the link between neurons is discovered. Then, using the trained samples, a new image is created by segmenting the trained image [32].

4 Biomedical Image Classification Framework

Most medical image analysis frameworks consist of a set of basic steps to ensure better and more effective performance, including dataset collection, preprocessing phase, segmentation phase, feature extraction, feature selection, and filly classification phase, as described in Fig. 4. The first step in medical image analysis entails gathering a large collection of medical images connected to a specific target disease. The impact of medical imaging on medicine is expanding quickly due to the recent growth of advanced imaging methods such as computed tomography (CT) for the diagnosis of many types of cancer (including lung, liver, and pancreatic cancer), magnetic resonance imaging (MRI) to distinguish between healthy and diseased soft tissues of the body, functional magnetic resonance imaging (fMRI), ultrasound, single photon emission computed tomography (SPECT), X-ray, and positron emission tomography (PET) to understand the functioning of these. In order to better train the proposed system, the medical images obtained from various imaging techniques are maintained in the medical database for future study by various research centers [65].

4.1 *Image Preprocessing*

Medical images usually suffer from low contrast, blur, noise, artifacts, and diminished colors. So, preprocessing has crucial significance. It is one of the preliminary processes needed to obtain highly accurate steps [66]. Therefore, there are many preprocessing methods to enhance and remove noise from medical images. For example, and not as a limitation, histogram equalization is an image processing method used to enhance contrast in images. This is accomplished by effectively spreading out the most prevalent intensity levels, thereby expanding the image's intensity range. This technique often results in more visually stunning images overall by closely representing an image's valuable information in contrast levels. Because of this, regions with less local contrast can acquire more contrast [67]. A mean filter, also known as an average filter, is a straightforward linear filter that is a quick and

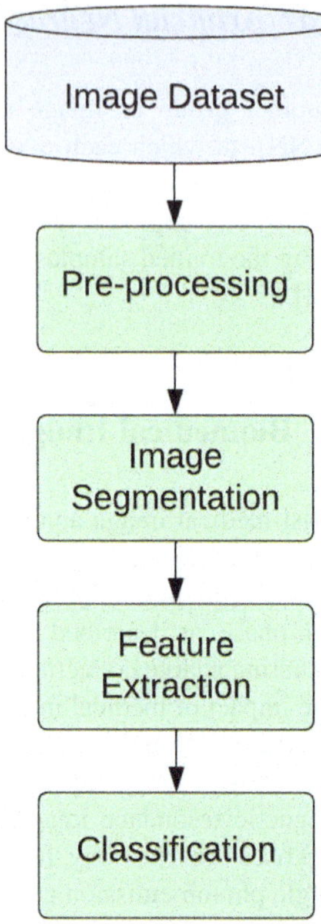

Fig. 4 Stages for biomedical image classification framework

simple way of smoothing images. A common technique for lowering noise and minimizing pixel-to-pixel intensity variance is the average filter. The basic idea behind a filter is for any element of the signal (image) to take an average across its neighborhood [68]. Since the median filter is well recognized for keeping edges during noise removal, even though it is a filtering technique for signal and image reduction of noise, it is crucial in the field of image processing [69]. Also, the Wiener filter is used to lower signal noise. Inverse filtering can be used to restore an image that has been blurred by a well-known low-pass filter. However, inverse filtering is extremely susceptible to additive noise. Wiener filtering achieves the best trade-off between noise smoothing and inverse filtering [70].

4.2 Image Segmentation

Medical image segmentation is essential to medical research as well as computer-aided diagnosis and is a crucial stage in medical image analysis. With the recent sustained progress of computer vision [71], it is the process of dividing an image into several regions using various methods [72], explained above in Sect. 3.

4.3 Feature Extraction

In image processing, feature extraction is an approach for reducing the size of vast amounts of duplicated data. It is the process of extracting a set of features from the input data. Medical images are analyzed using feature extraction approaches to identify the most prominent aspects that constitute the primary representation of the image [73].

4.4 Classification

The classification of medical images into several categories is one of the most scientific methods in image recognition. It aids clinicians in the diagnosis of diseases or for future research. After extracting effective features using feature extraction methods, use these features to build models that classify the image dataset [74]. Since soft computing offers a variety of general classification techniques that can be used to classify a wide range of data. As a result, it might be difficult to determine which categorization approach would be best for a certain study. However, by taking into consideration a number of variables and prior information, classification methods that are tailored to specific applications frequently perform better in terms of accuracy and time complexity. According to the relative weight performance measurements, no single classification algorithm was able to answer every one of the hundred classification tasks with the highest level of performance across all possible experimental configurations. Furthermore, data mining faces hitherto unimagined difficulties as massive datasets spread across numerous sectors. Despite several evaluations of various classification techniques, no one technique has emerged as the best across all datasets. As a result, the subject is still being researched [8]. Many classification methods have been presented in the past, including nearest-neighbor, decision tree induction, error backpropagation, lazy learning, rule-based learning, and statistical learning, which is a relatively new addition. The performance of these methods, however, is much below the practical norm, and their recent development has been relatively gradual [75]. Furthermore, selecting and extracting features takes time and varies depending on the object. Since 2012, deep neural networks (DNNs), particularly convolutional neural networks (CNNs), have significantly improved performance in changing image classification tasks. CNN has conducted some medical image classification research with results that are competitive with those of human professionals [76]. According to Scopus databases, Fig. 3 illustrates the statistics for CNN and biomedical imaging classification and detection studies from 2013 to 2022, and Fig. 5, shows the distribution of CNN for the biomedical imaging classification research area.

According to the work in [77], a technique for detecting brain tumors in two-dimensional magnetic resonance imaging (MRI) was developed using a CNN. The investigation used a real-time dataset that contained photographs of various tumor parameters, such as image shape, intensity, and size. The study's goal was to categorize the MRI images of the brain into benign and malignant brain tissues, which is essentially a two-way classification. Experimental results of the proposed CNN model archived at a rate of 97.5% accuracy. A 3D Inception-based CNN was utilized by Khvostikov et al. [78] to diagnose AD. The technique was based on the combination of diffusion tensor imaging (DTI) and structural magnetic resonance imaging (SMRI) modalities on hippocampus regions of interest (RoI). They contrasted that strategy's performance with that of the network built using AlexNet. 3D Inception reported better performance than AlexNet. While in [79], the authors presented two techniques for classifying medical images and identifying Alzheimer's disease. The first method processes structural brain scans in 2D and 3D from the Alzheimer's Disease Neuroimaging Initiative (ADNI) dataset using simple CNN architectures based on convolution in both 2D and 3D. The transfer learning principle is used in the second method, which makes use of the VGG19 model and other previously trained medical image classification models. The experimental results show that the suggested architectures are adequate, straightforward structures that minimize computational complexity, memory needs, overfitting, and reasonable time. In addition, they reach accuracy levels of 93.61% and 95.17% for 2D and 3D multi-class AD stage classifications, respectively. The VGG19 pre-trained model has been improved, and it now achieves a multi-class AD stage classification accuracy of 97%. In addition, to categorize the ISIC dataset for skin cancer, simple convolutional neural networks and DenseNet architectures were suggested [80]. Adopting a transfer learning strategy increases classification accuracy. An accuracy of 95% in classifying objects has been discovered.

In terms of breast cancer classification and detection, Wang et al. [81] suggested a CAD system based on CNN that classified breast lesions into benign and malignant categories. In order to provide effective feature extraction and extract multi-view

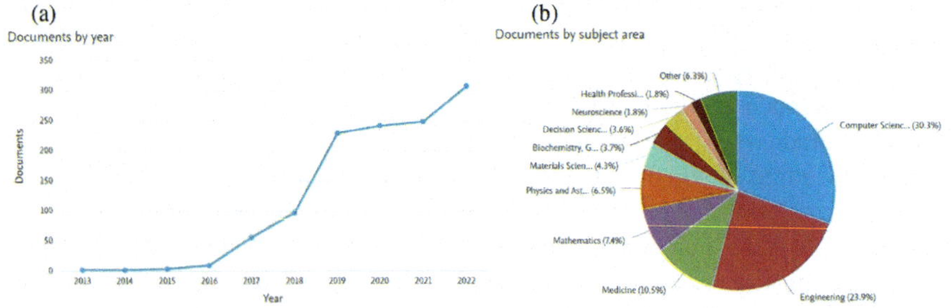

Fig. 5 Research Impact using Scopus from 2013 to 2022 on biomedical classification using C: **a** Histogram, **b** Distribution for biomedical image research areas

features from both views, the proposed CNN adopts a modified Inception-v3 architecture. A total of 316 breast lesions were used to train the proposed CNN. The sensitivity and specificity were 0.886 and 0.876, respectively, while the AUC value was 0.9468. Also, Byra et al. [82] developed a CNN system that used transfer learning to identify between benign and malignant breast lesions, and they achieved an AUC of 93.6% on a test set of 150 instances. They configured the CNN VGG19 model for 882 images of ultrasound breast masses after it had been trained on the ImageNet dataset. In [83], the scientists developed a deep convolutional neural network-based computer-assisted diagnostic system using three of the most magnificent CNN models: VGG16, ResNet50, and Inception v3. Better results can be achieved by switching to learning with pre-trained weights rather than random initialization. On the DDSM database, the suggested system achieved 97.4% accuracy and 0.99 AUC; on the INbreast database, 95.5% accuracy and 0.97 AUC; and on the BCDR database, 96.60% accuracy and 0.96 AUC. After all, the authors combined all the datasets to produce a large collection of images and used them to fine-tune the CNNs. The extracted regions of interest (ROIs) from the full mammograms were pre-processed and standardized. The accuracy rate is 98.94%. An approach based on convolutional neural networks was developed by Baffa and Lattari [84]. Both static and dynamic infrared images were employed. They ran a CNN architecture test on the data set. Applying distinct techniques to IR images for data augmentation, mean subtraction, and grey to RGB conversion. They reported 97% sensitivity, 100% specificity, and 98% accuracy for static procedures.

In [90], a VGG-16 network-based Faster Region Framework (Faster R-CNN) was introduced for the diagnosis of COVID-19 using X-ray images from convolutional neural networks. The authors created a customized data set of 5,450 pictures from 2,500 patients. The information was used for training and assessment. The authors combine and update two publicly accessible data sets to produce the unique experiment dataset. The accuracy, sensitivity, and precision of the authors' binary classification were 97.36%, 97.65%, and 99.28%, respectively. In [91], CoroNet is a model of the Deep Convolutional Neural Network developed by Khan et al. to autonomously diagnose infection in the chest using X-ray imaging. The proposed model is based on the Xception architecture and was pretrained using the ImageNet data set before being trained on the Covid-19 dataset and extra chest pneumonia X-ray images from two publically available sources. The experimental data show that the proposed model has an overall recall rate of 93% for COVID-19 cases, an accuracy rate of 89.6%, and a rate of 98.2% for four-class instances. COVID-19, pneumonia (viral and bacterial), and normal patients are the classification classes. Table 2 summarizes a few prominent state-of-the-art research works in biomedical image classification using CNN.

Table 2 Summary of recent related work from 2018 to 2021 on CNN classifier for different biomedical images

References	Dataset	Year	Contraption	Accuracy
[85]	– HAM10000	– 2021	– Skin cancer classification	– 91.43%
[86]	– Dermquest – DermIS	– 2020	– Detection of skin cancer	– 91%
[81]	– ABUS imaging	– 2020	– Breast cancer classification	– 88%
[87]	– ISIC database	– 2021	– Skin cancer classification	– 86%
[80]	– ISIC database	– 2021	– Skin cancer classification	– 95%
[83]	– DDSM	– 2018	– Breast cancer classification	– 97.4%
	– Inbreast			– 95.5%
	– BCDR			– 96.60%
[77]	– BRATS	– 2018	– Brain tumor classification	– 97.5%
[79]	– ADNI	– 2021	– Detection of Alzheimer's disease	– 93.61% for 2D – 95.17% for 3D
[88]	– 3895 thermal breast images	– 2020	– Detection of breast cancer	– 98.95%
[89]	– DMR	– 2020	– Detection of breast cancer	– 94.3%
[84]	– DMR	– 2018	– Breast cancer classification	– 98%
[90]	– COVIDx	– 2020	– COVID-19 classification	– 97.36%
[91]	– Collected chest X-ray images	– 2020	– COVID-19 detection	– 89.6%

5 Conclusion and Future Research Issues

Soft computing techniques have attracted growing attention over the past several years due to their applicability for problem-solving in the processing and evaluation of medical images. Processing of medical images includes two major phases called segmentation and classification. Image segmentation is the process of separating an image into a number of distinct classes depending on the characteristics of the grayscale or color image, such as texture, shape, size, or position, to extract regions of interest. Segmentation is a critical preprocessing phase in any computer-aided diagnosis (CAT) system. The problem of image segmentation can be solved using a variety of techniques. The most crucial image segmentation approaches are threshold-based techniques, clustering-based techniques, edge-based techniques, region-based techniques, watershed-based techniques, partial differential equation-based techniques, and artificial neural network-based techniques. Machine learning and its subcategories, such as deep learning and neural networks, have gathered tremendous speed in recent years as a result of the increasing volatility, necessity, and applicability of AI. Training requires software and tools such as classifiers, which feed enormous amounts of data, analyze it, and extract useful features. Therefore, medical image classification is one of the most important challenges in the field of image recognition because it aims to categorize medical images into different groups to aid doctors in the diagnosis of diseases or in further research. Many classification techniques, including machine learning techniques, have been proposed in the past, but their effectiveness falls far short of the norm in real-world applications, and their recent development has been quite gradual. Additionally, the time required to choose and extract characteristics varies according to the object. Deep neural networks (DNNs), in particular convolutional neural networks (CNNs), have gained significant traction in the field of medical image classification. It is expected in the future that the use of soft computing methods will increase in many fields, such as biomedical images, health care systems, and computer-aided diagnosis systems, due to the efficiency of these methods in solving complex problems. Many nature-inspired optimization algorithms (NIOAs) have not been applied yet, and hyper-inspired algorithms will be used to solve the segmentation problem, especially in thresholding method. Also, there are many NIOAs that will be widely used in the scope of image classification to optimize the hyper-parameters of CNN in order to improve the classification efficiency and accuracy.

References

1. Zadeh LA (1994) Fuzzy logic, neural networks, and soft computing. Commun ACM 37(3):77–84
2. Huang Y, Lan Y, Thomson SJ, Fang A, Hoffmann WC, Lacey RE (2010) Development of soft computing and applications in agricultural and biological engineering. Comput Electron Agric 71(2):107–127
3. Devi M, Singh S, Tiwari S, Chandra Patel S, Ayana MT (2021) A survey of soft computing approaches in biomedical imaging. J Healthcare Eng 2021
4. Houssein EH, Helmy BE-D, Oliva D, Jangir P, Premkumar M, Elngar AA, Shaban H (2022) An efficient multi-thresholding based covid-19 ct images segmentation approach using an improved equilibrium optimizer. Biomed Signal Process Control 73:103401
5. Houssein EH, Helmy BE-D, Oliva D, Elngar AA, Shaban H (2021) A novel black widow optimization algorithm for multilevel thresholding image segmentation. Expert Syst Appl 167:114159
6. Houssein EH, Abdelkareem DA, Emam MM, Hameed MA, Younan M (2022) An efficient image segmentation method for skin cancer imaging using improved golden jackal optimization algorithm. Comput Biol Med 149:106075
7. Dhal KG, Das A, Ray S, Gálvez J, Das S (2020) "Nature-inspired optimization algorithms and their application in multi-thresholding image segmentation. Archives of computational methods in engineering, vol 27, pp 855–888
8. Lashari SA, Ibrahim R (2013) A framework for medical images classification using soft set. Procedia Technol 11:548–556
9. Jain AK, Duin RPW, Mao J (2000) Statistical pattern recognition: a review. IEEE Trans Pattern Anal Mach Intell 22(1):4–37
10. Falcone R, Lima C, Martinelli E (2020) Soft computing techniques in structural and earthquake engineering: a literature review. Eng Struct 207:110269
11. Kamble AJ, Rewaskar RP (2020) Soft computing-fuzzy logic: an overview. Int J Fuzzy Math Arch 18(1):45–52

12. Kambalimath S, Deka PC (2020) A basic review of fuzzy logic applications in hydrology and water resources. Appl Water Sci 10(8):1–14
13. Harirchian E, Hosseini SEA, Jadhav K, Kumari V, Rasulzade S, Sık I, Wasif M, Lahmer T (2021) A review on application of soft computing techniques for the rapid visual safety evaluation and damage classification of existing buildings. J Build Eng 43:102536
14. Zhang W, Zhang R, Wu C, Goh ATC, Lacasse S, Liu Z, Liu H (2020) Stateof-the-art review of soft computing applications in underground excavations. Geosci Front 11(4):1095–1106
15. Sharkawy A-N (2020) Principle of neural network and its main types. J Adv Appl Comput Math 7:8–19
16. Jumani TA, Mustafa MW, Alghamdi AS, Rasid MM, Alamgir A, Awan AB (2020) Swarm intelligence-based optimization techniques for dynamic response and power quality enhancement of ac microgrids: a comprehensive review. IEEE Access 8:75986–76001
17. Houssein EH, Sayed A (2023) Dynamic candidate solution boosted beluga whale optimization algorithm for biomedical classification. Mathematics 11(3):707
18. Sharma A, Sharma A, Pandey JK, Ram M (2022) Swarm intelligence: foundation, principles, and engineering applications
19. Beni G, Wang J (1993) Swarm intelligence in cellular robotic systems. In: Robots and biological systems: towards a new bionics? Springer, pp 703–712
20. Chan F, Tiwari M (2007) Swarm Intelligence: focus on ant and particle swarm optimization. BoD–Books on Demand
21. Kennedy J, Eberhart R (1995) Particle swarm optimization. In: Proceedings of ICNN'95-international conference on neural networks, vol 4. IEEE, pp 1942–1948
22. Karaboga D et al. (2005) An idea based on honey bee swarm for numerical optimization. Technical report-tr06, Erciyes university, engineering faculty, computer ..., Tech. Rep.
23. Dorigo M (1992) Optimization, learning and natural algorithms. PhD Thesis, Politecnico di Milano
24. Passino KM (2002) Biomimicry of bacterial foraging for distributed optimization and control. IEEE Control Syst Mag 22(3):52–67
25. Yang X-S (2009) "Firefly algorithms for multimodal optimization. In: Stochastic algorithms: foundations and applications: 5th international symposium, SAGA 2009, Sapporo, Japan, October 26-28, 2009. Proceedings 5. Springer, pp 169–178
26. Bansal JC, Sharma H, Jadon SS, Clerc M (2014) Spider monkey optimization algorithm for numerical optimization. Memetic Comput 6:31–47
27. Telikani A, Gandomi AH, Shahbahrami A (2020) A survey of evolutionary computation for association rule mining. Inf Sci 524:318–352
28. Banzhaf W, Nordin P, Keller RE, Francone FD (1998) Genetic programming: an introduction: on the automatic evolution of computer programs and its applications. Morgan Kaufmann Publishers Inc.
29. Goldberg DE, Korb B, Deb K (1989) Messy genetic algorithms: Motivation, analysis, and first results. Complex Systems 3(5):493–530
30. Fogel DB (1998) Artificial intelligence through simulated evolution. Wiley-IEEE Press
31. Beyer H-G, Schwefel H-P (2002) Evolution strategies–a comprehensive introduction. Nat Comput 1:3–52
32. Houssein EH, Helmy E-D, Oliva D, Elngar AA, Shaban H et al (2021) Multilevel thresholding image segmentation based on nature-inspired optimization algorithms: a comprehensive review. Metaheuristics in machine learning: theory and applications, pp 239–265
33. Abualigah L, Al-Okbi NK, Elaziz MA, Houssein EH (2022) Boosting marine predators algorithm by salp swarm algorithm for multilevel thresholding image segmentation. Multimedia Tools Appl 81(12):16707–16742
34. Anitha J, Pandian SIA, Agnes SA (2021) An efficient multilevel color image thresholding based on modified whale optimization algorithm. Expert Syst Appl 178:115003
35. Rodríguez-Esparza E, Zanella-Calzada LA, Oliva D, Heidari AA, Zaldivar D, Pérez-Cisneros M, Foong LK (2020) An efficient harris hawks-inspired image segmentation method. Expert systems with applications, vol 155, p 113428

36. Abd El Aziz M, Ewees AA, Hassanien AE (2017) Whale optimization algorithm and moth-flame optimization for multilevel thresholding image segmentation. Expert Syst Appl 83:242–256
37. Otsu N (1979) A threshold selection method from gray-level histograms. IEEE Trans Syst Man Cybern 9(1):62–66
38. Kapur JN, Sahoo PK, Wong AK (1985) A new method for gray-level picture thresholding using the entropy of the histogram. Comput Vis Graph Image Process 29(3):273–285
39. Ben Hamza A (2006) Nonextensive information-theoretic measure for image edge detection. J Electron Imaging 15(1):013011–013011
40. Bhargavi K, Jyothi S (2014) A survey on threshold based segmentation technique in image processing. Int J Innov Res Dev 3(12):234–239
41. Reshma G, Al-Atroshi C, Nassa VK, Geetha B, Sunitha G, Galety MG, Neelakandan S (2022) Deep learning-based skin lesion diagnosis model using dermoscopic images. Intell Autom Soft Comput 31(1)
42. Uplaonkar DS, Virupakshappa, Patil N (2022) Modified otsu thresholding based level set and local directional ternary pattern technique for liver tumor segmentation. Int J Syst Assur Eng Manag, 1–11
43. Hosny KM, Khalid AM, Hamza HM, Mirjalili S (2022) Multilevel segmentation of 2d and volumetric medical images using hybrid coronavirus optimization algorithm. Comput Biol Med 150:106003
44. Su H, Zhao D, Elmannai H, Heidari AA, Bourouis S, Wu Z, Cai Z, Gui W, Chen M (2022) Multilevel threshold image segmentation for covid-19 chest radiography: a framework using horizontal and vertical multiverse optimization. Comput Biol Med 146:105618
45. Panda R, Agrawal S, Samantaray L, Abraham A (2017) An evolutionary gray gradient algorithm for multilevel thresholding of brain mr images using soft computing techniques. Appl Soft Comput 50:94–108
46. Aranguren I, Valdivia A, Morales-Castan˜eda B, Oliva D, Abd Elaziz M, Perez- Cisneros M (2021) Improving the segmentation of magnetic resonance brain images using the lshade optimization algorithm. Biomedical signal processing and control, vol 64, p 102259
47. Wang R, Zhou Y, Zhao C, Wu H (2015) A hybrid flower pollination algorithm based modified randomized location for multi-threshold medical image segmentation. Bio- Med Mater Eng 26(s1):S1345–S1351
48. Abdel-Basset M, Chang V, Mohamed R (2020) Hsma woa: a hybrid novel slime mould algorithm with whale optimization algorithm for tackling the image segmentation problem of chest x-ray images. Appl Soft Comput 95:106642
49. Díaz-Cortés M-A, Ortega-Sánchez N, Hinojosa S, Oliva D, Cuevas E, Rojas R, Demin A (2018) A multi-level thresholding method for breast thermograms analysis using dragonfly algorithm. Infrared Phys Technol 93:346–361
50. Mahdy LN, Ezzat KA, Torad M, Hassanien AE (2020) Automatic segmentation system for liver tumors based on the multilevel thresholding and electromagnetism optimization algorithm. Int J Imaging Syst Technol 30(4):1256–1270
51. Naik MK, Panda R, Samantaray L, Abraham A (2022) A novel threshold score based multiclass segmentation technique for brain magnetic resonance images using adaptive opposition slime mold algorithm. Int J Imaging Syst Technol 32(4):1397–1413
52. Jena B, Naik MK, Panda R, Abraham A (2022) A novel minimum generalized cross entropy-based multilevel segmentation technique for the brain mri/dermoscopic images. Comput Biol Med 151:106214
53. Pitoya PA, Suputraa I (2021) Dermoscopy image segmentation in melanoma skin cancer using otsu thresholding method. J. Elektronik Ilmu Komputer Udayana P-ISSN 2301:5373
54. Erukala Mahender D, Babu CR, Kumar KS (2021) Multi-level thresholding for image segmentation on medical images using multi otsu and sine cosine optimization algorithm. Annals of the romanian society for cell biology, pp 12305–12316
55. Houssein EH, Emam MM, Ali AA (2021) An efficient multilevel thresholding segmentation method for thermography breast cancer imaging based on improved chimp optimization algorithm. Expert Syst Appl 185:115651

56. Liu L, Zhao D, Yu F, Heidari AA, Ru J, Chen H, Mafarja M, Turabieh H, Pan Z (2021) Performance optimization of differential evolution with slime mould algorithm for multilevel breast cancer image segmentation. Comput Biol Med 138:104910
57. Al-Rahlawee ATH, Rahebi J (2021) Multilevel thresholding of images with improved otsu thresholding by black widow optimization algorithm. Multimedia Tools Appl 80(18):28217–28243
58. Houssein EH, Hussain K, Abualigah L, Abd Elaziz M, Alomoush W, Dhiman G, Djenouri Y, Cuevas E (2021) An improved opposition-based marine predators algorithm for global optimization and multilevel thresholding image segmentation. Knowl Based Syst 229:107348
59. Anter AM, Hassenian AE, Oliva D (2019) An improved fast fuzzy c-means using crow search optimization algorithm for crop identification in agricultural. Expert Syst Appl 118:340–354
60. Kamil MY, Salih AM (2019) Mammography images segmentation via fuzzy cmean and k-mean. Int J Intell Eng Syst 12(1):22–29
61. Kaur D, Kaur Y (2014) Various image segmentation techniques: a review. Int J Comput Sci Mob Comput 3(5):809–814
62. Solomon C, Breckon T (2011) Fundamentals of digital image processing: a practical approach with examples in Matlab. Wiley
63. Liang Y, Zhang M, Browne WN (2014) Image segmentation: a survey of methods based on evolutionary computation. In: Simulated evolution and learning: 10th international conference, SEAL 2014, Dunedin, New Zealand, December 15-18, 2014. Proceedings 10. Springer, pp 847–859
64. Khan W (2013) Image segmentation techniques: a survey. J Image Graph 1(4):166–170
65. Amirjahan M, Sujatha DN (2016) Comparative analysis of various classification algorithms for skin cancer detection. PG & Research Department of Computer Science, Raja Doraisingam Govt. Art College, Sivagangai, Tamil Nadu, India, pp 199–205
66. Jeyavathana RB, Balasubramanian R, Pandian AA (2016) A survey: analysis on pre- processing and segmentation techniques for medical images. Int J Res Scientific Innov (IJRSI) 3(6):113–120
67. Shankar KP, Shyry SP (2021) A survey of image pre-processing techniques for medical images. In Journal of physics: conference series, vol 1911, no 1. IOP Publishing, p 012003
68. Shetti P, Patil A (2017) Performance comparison of mean, median and wiener filter in mri image de-noising. Int J Res Trends Innov 2:371–375
69. Sagayam KM, Bruntha PM, Sridevi M, Sam MR, Kose U, Deperlioglu O (2021) A cognitive perception on content-based image retrieval using an advanced soft computing paradigm. In: Advanced machine vision paradigms for medical image analysis. Elsevier, pp 189–211
70. Jadwa D (2018) Wiener filter based medical image de-noising. Int J Sci Eng Appl 7(9):318–323
71. Xu Y, He X, Xu G, Qi G, Yu K, Yin L, Yang P, Yin Y, Chen H (2022) A medical image segmentation method based on multi-dimensional statistical features. Front Neurosci 16
72. Emam MM, Houssein EH, Ghoniem RM (2023) A modified reptile search algorithm for global optimization and image segmentation: case study brain mri images. Comput Biol Med 152:106404
73. Nagarajan G, Minu R, Muthukumar B, Vedanarayanan V, Sundarsingh S (2016) Hybrid genetic algorithm for medical image feature extraction and selection. Procedia Comput Sci 85:455–462
74. Houssein EH, Abohashima Z, Elhoseny M, Mohamed WM (2022) Hybrid quantum- classical convolutional neural network model for covid-19 prediction using chest x-ray images. J Comput Design Eng 9(2):343–363
75. Houssein EH, Saber E, Ali AA, Wazery YM (2022) Centroid mutation-based search and rescue optimization algorithm for feature selection and classification. Expert Syst Appl 191:116235
76. Yadav SS, Jadhav SM (2019) Deep convolutional neural network based medical image classification for disease diagnosis. J Big Data 6(1):1–18
77. Seetha J, Raja SS (2018) Brain tumor classification using convolutional neural networks. Biomed Pharmacol J 11(3):1457
78. Khvostikov A, Aderghal K, Krylov A, Catheline G, Benois-Pineau J (2018) 3d inception- based cnn with smri and md-dti data fusion for alzheimer's disease diagnostics. arXiv preprint arXiv: 1809.03972

79. Helaly HA, Badawy M, Haikal AY (2021) Deep learning approach for early detection of alzheimer's disease. Cognitive computation, pp 1–17
80. Alagu S, Bagan KB (2021) "Skin cancer classification in dermoscopy images using convolutional neural network. In: AIP conference proceedings, vol 2336, no 1. AIP Publishing LLC, 2021, p 040013
81. Wang Y, Choi EJ, Choi Y, Zhang H, Jin GY, Ko S-B (2020) Breast cancer classification in automated breast ultrasound using multiview convolutional neural network with transfer learning. Ultrasound Med Biol 46(5):1119–1132
82. Byra M, Galperin M, Ojeda-Fournier H, Olson L, O'Boyle M, Comstock C, Andre M (2019) Breast mass classification in sonography with transfer learning using a deep convolutional neural network and color conversion. Med Phys 46(2):746–755
83. Chougrad H, Zouaki H, Alheyane O (2018) Deep convolutional neural networks for breast cancer screening. Comput Methods Programs Biomed 157:19–30
84. Baffa MDFO, Lattari LG (2018) Convolutional neural networks for static and dynamic breast infrared imaging classification. In: 2018 31st SIBGRAPI conference on graphics, patterns and images (SIBGRAPI). IEEE, pp 174–181
85. Ali MS, Miah MS, Haque J, Rahman MM, Islam MK (2021) An enhanced technique of skin cancer classification using deep convolutional neural network with transfer learning models. Mach Learn Appl 5:100036
86. Zhang L, Gao HJ, Zhang J, Badami B (2020) Optimization of the convolutional neural networks for automatic detection of skin cancer. Open Med 15(1):27–37
87. Mijwil MM (2021) Skin cancer disease images classification using deep learning solutions. Multimedia Tools Appl 80(17):26255–26271
88. Ekici S, Jawzal H (2020) Breast cancer diagnosis using thermography and convolutional neural networks. Med Hypotheses 137:109542
89. . Cabıog˘lu C, Og˘ul H (2020) Computer-aided breast cancer diagnosis from thermalimages using transfer learning. In: Bioinformatics and biomedical engineering: 8th international work-conference, IWBBIO 2020, Granada, Spain, May 6–8, 2020, Proceedings 8. Springer, pp 716–726
90. Shibly KH, Dey SK, Islam MT-U, Rahman MM (2020) Covid faster r–cnn: a novel framework to diagnose novel coronavirus disease (covid-19) in x-ray images. Inf Med Unlocked 20:100405
91. Khan AI, Shah JL, Bhat MM (2020) Coronet: a deep neural network for detection and diagnosis of covid-19 from chest x-ray images. Comput Methods Programs Biomed 196:105581

Chapter 2
Optimization Technique Used in Biomedical for Qualitative Sleep Analysis

Hasina Adil, Mustafa Adil, Santosh S. Raghuwanshi, and Alok Gupta

Abstract A good sleep is as important as good nutrition. The intangible nature of sleep makes it difficult to measure and can only be assessed on individual's experience. The accurate qualitative measurement of sleep is a challenging engineering and clinical problem. Qualitative measurement of sleep is a critical process that uses a huge number of parameters for assessing sleep stages and sleep cycles. The clinical sleep analysis uses Polysomnography (PSG) as Gold Standard method for monitoring and diagnosis. In the PSG process, a huge number of sensors are attached on user, which makes this process critical, costly, and uncomfortable. The available commercial smart wearable devices for sleep monitoring have limitations that the information from such devices cannot be used by medical professionals for analysis and diagnosis purposes. This chapter presents parameters involved in the qualitative measurement of sleep in clinical and commercial monitoring systems. The parameter optimization method for qualitative sleep analysis is based on reduction of parameters by finding independent parameters and their derived dependent parameters. The optimization technique in biomedical application leads the medical practitioners to select the most important key parameters needed to be monitored for the sleep analysis of person in a given condition among all vital parameters.

Keywords Gold standard Polysomnography · Electrocardiogram · Electrooculogram · Electroencephalogram · Cerebral blood flow · Non-rapid eye

---

H. Adil (✉)
Department of Basic Science, Oriental University, Indore, Madhya Pradesh, India
e-mail: dr.hasinaadil@orientaluniversity.in

M. Adil
System Engineer, Tata Consultancy Services (TCS), Indore, Madhya Pradesh, India

S. S. Raghuwanshi
Department of Electrical Engineering, Medi-Caps University, Indore, India

A. Gupta
Research and Development, 3VTechnix, Indore, Madhya Pradesh, India
e-mail: alok@3vthechnix.com

U. N. Dulhare and E. H. Houssein (eds.), *Machine Learning and Metaheuristics: Methods and Analysis*, Algorithms for Intelligent Systems,
https://doi.org/10.1007/978-981-99-6645-5_2

movement · Rapid eye movement · Sleep parameter allocation · Parameter optimization

Abbreviations

BCG	Ballistocardiography
BP	Blood pressure
BPM	Beats per minute
CBF	Cerebral blood flow
CPAP	Continuous positive airway pressure
CTT	Continuous positive airway pressure titration test
EB	Eye blinking
ECG	Electrocardiogram
EDR	Electrocardiogram derived respiration
EEG	Electroencephalogram
EMG	Electromyography
EOG	Electro-oculogram
ESS	Epworth sleepiness scale
GERD	Gastroesophageal reflux disease
HR	Heart rate
MSLT	Multiple sleep latency test
MWT	Maintenance of wakefulness test
NREM	Non-rapid eye movement
PAP	Positive airway pressure
PLMS	Periodic leg movements of sleep
PPG	Photoplethysmogram
PSG	Polysomnography
PSQI	Pittsburgh Sleep Quality Index
REM	Rapid eye movement
RLS	Restless Leg Syndrome
RR	Respiration Rate

1 Introduction

One of the most important and regular tasks of human being for which they spend their one third of life is sleep. Sleep is rejuvenation process for all living organisms for their growth. A sound sleep provides freshness and energy to all living beings [1]. Inadequate and improper quality of sleep invites diseases like stress, obesity, depression, thought of ending life and breast cancer [2]. The average sleep hours vary according to age, from 14 to 8 h depending upon number of sleep hours and sleep habits of an individual. To live healthy life, sleep clinics and smart wearable devices are available for sleep analysis. The monitoring system for measuring sleep has to meet certain stringent specifications, sensitivity, reliability, and stability to monitor brain waves, heart rate, and other body parameters throughout the sleep hours. The monitoring of sleep quality involves the identification of the primary stages of sleep and their respective durations, which are influenced by changes in brain activity and physiological functions of the body during sleep. The standard method measures all vital parameters of body like heartbeats, oxygen saturation level, muscle movements, and eye movement [3].

The Polysomnography (PSG) parameters are the bench mark for qualitative sleep measurement. Only in hospitals, it is possible to measure these parameters under the supervision of medical expert, which makes it very expensive and cannot be affordable by general users [4]. On the other hand, commercial sleep monitor's information cannot be directly used by doctors for further diagnosis [5]. There is a genuine need to reduce the process complexity by finding optimized parameters with the help of optimization techniques, which will provide accurate information about quality of sleep. There are various optimization techniques used in bio-signal processing. The most commonly used optimization techniques include filtering, feature extraction, signal segmentation, data fusion and optimization based on correlation factor. These techniques help to enhance signal processing, extract relevant features, and improve the accuracy of qualitative analysis. Recently, an optimal wavelet filter technique has been used to assess the quality of sleep in healthy and sleep disorder patients [6]. The accuracy of automated sleep stage detection has been improved by employing principal component analysis (PCA) technique [7]. The independent component analysis (ICA) technique is also widely utilized to extract meaningful characteristics and relevant features that represent specific information in the bio-signal [8].

The signal segmentation optimization technique is another tool to divide a continuous bio-signal into meaningful segments or epochs to identify different patterns or events within the signal for example dynamic time warping (DTW) approach is used in biomedical signal processing [9] and hidden Markov models (HMM) for the estimation of sleep wake cycle [10], whereas, optimized clustering techniques are commonly used for biomedical data set [11]. On the other hand, data fusion combines information from multiple sensors or data sources to improve overall analysis. Optimization techniques like sensor fusion [12], Bayesian inference [13], and Kalman filtering [14] are employed to integrate data from different modalities or sources, providing a more comprehensive understanding of the underlying bio-signals. The analysis of sleep is a complex process and needs more parameters to provide qualitative information [15].

Various physiological parameters are involved in the qualitative assessment of sleep [16]. There is a significant change in the vital parameters of life during the sleep. The major parameters are variation in heart rate, breathing, activity of brain, muscle activity, body position, and movement in eyes. These parameters are considered as sleep parameters and are allocated to retrieve the information about quality of sleep. Several techniques have been developed either by using this single parameter or in combination of parameters. The limitation with this type of monitor is that they are based on specific parameter, though the functionality of every organ in human body shows change during sleep. The brain plays an important role during the sleep generating various frequency signals, which provide clear information of sleep stages. The movement of eyes also helps in the identification of various sleep stages. Muscle activity, heart rate and respiration rate provide important information about a particular sleep stage. Individual sleep parameter will not provide complete information about quality of sleep. The medical experts in sleep clinics monitor number of parameters to find out sleep quality and related disorder. The sleep trackers monitor awake-sleep time and provide information about sleep quality based on total time spend in bed and actual sleep time computation, which is not so accurate since no technical raw data about sleep parameter for analysis are available. For human suffering from various diseases, the single parameter information does not meet the requirement of sleep assessment. There is a basic need to allocate and optimize such sleep parameters, which have direct involvement in whole sleep process to provide complete information about sleep. The allocated and optimized parameters used in qualitative sleep monitoring ensure the subject that their sleep assessment meets the standard parametric measurement along with raw data.

The sleep quality is analyzed by monitoring different sleep stages and their durations [17]. Sleep quality means total sleep time without fragmentation from wakefulness to deep sleep. It tells how well a person sleeps and not wake up whole night. It provides information of refreshment upon awakening, time awake, hours of sleep, number of arousals, and sound sleep without restless body movement. The qualitative sleep analysis gives us sleep efficiency.

Sleep Physiology is the study of different stages of sleep in living organisms. By monitoring and analyzing these sleep stages; one can detect the sleep quality. Normally sleep is divided into Rapid Eye Movement (REM) and Non-Rapid Eye Movement (NREM) sleep. The alternate NREM and REM stage completes one sleep cycle. In the absence of both sleeps, humans experience wakefulness stage. Hypnogram is used to represent various stages of sleep as a function of time and helps in qualitative analysis of sleep [18].

The qualitative analysis of sleep is performed by different methods based on various physiological parameters of body. By applying optimization technique, optimized sleep parameters can be found out so that all optimized key parameters help in the accurate qualitative sleep analysis [19]. The advantage of using optimization technique results in less number of parameters means less number of electrodes attached on the body, which make the process of assessment easy, unobtrusive, and cost-effective.

2 Qualitative Sleep Analysis Methods

A variety of sleep assessment techniques have been developed in an effort to meet the requirements of an ideal sleep analyzing method [20]. Various standard sleep scales based on self-administered questionnaire like Pittsburgh Sleep Quality Index

(PSQI), Epworth Sleepiness Scale (ESS) have been used by medical professionals for qualitative sleep analysis and polysomnography, ballistocardiography are clinical setups to diagnose the type of sleep disorders. Many indirect wearable and non-wearable methods are proposed by researchers in recent years to measure daily sleep quality. These techniques are based on vital parameters like heart rate, respiration rate, electro-oculogram, electromyography and actigraphy, etc. However, most of the commercially available methods do not provide brain wave information as sleep is a function of brain. This will result in limiting the performance of sleep monitoring system. The qualitative sleep can be assessed by using subjective, clinical, and commercial methods. The standard methods used for qualitative sleep analysis are:

2.1 Subjective Sleep Quality Assessment

It is a method of assessing person through questions and filling diaries individually. The two methods PSQI and ESS are very common for this assessment.

2.1.1 The Pittsburgh Sleep Quality Index (PSQI)

In this method, seven features of sleep are monitored for quality of sleep at night. It is comprised of 19 self-administered questions having score ranging from 0 to 21. A lower scale value indicates good sleep while a higher scale value implies the sleep deprivation [21].

2.1.2 The Epworth Sleepiness Scale (ESS)

It monitors eight features of sleep during day time having score range 0–24. The ESS scale results are close to clinical diagnosis done by MSLT and MSW test [22].

2.2 Clinical Sleep Quality Assessment

2.2.1 Polysomnography (PSG)

PSG Gold-standard clinical method executed under the guidance of medical professionals for monitoring multiple physiological parameters. These are Electroencephalography, Electrooculogram, Electrocardiogram, Blood Pressure, Electromyography, Restless Leg Syndrome (RLS), Periodic Leg Movements of Sleep (PLMS), Pulse Oximetry, Photoplethysmogram (PPG), Respiration Effort, Capnometry, Esophageal Tests, Nasal and Oral Airflow, Gastroesophageal Reflux Disease

(GERD), and Actigraphy. In PSG, person has to stay in hospital overall night wearing large number of various type of electrodes to record complex bioelectrical and mechanical signals from different parts of body which feels obtrusive in sleep. The recorded sleep pattern is used to find the type of sleep disorder. Hospital or clinic set up is required for this complex procedure. It is an expensive test and cannot be used frequently by patient for sleep monitoring and analyzing the sleep.

2.2.2 Ballistocardiography (BCG)

It is non-invasive continuous sleep monitoring method used to detect heartbeats along respiration rates due to vibrational activity of heart of patient. It is equivalent to electrocardiograph in terms of waves obtained in patterns but more expensive than ECG setup and has to perform in clinical environment under expert medical guidance [23].

2.2.3 Multiple Sleep Latency Test (MSLT)

This test is performed in sleep clinics in day time to check excessive sleepiness during the day. The medical professionals scheduled five naps to check the time of fall asleep and how much long you remain in bed during this nap. The sensors are attached on the face, chin, and head to monitor the time of sleep stages and video camera captured your sleep habit, behavior, and activities to help in understanding the sleep pattern. The recorded pattern will be examined by sleep team members, and the result of test will be handed over within 2 weeks. Doctors then prescribe the treatment according to the type of sleep disorder. This test is also expensive and cannot be used on regular basis [24].

2.2.4 Maintenance of Wakefulness Test (MWT)

This test is performed to diagnose sleep disorder to check your alertness during the day time in quiet dark room, and how much period you remain awake. This test is conducted in five sleep trials by putting sensors on your chin, face, and head to record your wake and sleep patterns. The patient remains awake for more than forty minutes in this 2-h sleep trial, the test ended and will start after a break under medical guidance. The medical setup is of same type as MSLT but it checks patient's alertness [25].

2.2.5 Continuous Positive Airway Pressure (CPAP) Titration Test

The Continuous Positive Airway Pressure (CPAP) titration test, also known as CTT, is a monitoring technique used to assess sleep. This test involves the use of a nasal mask that periodically adjusts the air pressure, while various sensors are employed to measure oxygen levels, breathing patterns, heart rate, brain waves, and movements of the limbs. Its purpose is to calibrate the optimal levels of continuous positive airway pressure (CPAP) for obstructive sleep apnea, and bi-level positive airway pressure (BIPAP) for central sleep apnea treatments. This determination is crucial in preventing blockage of the upper airway [26].

2.3 *Commercial Sleep Quality Assessment*

2.3.1 Videosomnography

In this sleep monitoring method, cameras are used to record and study the sleep behavior as well as sleep patterns of a person in its natural environment. The limitations with this method are installation of camera, proper positioning of camera to capture subject night sleep activities, and interference of signals. Women especially feel that this method does not maintain their privacy. The medical assistance is needed to analyze sleep behavior [27].

2.3.2 Actigraphy

In this method, sleep pattern is obtained from body movements due to sensors worn on wrist and ankle for prolonged period of time. This method is comfortable for all age group people and can be performed in clinic as well as home under the observation of medical expert in setting of device and understanding of actigraph to assess sleep measurement [28].

2.3.3 Interferometry

This method used laser to measure the variation in chest displacement during heartbeats and respiration and further processed in BCG manner to analyze the sleep pattern for sleep monitoring [29].

2.3.4 Force Sensor Measurements

In this method, unnoticeable force sensors are installed in the bed, pillow, t-shirts to measure heartbeats, respiration, and movements of body to analyze the quality and quantity of sleep of the user. The limitation is to take care of sensors attached to the sensing items for long-term use, required Wi-Fi arrangements to record sleep patterns and needed medical expert for data interpretation [30].

The recorded sleep patterns by different monitors help in identification of sleep disorders and their type. The sleep stages and it's time for people depend on their age and work nature. Monitoring the quality of sleep allows individuals to assess and enhance their overall fitness and well-being.

3 Sleep Parameter Allocation

Sleep is the function of entire body and whole body goes in resting process showing change in the signals generated during this resting mode. The monitoring of sleep can be possible by capturing signal from any part of the body whole night. The analysis of the signal and understanding the change in signal with the change in sleep cycle helps to give quality of sleep. Researchers of medical field studied a lot about sleep process and found certain important parameters to assess the quality and way to find sleep disorder by a laboratory test called PSG. In medical assistance method, this test and the parameters included in measuring the signals are known as gold standard. It is very complex clinical process of monitoring multiple parameters at a time by different channels using electrodes attached on the body of subject and at the same time quite difficult too for common user to understand and analyze the result obtained from PSG.

The PSG test includes the study of two things, first wake-sleep detection of an individual and second the study of sleep stages, which includes the duration of particular stage of sleep and changes that occur in these stages. The complementary information of individual subject is also obtained from the testing and measurement of multiple parameters. On the basis of this study, quantity, quality, and disorder of sleep are identified. PSG is time-consuming process, computational cost is also high and dependent on experts for interpretation of the results. This limitation is overcome by using parameter optimization method.

The sleep assessment techniques developed for qualitative sleep measurement works on either two, three, or five sleep stages. These are NREM1, NREM2, NREM3, REM, and awake stage. These techniques have selected PSG sleep parameters, namely, EEG, sometime in combination with EMG, EOG, ECG, and respiration rates to get information of sleep quality. Some of the most important sleep parameters allocated in the sleep stage classification are summarized [31] in Table 1. This system used mainly EEG in combination with EMG, EOG, or ECG and respiratory but the comparisons between these systems cannot be done since they differ in their state of the art, validation, and recording method.

The sleep sensing system can be broadly classified into two types; diagnostic and daily tracking system, which is further classified as contact or contactless techniques. The sleep detection method used by researchers differs in its technology; some have selected a single sleep parameter or a number of sleep parameters from PSG parameters to assess information of sleep quality. The main parameters allocated in diagnostic contact method are EEG, ECG, EMG, EOG, respiration, and snoring. In daily tracking contact system, parameters allocated are EEG, ECG, breathing rate,

Table 1 Allocation of important sleep parameters in classification of sleep stages

Serial Number	Sleep Parameters	Researchers/Year	Sleep Stages
1	EEG, EMG	Zoubek et al. / 2007	Wake, NREM1, NREM2, Slow wave sleep and REM
2	EEG (C3-A2), EMG and EOG	Tang et al. /2007	Awake, NREM1, NREM2, NREM3, NREM4, and REM
3	Single EEG (C3-A2),	Jo et al. /2010	Wakefulness, shallow sleep, deep sleep, and REM
4	EEG	Gunes et al. /2010	Wake, NREM1, NREM2, NREM3, and REM
5	Single EEG	Fraiwan et al., 2010	Wake, NREM1, NREM2, NREM3, NREM4, and REM
6	EEG, ECG, and respiratory signals	Helland et al. / 2010	Wake, NREM1, NREM2, NREM3, and REM
7	EEG (C3-A2), EMG, and EOG	Tagluk et al. / 2010	NREM1, NREM2, NREM3, NREM4, and REM
8	EEG, EMG	Chapotot and Becq/2010	Wake, NREM1, shallow NREM2, deep NREM3, REM
9	Single EEG (C3-A1)	Fraiwan et al. / 2011	Wake, NREM1, NREM2, NREM3, and REM
10	EEG (C3-A2and C4-A1), EMG, and EOG	Álvarez-Estévez et al. /2013	Awake, NREM1, NREM2, NREM3, NREM4, and REM
11	EEG, EOG, and EMG (F3-A2, C3-A2, O1-A2, F4-A1, C4-A1, O2-A1, ROC-A1, LOC-A2)	S. Khalighi et al./ 2013	Sleep–wake, multiclass sleep staging

and body motion while contactless system includes radar, accelerometer sensor for measuring breathing, heart rate, and body motion. The snoring parameter is also considered in contactless sleep assessment method. The clinical as well as commercial state of art techniques consider and select few parameters among sleep parameters of PSG method and compute sleep-related results.

The following sleep parameters of PSG procedure are considered as primary, for parameter allocation.

- Brain waves (EEG),

- Eye movements (EOG),
- Heart rate (ECG),
- Breathing pattern (capnometry),
- Blood oxygen level (pulse oximetry),
- Chest and abdominal movement (PPG respiratory effort),
- Body position (video and actigraphy),
- Limb movement (EMG and actigraphy),
- Snoring and other noise (audio).

These are our allocated parameters for optimization process. The number of parameters and sensors is large in PSG monitoring method, which makes this process complicated, uncomfortable, and costly. To reduce the process complexity and cost of qualitative sleep analysis, it is required to optimize parameters and digging out those which are having direct involvement in the sleep process. The aim is to eliminate parameters of PSG, which are correlated with others; the technique used is termed as elimination technique. The parameters obtained are optimized parameters from standard sleep parameters with less complexity of measurement.

4 Parameter Optimization Technique

Parameter optimization means to select best parameters from the set of available parameters. It saves the time by reducing error and cost. It provides safety and efficacy. Parameter of optimization is broadly classified as problem type and variable type. Mathematically, the variable type is further divided into independent or primary variables and dependent or secondary variables. Greater the number of variables in a given problem, greater will be the complications in the process of optimization. There will always be a relationship between these given variables of a specific problem. Once the relationship will be known then optimization techniques lead to best result of given complicated problem that's why parameter optimization technique is widely used in biomedical processes.

The qualitative analysis of sleep involves a greater number of parameters. Medical experts need to obtain information from various data sets. It is a complex task to perform without appropriate tools. Optimization techniques in biomedical analysis of signals become boon for practitioners and individual as it makes the analysis of only those signals which are necessary for the assessment of sleep quality.

Recently, we employed a novel parameter optimization technique for sleep analysis in which correlation analysis is performed between the various parameters of sleep [32]. This technique reliably finds positive and negative correlations between all dependent and independent variables involved in the sleep process. Such correlations are used to optimize the parameters from the given variable set. It could be shown that an optimized set of parameters is able to analyze the sleep quality in comparison to our PSG analyses.

The parameters directly involved in qualitative sleep assessment are first allocated and optimized by applying optimization technique. In optimization process, all parameters are considered as independent parameters. Among all, some parameters are independent key parameters while others are dependent parameters. The independent parameter cannot be dropped out in any case and value of dependent parameters can be derived from independent one. It shows that dependent parameter has close correlation with independent parameter. The dependent parameters are important but can be measured indirectly through independent parameter. The qualitative analysis parameter can be optimized by eliminating derived parameters (i.e. dependent one) from all sleep analysis parameters, the remaining independent parameters are termed as key parameter. This would result in simplified biomedical signal processing. These key parameters, which have been reduced through optimization, and the process of eliminating dependent parameters are collectively termed as the optimization method. The elimination technique based on their correlation helps to reduce measurement complexity for assessing the quality of sleep with less disturbances.

The flowchart below in Fig. 1 explains the process of optimization by finding correlation among parameters.

The method of optimization includes certain steps:

i. First the selection of widely used and accepted parameters by medical experts, researchers, and commercial developers to obtain sleep information.
ii. Secondly, this parameter directly involved in sleep quality measurement by providing clear information of sleep stages and its changes in whole night.
iii. Third, how many other parameters' information can be related and obtained from these selected parameters in previous step.
iv. Fourth to find out the correlation factor among selected and other parameters of sleep.
v. In the, last, fifth step, that parameter is considered as optimized key parameter for sleep quality assessment, which has close correlation value with most of the other ones and is used in qualitative sleep analysis method.

On the basis of the above steps, literature reveals that EEG, ECG, EOG, and EMG are major parameters that have been widely used by medical experts and researchers to study the quality of sleep of a person. Temperature, blood flow, and actigraphy are also used in the combination with major parameters to assess the sleep quality. The primary concern is to find those parameters which have direct involvement in the identification of sleep stages. Here, the two parameters, temperature and actigraphy, do not show direct relation with sleep. These parameters can be eliminated from the list of primary parameters on the basis that temperature changes according to the climatic conditions so body temperature shows variation and on the other hand actigraphy works on body movement also not give any direct information of sleep. In the optimization process, temperature and actigraphy parameters are eliminated, and the other parameters EEG, ECG, EOG, EMG, and blood flow be considered as primary parameters to provide direct information of sleep.

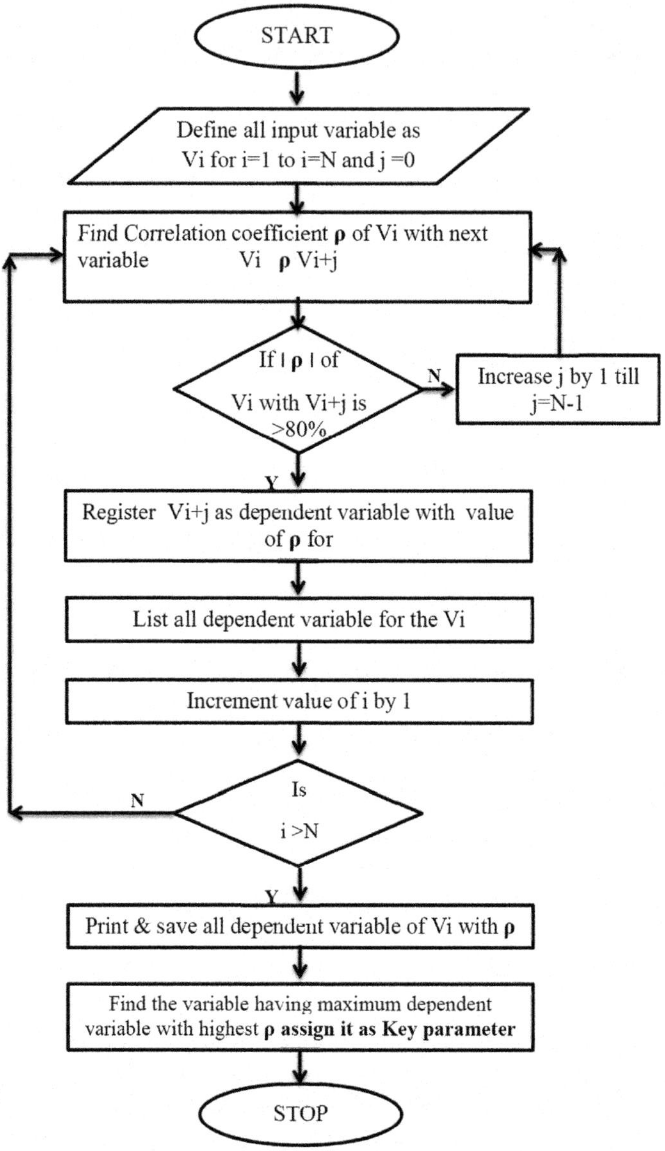

Fig. 1 The flowchart of optimization technique to find the correlation among various parameters

The human brain produces electrical activity across a wide range of frequencies, which can be measured using EEG. Brain waves are electrical impulses generated by the synchronized activity of neurons in the brain. They can be categorized into different frequency bands, namely beta, alpha, theta, and delta. Understanding these brain wave signals is crucial for various applications such as sleep diagnosis and analysis. Beta waves are high-frequency brain waves that typically range from 13 to 30 Hz. They are predominantly observed in awake and alert states. Alpha waves are relatively slower brain waves ranging from 8 to 13 Hz. They are commonly observed when an individual is awake but relaxed, with closed eyes. Theta waves are slower brain waves with a frequency range of 4–8 Hz. They are predominantly observed during states of deep relaxation, daydreaming, and light sleep. Delta waves are the slowest brain waves, typically ranging from 0.5 to 4 Hz. They are most commonly observed during deep sleep and unconscious states. Understanding the characteristics and functions of beta, alpha, theta, and delta brain waves has significant implications across various domains. The comprehensive understanding of these brain wave signals can contribute to advancements in neuroscience research, clinical diagnosis, and therapeutic interventions.

Our study of parameter allocation concludes that EEG is an important parameter, as sleep is defined as function of brain and is also used by researchers in development of new sleep detection tool either single or in combination with other bio-signals. In the first step of optimization, EEG is considered as key independent parameter. This parameter is significantly used for identification of sleep stages in PSG as it directly provides the information of various sleep stages by capturing beta, alpha, theta, and delta brain wave signals. The main information of various sleep stages along with sleep–awake state is obtained by EEG waveforms. Secondly, all the activity of body is controlled by nervous system so EEG signals are not pure and have information of other signals referred as physiological artifacts as shown in Fig. 2.

These artifacts are muscle activity like EMG and ECG, eye movement, blinking signals, and movement of head. This means that recorded EEG signal of subject shown in the above figure not only contains the pure EEG signals but also the information of other bio-signals like EOG, EMG and ECG too. The EEG waveform is a complex signal that also contains the information about the movement of limbs, snoring signals, and changes in body positions during the sleep. The body movement during sleep shows the change in both EEG signal as well as EMG signals, which is

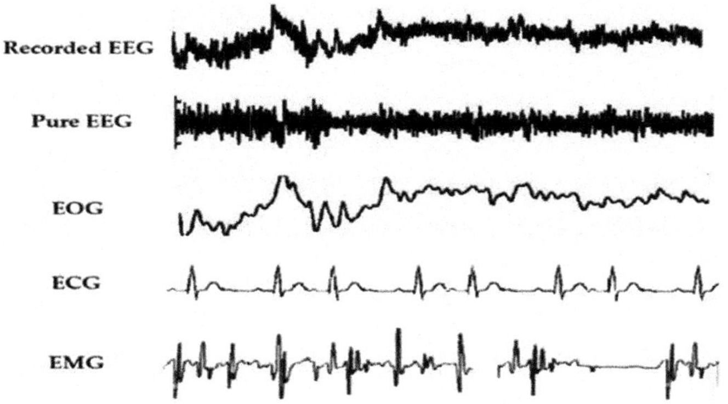

Fig. 2 Physiological artifacts in recorded EEG signals

Table 2 Relationship of EEG primary parameter with derived parameter

Parameter	Electromyography (EMG)	Snoring
EEG	√	√

known as motion artifact. The EEG signals are in well correlation with EMG signals [33] as well as that of snoring signals [34]. The EEG is our first optimized parameter. The EMG and snoring information can be obtained from EEG signals. The parameter EEG has relationship with EMG and snoring as shown in Table 2.

The first optimized parameter is EEG that has direct involvement in sleep analysis.

The second important independent parameter is EOG. There is complete involvement of eyes in sleep process. The person can be stated clearly in awake or sleep state by observing the physical status of eyes. Eyes have direct relation with sleep stages. The NREM and REM information can be obtained by observing the movement of eyes of a person. There is rapid eye movement in dreaming stage and no movement in deep sleep stage. The movement of eyes comes under muscular activity. EOG is a subpart of EMG and shows close correlation with EMG. Confirmative information of non-rapid eye movement (NREM) and rapid eye movement (REM) will be obtained by EOG signals. The second independent and optimized key parameter is EOG for sleep analysis. The parameter EOG has relationship with only EMG as shown in Table 3.

Heart, lungs, and circulatory system of a human body refer as cardiovascular system. The rate and regularity of heartbeats are measured by ECG signal. The third important key parameter is ECG as heart rate gets changed during different sleep stages. Sleep physiology shows that heart rate becomes slower as person entered in the deep sleep stage. The blood pressure (BP), respiration rate (RR), heart rate, and oxygen saturation are in close relation with ECG signals. There is inverse relation between oxygen saturation and heart rate. On the other hand, respiration rate is in direct relation with heart rate [35]. The experimental study of various clinical group of Anoxia shows that the fall in arterial oxygen saturation level increases the heart rate as shown in Fig. 3 [36]. This means that as the percentage of oxygen decreases in the air, the heart rate of subjects under study increases, which results in increase in respiration rate of patients.

Respiration rate and heart rate are in direct relationship with each other. When the oxygen saturation level in the body decreases, it demands more oxygen, which in turn increases the heart rate and results in an increase in respiration rate. The experimental study of heart rate obtained by general oximeter as well as by standard ECG method shows significantly correlation with each other as represented in Fig. 4 [37]. It shows positive correlation with high correlation factor value. It means that

Table 3 Relationship of EOG primary parameter with derived parameter

Parameter	Electromyography (EMG)
EOG	√

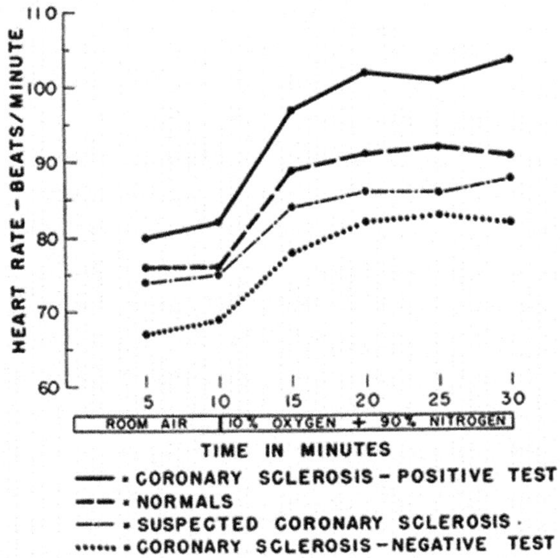

Fig. 3 Fall in the arterial oxygen saturation in various clinical group of Anoxia shows an increase value of heart rate [36]

heart rate value measured by oximeter results same as that of measured by ECG signals.

The respiratory rates obtained from chest strap are compared with indirect measurement of respiration rate obtained from the electrocardiogram (ECG) as shown in Fig. 5 [38]. This shows close linear correlation between respiratory rates estimated from ECG and chest strap by using autocorrelation and zero-crossing method. This implies that respiration rate can also be measured by ECG signals.

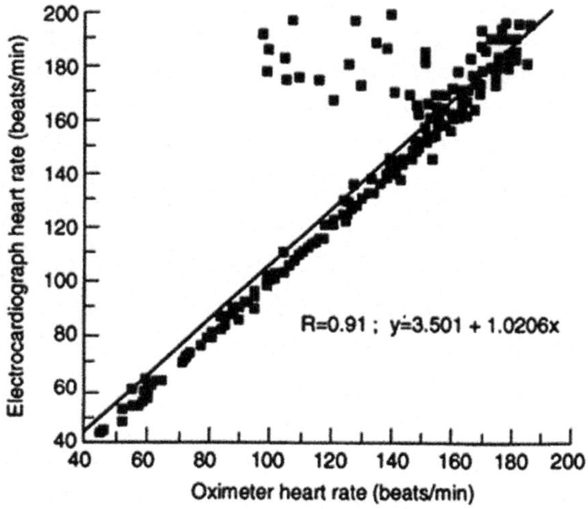

Fig. 4 The relation of ECG reading of heart rate to the general oximeter reading [37]

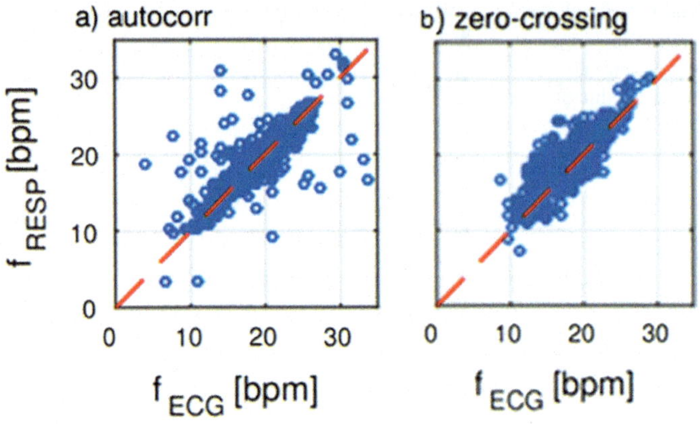

Fig. 5 The respiration-derived respiratory rate and ECG-derived estimated respiratory rate correlation curve [38]

The parameter ECG has relationship with blood pressure, oxygen saturation level, respiration rate, and heart rate as shown in Table 4.

According to the hemodynamic relationship [39].

$$F = \Delta P / R$$

where,

F = flow of blood in a blood vessel

ΔP = driving pressure, perfusion pressure, or pressure gradient

R = the resistance offered by the blood vessel.

The above relation shows that blood flow is directly related with blood pressure (BP). It is difficult to get direct measurement of blood pressure (BP) from ECG signal. ECG Derived Respiration (EDR) method is used to measure BP and pulse transit time [40]. The oximetry, BP, and blood flow are in close correlation with each other. Further, blood flow has direct relation with brain activity. The study of sleep shows that during sleep stage II cerebral blood flow (CBF) is 3–10% below the level compared to awake state, whereas 20–45% fall is shown during sleep stage III and IV [41]. On the other hand, as shown in Fig. 6, there is an increase in blood flow and blood pressure when there is extreme effect of oxygen deficiency in the body. There is a linear relationship between blood flow and perfusion pressure. BP increases, blood flow also increases but at the same time there is a decrease in oxygen saturation

Table 4 Relationship of ECG primary parameter with derived parameter

Parameter	Heart Rate (HR)	Respiration Rate (RR)	Oxygen saturation level (O_2)	Blood Pressure (BP)
ECG	√	√	√	√

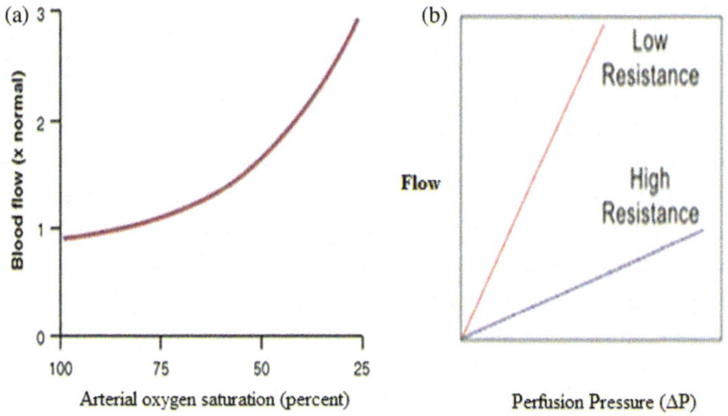

Fig. 6 **a** Relation between blood flow and oxygen saturation and **b** Relation between blood flow and perfusion pressure

Table 5 Relationship of CBF primary parameter with derived parameter

Parameter	Oxygen saturation level (O_2)	Blood Pressure (BP)
CBF	√	√

level. These parameters can also be calculated by measuring any one parameter among them.

The CBF is in inverse relationship with cortical oxygenation whereas mean arterial pressure is directly proportional with CBF [42]. The accurate measurement of BP can be carried out with the help of both ECG and cerebral blood flow (CBF) signals. The parameter CBF has relationship with BP and $\mathbf{O_2}$(Oxygen saturation level) as shown in Table 5.

In this way, each and every parameter has been studied and checked as independent or dependent parameter. The relationship of individual parameter with one or more parameters has been found. These key parameters have direct or indirect relationship with dependent parameters that have been studied so far, based on their correlation values with one another. The most significant parameters that have relation with derived parameters are selected and called as optimized parameter.

5 Result Analysis

The proposed technique for parameter optimization aims to reduce the number of parameters by identifying key parameters and eliminating dependent parameters that exhibit correlation. The qualitative sleep assessment by this method differs from both, commercially available devices based on dependent and dynamic parameter sensing and clinical complex (PSG) in hospitals which uses multiple parameter monitoring.

The PSG method and its parameters are gold standard for qualitative sleep analysis. In the PSG process, multiple parameters are monitored, and a number of sensors are connected making the assessment procedure complicated and costly. The optimized parameters that show direct involvement in the sleep process provide accurate information about sleep and its stages, while also simplifying the complexity of the process. In the proposed optimization technique, PSG parameters are used and optimized to get information about sleep. Every sleep parameter of PSG is treated as variable and checked to find it as a primary or derived parameter. The optimized parameters obtained from the experimental values of the correlation factor among various PSG parameters include:

- Electroencephalography (EEG)
- Electrooculogram (EOG),
- Electrocardiogram (ECG),
- Cerebral blood flow (CBF).

The above parameters show strong correlation with other derived parameters, and hence the derived one can be eliminated from measurement process to make system unobtrusive. This reduces complexity in biomedical signal processing for qualitative sleep analysis.

The CBF can be used to provide information about blood pressure and oxygen saturation. The heart rate, respiration rate, oxygen saturation information, and blood pressure can be found out with ECG parameter. EOG gives the information of eye muscle activity. EMG and snoring information can be obtained from EEG parameter.

During sleep, brain activity changes and easily noticed by EEG waveforms in finding sleep stages. The confirmation of particular sleep stage will be obtained by EOG, ECG, and CBF [43]. These parameters show an increase or decrease in their value during sleep cycle and help in providing information regarding different stages of sleep. The above four significant parameters for sleep assessment can provide the accurate results that are close to the qualitative sleep analysis at hospitals for sleep monitoring. Monitoring system on these optimized parameters is more precise as compared to other commercial monitors that are based on heart rate, actigraphy, body temperature, respiration rate signals, and EEG signals only. The parameter optimization results are consolidated in Table 6 showing close relation of key parameter with dependent parameters.

Table 6 Relationship of optimized key parameter with dependent parameter

Parameter	Heart Rate (HR)	Respiration Rate (RR)	Oxygen saturation level (O_2)	Electromyography (EMG)	Snoring	Blood Pressure (BP)
EEG	X	X	X	√	√	X
EOG	X	X	X	√	X	X
ECG	√	√	√	X	X	√
CBF	X	X	√	X	X	√

Table 7 Key parameter and derived parameter with correlation factor value

Parameter	Respiration Rate (RR)	Snoring	Electromyography (EMG)	Heart Rate (HR)	Blood Pressure (BP)	Oxygen saturation level (O_2)
EEG	X	0.8	0.8	X	X	X
EOG	X	X	0.9	X	X	X
ECG	0.9	X	X	0.9	0.9	−0.9
CBF	X	X	X	X	0.9	−0.8

The optimized key parameter with dependent parameter correlation factor value is shown in Table 7 [44]. It provides information that selected parameters, EEG, EOG ECG, and cerebral blood flow, are key parameters with the help of these we can derive the values of dependent parameters.

6 Conclusion

The main aim of our chapter was to optimize the sleep parameters used in sleep clinics and daily sleep monitoring system by using parameter optimization technique based on correlation analysis of primary and dependent parameters. In this method, PSG sleep parameters for analysis of sleep are considered for optimization. Using optimization technique, we obtained four optimized parameters from the set of 10 parameters used in standard sleep analysis method. The first step towards the qualitative sleep assessment is to allocate the sleep parameters involved in the method of sleep measurement. The second step is to establish the correlation based on experimental results obtained for various sleep parameters. These optimized parameters make sleep analysis system simple and unobtrusive. We obtained Electroencephalogram (EEG), Electrooculogram (EOG), Electrocardiogram (ECG), and Cerebral Blood Flow (CBF) as four key parameters in order to develop sleep monitor while derived parameters are eliminated to reduce complexity of the system. These optimized parameters are significant and provide accurate qualitative sleep analysis as carried out during clinical sleep monitoring. When comparing the assessment derived from this technique to other commercially available daily sleep monitors that rely on one or two parameters, it becomes evident that the former offers a higher level of accuracy.

References

1. Adil H, Koser AA, Gupta A (2019) Importance of sleep for medical and general wellness. Res J Pharmacy Technol 12(12):5827–5830. https://doi.org/10.5958/0974-360X.2019.01009.6
2. Partinen M (2021) Sleep research in 2020: COVID-19-related sleep disorders. Lancet Neurol 20(1):15–17. https://doi.org/10.1016/S1474-4422(20)30456-7

3. Ibáñez V, Silva J, Cauli O (2018) A survey on sleep assessment methods. PeerJ 6:e4849
4. Markun LC, Sampat A (2020) Clinician-focused overview and developments in polysomnography. Curr Sleep Med Rep 6:309–321. https://doi.org/10.1007/s40675-020-00197-5)
5. Siyanbade J, Abdulrazak B, Sadek I (2022) Unobtrusive monitoring of sleep cycles: a technical review. BioMedInformatics 2(1):204–216. https://doi.org/10.3390/biomedinformatics2010013
6. Sharma M, Jainendra T, RajendraAcharya U (2021) Automatic sleep-stage scoring in healthy and sleep disorder patients using optimal wavelet filter bank technique with EEG signals. Int J Environ Res Public Health 18(6):3087. https://doi.org/10.3390/ijerph18063087
7. Tăuţan AM, Rossi AC, de Francisco R, Ionescu B (2020) Dimensionality reduction for EEG-based sleep stage detection: comparison of autoencoders, principal component analysis and factor analysis. Biomed Tech (Berl) 66(2):125–136. https://doi.org/10.1515/bmt-2020-0139
8. Shehnas S, Anas AS (2022) Independent component analysis for biomedical signal separation using lambert's continued fraction. Int J Eng Res Technol 11(6)
9. Adhikary S, Ghosh A (2022) Dynamic time warping approach for optimized locomotor impairment detection using biomedical signal processing 72:103321. https://doi.org/10.1016/j.bspc.2021.103321
10. Semhar B, Ogbagaber YC, Li K, Iannotti RJ, Albert PS (2022) A hidden Markov modeling approach combining objective measure of activity and subjective measure of self-reported sleep to estimate the sleep-wake cycle. J Appl Stat. https://doi.org/10.1080/02664763.2022.2151576
11. Venkatesan A (2017) Optimized clustering techniques with special focus to biomedical datasets. https://doi.org/10.4018/978-1-5225-0660-7.ch015
12. Kavita P, Alli DR, Annepu BR (2022) Study of image fusion optimization techniques for medical applications. Int J Cogn Comput Eng 3:136–143.https://doi.org/10.1016/j.ijcce.2022.05.002
13. Li H, Lin Z, An Z, Zuo S, Zhu W, Zhang Z, Mu Y, Cao L, PradesGarcía JD (2022) Automatic electrocardiogram detection and classification using bidirectional long short-term memory network improved by Bayesian optimization. Biomed Signal Process Control 73:103424. https://doi.org/10.1016/j.bspc.2021.103424
14. Akram MA, Liu P, Tahir MO, Ali W, Wang Y (2019) A state optimization model based on Kalman filtering and robust estimation theory for fusion of multi-source information in highly non-linear systems. Sensors 19(7):1687. https://doi.org/10.3390/s19071687
15. Adil H, Koser AA, Gupta A (2021) A qualitative study on commercially available: sleep monitoring systems and sleep analysis tools. Res J Pharmacy Technol 14(1):413–419. https://doi.org/10.5958/0974-360X.2021.00075
16. Pan Q, Brulin D, Campo E (2020) Current status and future challenges of sleep monitoring systems: systematic review. JMIR Biomed Eng 5(1):e20921
17. Xue B, Deng B, Hong H, Wang Z, Zhu X, Feng DD (2020) Non-contact sleep stage detection using canonical correlation analysis of respiratory sound. IEEE J Biomed Health Inform 24(2):614–625. https://doi.org/10.1109/JBHI.2019.2910566
18. Van der Plas D, Verbraecken J, Willemen M, Meert W, Davis J (2021) Evaluation of automated Hypnogram analysis on multi-scored Polysomnographies. Front Digit Health 26(3):707589. https://doi.org/10.3389/fdgth.2021.707589
19. Adil H, Koser AA, Gupta A (2020) A unique unobtrusive intelligent sleep monitoring (ISM) method for sleep analysis. AIP conference proceedings, 2224 (060003), 060003–1–060003–4, https://doi.org/10.1063/5.0000653
20. Fabbri M, Beracci A, Martoni M, Meneo D, Tonetti L, Natale V (2021) Measuring subjective sleep quality: a review. Int J Environ Res Public Health 18(3):1082
21. Zitser J, Allen IE, Falgàs N, Le MM, Neylan TC, Kramer JH, Walsh CM (2022) Pittsburgh Sleep Quality Index (PSQI) responses are modulated by total sleep time and wake after sleep onset in healthy older adults. PLoS ONE 17(6):e0270095. https://doi.org/10.1371/journal.pone.0270095
22. Scharf MT (2022) Reliability and efficacy of the epworth sleepiness scale: is there still a place for it?. Nat Sci Sleep, 2151–2156.https://doi.org/10.2147/NSS.S340950

23. Mitsukura Y, Sumali B, Nagura M, Fukunaga K, Yasui M (2020) Sleep Stage Estimation from bed leg Ballistocardiogram sensors. Sensors 20(19):5688. https://doi.org/10.3390/s20195688
24. Goddard J, Tay G, Fry J, Davis M, Curtin D, Szollosi I (2021) Multiple sleep latency test: when are 4 naps enough? J Clin Sleep Med 17(3):491–497. https://doi.org/10.5664/jcsm.8976
25. Bijlenga D, Overeem S, Fronczek R, Lammers GJ (2022) Usefulness of the maintenance of wakefulness test in central disorders of hypersomnolence: a scoping review. Sleep 45(8):zsac091. https://doi.org/10.1093/sleep/zsac091. PMID: 35429267
26. Wilckens KA, Jeon B, Morris JL, Buysse DJ, Chasens ER (2022) Effects of continuous positive airway pressure treatment on sleep architecture in adults with obstructive sleep apnea and type 2 diabetes. Front Hum Neurosc 16:924069. https://doi.org/10.3389/fnhum.2022.924069
27. Lee J, Clark PC, Spratling R (2023) Transitioning a research protocol for Videosomnography to assess sleep and Nighttime caregiving activities in school-aged children with developmental disabilities during the COVID-19 pandemic. J Pediatr Health Care 37(2):133–136. https://doi.org/10.1016/j.pedhc.2022.11.001
28. Fekedulegn D, Andrew ME, Shi M, Violanti JM, Knox S, Innes KE (2020) Actigraphy-based assessment of sleep parameters. Ann Work Expo Healt 64(4):350–367. https://doi.org/10.1093/annweh/wxaa007
29. Jaros R, Nedoma J, Kepak J, Martinek R (2022) Fiber-optic interferometry-based heart rate monitoring. IEEE Trans Instrum Meas 71:1–15. https://doi.org/10.1109/TIM.2022.3178495
30. Haghi M, Asadov A, Boiko A, Ortega JA, Madrid NM, Seepold R (2023) Validating force sensitive resistor strip sensors for cardiorespiratory measurement during sleep: a preliminary study. Sensors 23(8):3973. https://doi.org/10.3390/s23083973
31. Khalighi S, Sousa T, Pires G, Nunes U (2013) Automatic sleep staging: a computer assisted approach for optimal combination of features and polysomnographic channels. Expert Syst Appl 40(17):7046–7059
32. Adil H, Koser AA, Qureshi SM, Gupta A (2021) Sleep quality assessment by parameter optimization. J Phys Conf Ser 2070(012013). https://doi.org/10.1088/1742-6596/2070/1/012013
33. Liang H, Yu Y, Mochida M, Liu C, Ueda N, Li P, Zhu C (2020) EEG-based EMG estimation of shoulder joint for the power augmentation system of upper limbs. Symmetry 12(11):1851
34. Alencar AM, Silva DGV, Oliveira CB, Vieira AP, Moriya HT, Filho GL (2013) Dynamics of snoring sounds and its connection with obstructive sleep apnea. Physica A: Stat Mech Appl 392 (1):271–77
35. Amare YE, Haile D (2020) Evaluation of pulmonary function tests among pregnant women of different trimesters in debreberhan referral hospital, shoa, Ethiopia. Int J Women's Health 12:1135–1143
36. Mathers JAL, Levy RL (1950) Correlation of the oxygen saturation of the blood and changes in the electrocardiogram, blood pressure and heart rate during the anoxemia test. Circulation AHA J 1(3):426–32
37. Iyriboz Y, Powers S, Morrow J, Ayers D, Landry G (1991) Accuracy of pulse oximeters in estimating heart rate at rest and during exercise. Br J Sports Med 25(3):162–164
38. Schrumpf F, Sturm M, Bausch G, Fuchs M (2016) Derivation of the respiratory rate from directly and indirectly measured respiratory signals using autocorrelation. Current Direct Biomed Eng 2(1):241–245
39. Richard E, Klabunde (2007) Cardiovascular physiology concepts 3rd edition text book published by Wolter Kluwer
40. He X, Goubran RA, Liu XP (2013) Evaluation of the correlation between blood pressure and pulse transit time. IEEE international symposium on medical measurements and applications (MeMeA). https://doi.org/10.1109/MeMeA.2013.6549697
41. Madsen PL, Vorstrup S (1991) Cerebral blood flow and metabolism during sleep. Cerebrovasc Brain Metab Rev 3(4):281–296
42. Lucas SJE, Tzeng YC, Galvin SD, Thomas KN, Ogoh S, Ainslie PN (2010) Inuence of changes in blood pressure on cerebral perfusion and oxygenation. Hypertension 55(3):698–705

43. Adil H, Koser AA, Gupta A (2020) Sleep monitoring using laser blood flowmeter. IEEE Xplore.https://doi.org/10.1109/SCEECS48394.2020.71
44. Marilyn J, Cipolla (2009) The cerebral circulation. Text book Morgan and Claypool Life Sciences Publishers

Chapter 3
Renewable Energy Optimization Solutions Using Meta-heuristics Methods

Santosh S. Raghuwanshi and Animesh Masih

Abstract Optimization is a systematic, mathematical, and technically developed technique of selecting the most feasible solution for a given problem under defined constraints and assumptions. Optimization handles the function with several variables of the prescribed problem for maximizing or minimizing it, subject to data, model constants, and parameters describing the objective function. Many design strategies are multiplex and are problematic to be solved by employing conventional techniques of optimization. An optimization algorithm produces the result in the form of an optimal solution if an optimal solution exists. The result obtained is interpreted and assessed, which may lead to modifications of the model and its pertinence of the optimal solution. For the purpose of choosing the best hybrid renewable energy system (HRES), this chapter introduces the metaheuristic-based grasshopper optimization algorithm (GOA) and results are validated by particle swarm optimization (PSO). PSO is a method of applying swarm intelligence while GOA is a method of applying population-based metaheuristic that can be off to find the solution to tough multiple problems. GOA and PSO can accomplish the search problem in obtaining an optimal solution by studying the communal deeds of the alive creatures. The HRES includes photovoltaic (PV) and wind turbine (WT) with storage, i.e., battery as a backup source. The four possible energy system configurations have been used for the selected area (22.58°N and 76.01°E). The four energy system configurations considered in this work are PV_battery, PV_WT, WT_battery, and PV_WT_battery. HRES, technical feasibility as well as economic feasibility is important. Therefore, optimization techniques considering both these aspects must be applied before actually implementing them. The cost of energy (COE) and loss of power supply reliability index (LPSRI) are minimal, this is the main objective function of this chapter. The results are validated by using genetic algorithm (GA) and hybrid optimization of multiple energy resources (HOMER) software. GA is based on evolutionary computation. The optimal methods reflect PV_WT_battery energy system configuration as an optimal solution of the located area. The results obtained from GOA are better

S. S. Raghuwanshi (✉) · A. Masih
Department of Electrical Engineering, Medi-Caps University, Indore, India
e-mail: santoshsraghuwanshi@gmail.com

U. N. Dulhare and E. H. Houssein (eds.), *Machine Learning and Metaheuristics: Methods and Analysis*, Algorithms for Intelligent Systems,
https://doi.org/10.1007/978-981-99-6645-5_3

for the optimal configuration hybrid PV_WT_battery energy system as compared to others.

Keywords GA · HOMER · Hybrid energy system · NPC · PSO · Reliability index

1 Introduction

While ease of use and availability are important factors in choosing an HRES for remote areas, it is equally important to conduct an economic feasibility assessment. Economic assessment involves identifying, calculating, and valuing the costs and benefits of a proposal to determine its overall value and compare it to other alternatives. Optimization techniques can also be used to minimize the economic cost of the system, analyze operating costs, calculate payback periods, and ensure long-term cost savings. Jahannoosh et al. [1] present that the selection of a suitable HRES for remote areas should not rely solely on ease of use and availability, but should also consider economic feasibility. This involves determining the costs and benefits of the proposed system, comparing it with alternatives, and using optimization techniques to minimize economic cost, analyze operating costs, and ensure long-term savings. One study used the hybrid grey wolf optimizer-sine cosine algorithm to evaluate the economic feasibility and reliability constraints of different system configurations [2, 3]. Another study proposed the Jaya Learning-Based Optimization algorithm to identify economic and reliability parameters for HRES. Additionally, the Manta ray foraging optimizer algorithm was used to evaluate the techno-economic parameters of an HES located in remote areas and compared with PSO and big-bang-big-crunch algorithm [4].

Oumaima et al. [5] presented a Numerical Heuristic Searching Algorithm, practiced in estimating parameters of photovoltaic segments. Hazara et al. [6] presented a GOA used in the cost-effective process of a hybrid energy system (windmill, thermal power). GOA selects the optimum number of energy systems, obligatory for the load. The outcomes of GOA were equated with the quantum-inspired PSO method. The outcomes of GOA are apposite for resolving partially shaded photovoltaic array reconfiguration course. Results were validated for the total cross-tied connection arrangement. Fathy et al. [7] presented an augmented photovoltaic module using a whale optimizing algorithm in standard and shading circumstances.

Hichem et al. [8] propose a binary version of GOA as a second stratagem-based element sub-set selection for COE. Various algorithms like binary grey wolf optimization approach, novel binary PSO, and binary dragonfly optimization were used to compare their results.

This chapter works to help with the selection of optimal energy configuration in terms of economic feasibility is done by applying GA, PSO, and GOA. The HOMER simulation tool is also applied for analyzing the economic feasibility of HRES.

2 Swarm Intelligence Optimization for Economic Feasibility Analysis

Swarm intelligence optimization draws inspiration from the joint behavior of birds, where a cluster of individuals work together to solve problems. In the context of computing, a swarm consists of a population of representative agents that collaborate with each other to generate coherent global patterns. This approach allows for decentralized problem-solving without relying on centralized control. Swarm intelligence optimization is broadly used in the economic optimization of HRES, where it helps to optimize the economic feasibility and reliability of different system configurations.

2.1 Objective Function

In order to achieve optimal sizing of HRES, the focus is on minimizing the objective function. The objective function of HRES to reduce the net present system cost (C_{NPSC}), COE (C_{COE}), and LPSRI.

$$\text{Minimizing}\, C_{NPSC}, C_{COE}, LPSRI \tag{1}$$

2.2 System Cost Optimization Model

The cost of the entire system comprises capital costs (C_C), operation and maintenance (C_{OM}), and replacement cost (C_{RC}) of each component which is computed as [9, 10].

$$C_{TSC} = C_C + C_{OM} + C_{RC} \tag{2}$$

NPSC is the ratio of total system cost to capital recovery factor (CRF). The NPC is as follows:

$$C_{NPSC} = \frac{C_{TSC}}{CRF(d, t)}$$

computing,

$$C_C = (N_{pv}C_{pv}) + (N_{wind}C_{wind}) + (N_{bat}C_{bat}) + (N_{inv}C_{inv}) + (N_{con}C_{con}) \tag{3}$$

$$C_{OM} = (N_{pv}C_{pv} + N_{wind}C_{wind} + N_{bat}C_{bat} + N_{inv}C_{inv} + N_{con}C_{con}) \sum_{t=1}^{LS} \left(\frac{1+i}{1+r}\right)^t \tag{4}$$

The life span of batteries, inverters, and converters is less and therefore need replacement during the life span of the system. Hence, these components are considered in the replacement segment and are computed as [11].

$$C_{RC} = (N_{bat}C_{bat} + N_{inv}C_{inv} + N_{con}C_{con})\sum_{t=1}^{LS}\frac{i}{\left((1+i)^t\right)-1} \quad (5)$$

CRF is as follows:

$$CRF(d,t) = \frac{d(1+d)^t}{(1+d)^t - 1}$$

COE is as follows [12]

$$C_{COE} = \frac{C_{TSC}}{L_{YLS}} \quad (6)$$

where

N_{pv}	No. of PV modules
N_{wind}	N0. of WT
N_{inv}	No. of Inverter
N_{con}	*No.of* converters
N_{bat}	No. of batteries
C_{pv}	Cost of PV module
C_{wind}	Cost of WT
C_{inv}	Cost of inverter
C_{con}	*Costof* converter
C_{bat}	Cost of battery
i	Inflation rate
r	Interest rate
d	Annual discount rate
L_{YLS}	Yearly load served by system
t	Component's life span (LS) (25 yrs is considered in this study)

2.3 Loss of Power Supply Reliability Index

LPSRI is the ratio of existing electricity to the possibility of letdown of HRES and its components. LPSRI is computed is [13, 14]

$$LPSRI = \frac{\sum_{t}^{8760}\left[P_l(t) - (P_{pv}(t) + P_{wind}(t) + P_{bat}(t))\right]}{\sum_{t}^{8760} P_l(t)} \tag{7}$$

where

P_l load power demand,
P_{pv} PV panel power,
P_{wind} WT power,
P_{bat} battery power.

2.4 Energy System Constraints

The desirable constraints for optimization are

$$0 \leq N_{wind} \leq N_{wind}^{max} \tag{8}$$

$$0 \leq N_{pv} \leq N_{pv}^{max} \tag{9}$$

$$0 \leq N_{bat} \leq N_{bat}^{max} \tag{10}$$

$$0 \leq N_{inv} \leq N_{inv}^{max} \tag{11}$$

$$0 \leq N_{con} \leq N_{con}^{max} \tag{12}$$

$$LPSRI \leq LPSRI_{desireable} \tag{13}$$

where

$N_{wind}^{max}, N_{pv}^{max}, N_{bat}^{max}, N_{inv}^{max}$ *and* N_{con}^{max} are a maximum number of WT, PV, batteries, inverter, and converter, respectively.

3 Particle Swarm Optimization

PSO is a speculative optimistic approach encouraged by nature, stimulated by communal deeds of bird congregation. The structure is reconfigured with a populace of random results and the quest for optimal goals by updating generations [15].

Fig. 1 Swarm of birds [16]

In PSO, the budding solution termed as particle moves over the problem space by succeeding in the present optimum units. If one perceives a needed pathway to move toward food or shield, the remaining swarm drives themselves to follow it rapidly even if some are on the reverse sideways of the swarm. This action is accomplished by the particle spaced in multi-dimension having a position and velocity. The particles hover through hyper-space with two crucial proficiencies: their best position owing to their memory and familiarity with their best swarm position. Affiliates of the swarm transfer the upright position to other swarms and they try to alter their position and velocity grounded on this upright location. The behavior exhibited by the swarm of birds is displayed in Fig. 1.

The PSO applies its basic concept of changing the position of each accelerating particle toward its p-best (fitness value) with g-best (location to neighbors) location. Accelerating is subjective to random span with discrete random statistics being created for accelerating toward p-best (fitness value) and g-best location [17].

3.1 Maneuvering Target Using PSO

Let us consider a swarm of units hovering in the space parameters while trying to attain its optimum center. Each particle is categorized by location vector as $x_i(t)$ and velocity vector as $v_i(t)$ [18], This is displayed in Fig. 2.

During the progression, individual particle drive has the information of p-best. It will also have the information of best location and social familiarity of g-best of its own neighborhood. The velocity constraint is updated by substituting the values in the formula [19],

$$v_i(t+1) = \alpha v_i + c_1 \times rand \times (pbest(t) - x_i(t)) + c_2 \times rand \times (gbest(t) - x_i(t)) \quad (14)$$

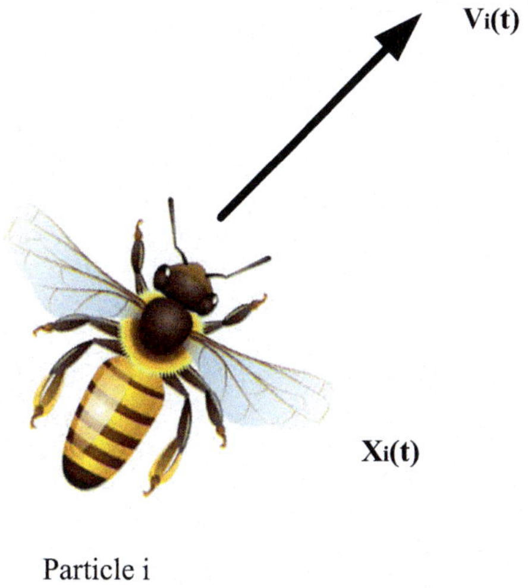

Fig. 2 Position and velocity vector of a particle [17, 18]

where

α	inertia weight that pedals the survey of the search space,
c_1	Cognition components,
c_2	Social components,
rand	a random number between 0 and 1.

The velocity updates of the particles built on the constraints of different parameters are displayed in Fig. 3.

The position updates of the particles are

$$x_i \times (t+1) = x_i \times (t) + v_i \times (t+1) \tag{15}$$

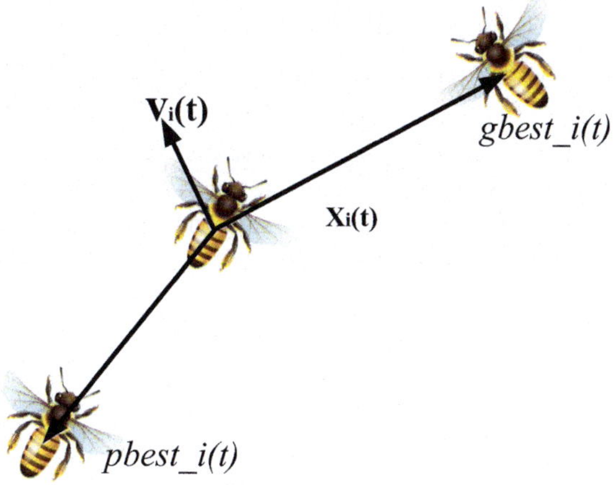

Fig. 3 Parameters for velocity update [19, 20]

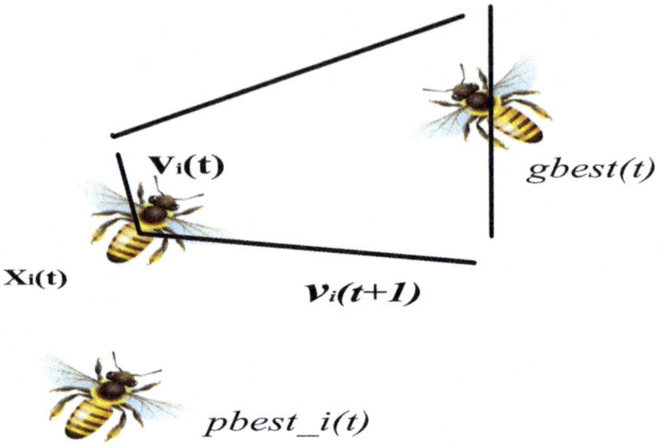

Fig. 4 Updated position using $x_i(t+1)$ *and* $v_i(t+1)$ [19]

where

$x_i(t)$	*Initial p* osition vector,
$x_i(t+1)$	*Updated p* osition vector,
$v_i(t+1)$	*Updated velocity* vector.

The position update process is shown in Fig. 4.

3.2 Flowchart of PSO

Figure 5 depicts the basic PSO flowchart.

3.3 Operation of PSO

The PSO employs the following procedures, starting with

Step 1: Initializing swarm from solution space.
Step 2: Different particles' fitness is assessed.
Step 3: Updating g-best, p-best, and velocity.
Step 4: A piece particle is moved to a new place.
Step 5: Go to Step 2 and replication all of the steps there until the convergence point requirement is met.

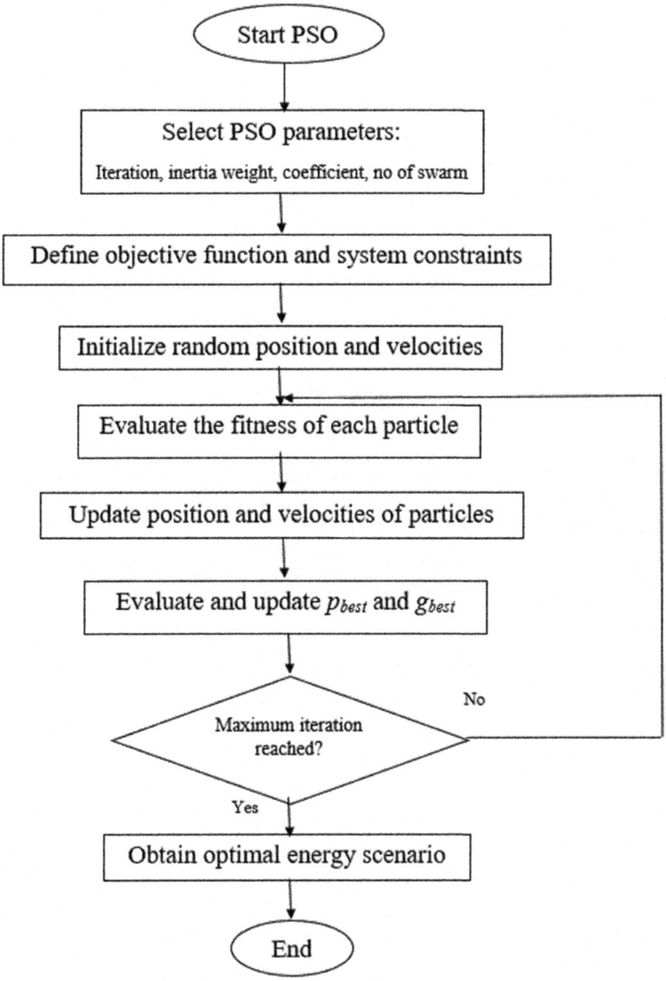

Fig. 5 PSO algorithm [20, 21]

4 Grasshopper Optimization Algorithm

GOA is employed, which emulates the performance of grasshoppers in their search for food. By simulating swarm behavior mathematically, GOA offers all possible solutions and moves over a search gap to find the optimal solution [22]:

$$Z_i^p(t+1) = \alpha\left(\sum_{j=1 j\neq i}^{N} \alpha \frac{up_p - lw_p}{2} k(r)\left(\left|X_j^p(t) - X_i^p(t)\right|\right)\frac{x_j(t) - x_i(t)}{f_{ij}}\right) + T_p \tag{16}$$

where

t Number of iterations,
N Quantity of grasshoppers,
up_p Upper bound dimension,
lw_p Lower bound dimension,
f_{ij} Distance of grasshopper amid ith and jth,
T_p p-dimensional position,
α Ease zone decreasing factor,
k Firmness of societal forces.

The firmness of societal forces is given by [6]

$$k(r) = s \times e^{\left(\frac{-d}{l_a}\right)} - e^{-d} \tag{17}$$

s and d are the strength of attraction and distance, respectively, l_a is attraction length.
The α is computed as [7, 8]

$$\alpha = \alpha_{\text{maxi}} - t\left\{\frac{\alpha_{\text{maxi}} - \alpha_{\text{mini}}}{t_{\text{maxi}}}\right\} \tag{18}$$

α_{maxi} Maximum value of ease zone decreasing factor,
α_{mini} Minimum value of ease zone decreasing factor,
t No. of iterations,
t_{maxi} max. no. of iterations.

Primarily, the quantity of grasshopper's group and the number of reiterations desired are entered for the best selection. The fitness function of the populace is assessed through an objective equation. The agent positions are updated by using Eqs. (16–18). The fitness of apiece grasshopper is determined by updating the finest agent. When the restraints are contented, the best number of the system is considered.

4.1 Flow Chart of GOA

The flow chart of GOA is shown in Fig. 6.

4.2 Operation of GOA

Step 1: Set some initial parameters of the grasshopper, the maximum and minimum alpha values (α_{maxi}, α_{mini}), and the iterations (i_{maxi}).
Step 2: Initialize the economic data, i.e., capital cost, annual cost (AC), fixed cost, etc.

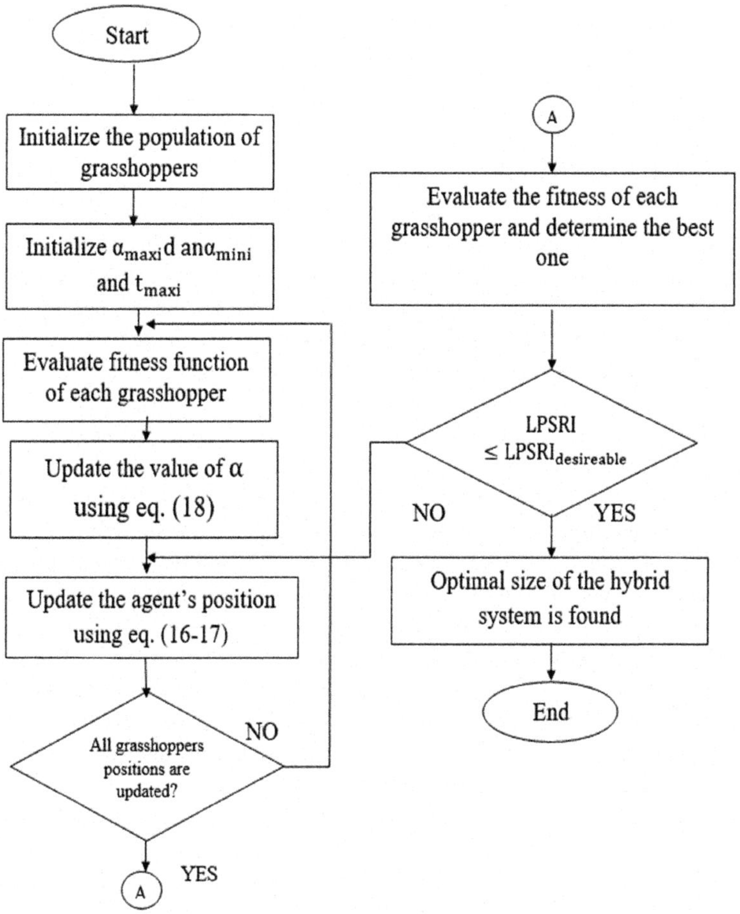

Fig. 6 Flow chart of GOA [23, 24]

Step 3: Assess the population's starting fitness level based on the goal function Eq. (16).

Step 4: Arrange the population in order of best to worst fitness function.

Step 5: Update the independent variables, which are the sizes of the PV, WT, and battery storage, using Eq. (16).

Step 6: Check if the updated independent variables satisfy the system constraints (Eqs. 8–13). If the constraints are met, calculate the fitness of the individual grasshoppers and save the optimal solution using Eq. (16).

Step 7: Increase the iteration count, i.e., $i = i + 1$.

Step 8: Return to Step 5 if the system restrictions are not fulfilled.

Step 9: Display the best solution agent and the ideal fitness function.

Figure 6 displays the flowchart illustrating the GOA in obtaining optimal energy system configurations.

5 Genetic Algorithms

The utmost prevalent method in evolutionary study is GA. This algorithm follows a cascade of phases for solving a problem. A GA uses genetic procedure as a base model in solving the problems. It is a quest technique for finding nearby solutions by optimization. Normally, an individual distinguishes all conceivable explanations equivalent to a precise query. The tradition of all the explanations that encounter this procedure establishes the search space. The problem entails in outcomes of the key that is accepted as the best solution, i.e., the one with the maximum payoff among all the probable results. If all the possible solutions to problems are summarized, the complications are not raised. But if the search space turns out to be bigger, the solution is not as simple as it appears. It will take a lot of time for a feasible solution. GA offers a solution to this time phase. Practically they follow a way by refashioning the elementary genetic coding to an algorithm mechanism [25].

GA grips a population to have a feasible solution. The individual result is characterized as a chromosome, which is just a conceptual illustration. The first part comprises coding the probable results into a chromosome. A collection of replicating operators has to be set on. Replicating operators can be used in a straight line on the chromosomes and are cast off to execute mutations and recombining them again and again over solutions to the difficulties [26].

The outcomes must be intelligent enough to compare each individual present in the population. Each individual is selected from their own value of fitness function. Every value of an individual chromosome resembles the fitness of the candidate's explanation. The finest solution must exploit the fitness function. The GA cycle is shown in Fig. 7.

The GA repeats its iteration method to evolve the population. GA selects the individual for reproduction. The individuals are randomly nominated based on the fitness of each individual with a probability of selecting the best for reproduction. After the selection of individuals with the best fitness, the succeeding step is to bring forth their offspring. The GA generates new sets of offspring chromosomes by using processes of recombination and mutation. Then the fitness function of the offspring chromosomes is evaluated. In the end, the weak chromosomes are eliminated and substituted by the new ones.

5.1 GA Flowchart

GA flowchart is shown in Fig. 8.

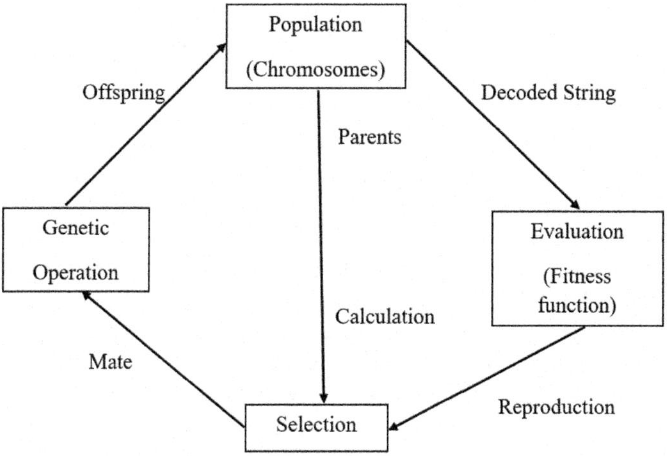

Fig. 7 GA cycle [26, 27]

Fig. 8 GA flowchart [27, 28]

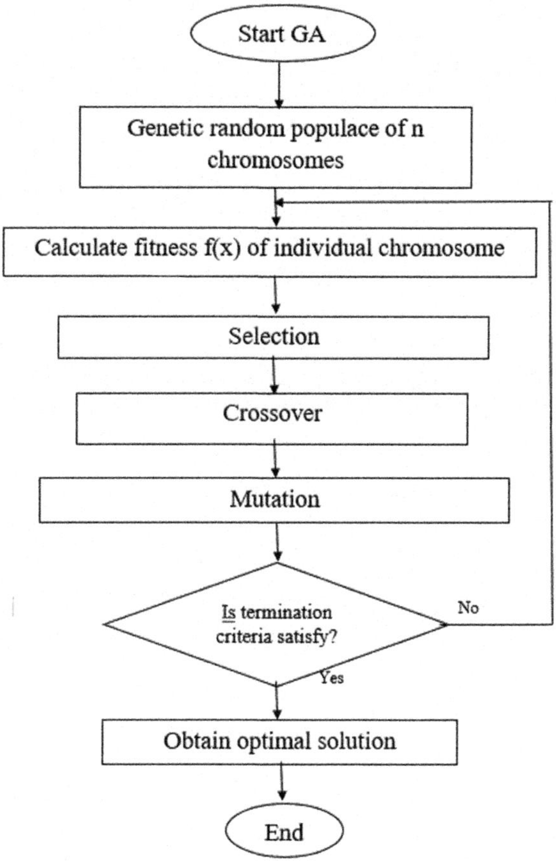

5.2 GA Operation

GA follows the following steps:

Step 1: Genetic random populace of n chromosomes is created.
Step 2: Fitness f(x) of individual chromosome x in populace is assessed.
Step 3: The new population is formed by reiterating selection, crossover, mutation, and accepting process.
Step 4: Newly created populace is replaced for an additional summation of procedure.
Step 5: If the conclusion form is fulfilled, stopover the procedure by retaining its finest result for the existing population.
Step 6: If the conclusion condition is not achieved repeat Step 2 for fitness calculation.

6 HOMER Pro Software

HOMER Pro is a software tool for designing and selecting the optimal energy solution for village, island, and utility link systems. Below, A flowchart for the HOMER Pro program is shown in Fig. 9. The steps are as follows:

Step 1: Select the location longitude and latitude of the selected area. Add the different types of load, e.g., domestic, commercial, industrial, etc.
Step 2: Add the power resource, e.g., renewable and conventional sources.
Step 3: Add and edit the resources capacity and cost per unit.
Step 4: Add the meteorology data of solar and wind.
Step 5: Connect the system components, e.g., converter, inverter, etc.
Step 6: Add the storage system, e.g., fuel cell, battery, etc.
Step 7: Calculate and optimize the system to meet the load demand.

6.1 HOMER Pro Flow Chart

See Fig. 9.

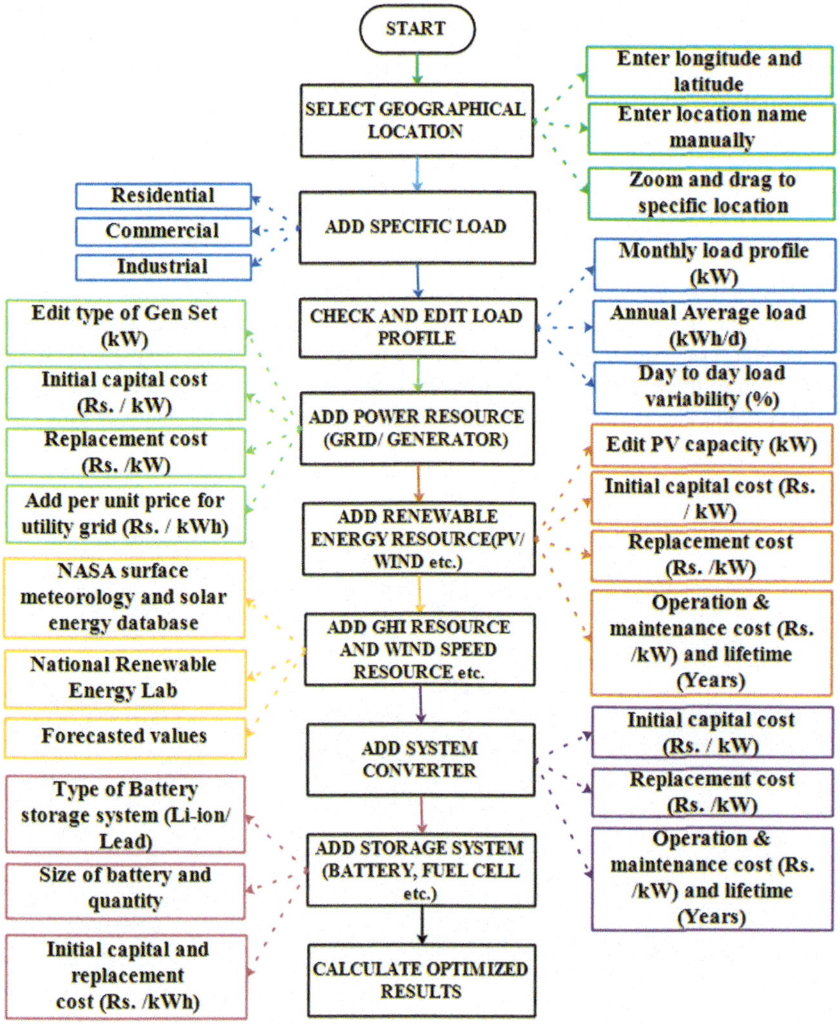

Fig. 9 Flowchart for the HOMER Pro program [29]

7 Application of HOMER, PSO, GOA, and GA for Techno-economic Optimization: A Case Study on Proposed System Architecture

Figure 10 depicts the Chhitkana township in the Indore region of the central Indian state of Madhya Pradesh, which is used as a work study. The latitude and longitude of the study region are 22.76°N and 76.017°E. There are 243 people living in the village overall (115 men and 128 women). The community has a 134-hectare total terrestrial area and roughly 30 homes.

7.1 *Proposed Hybrid Renewable Energy System*

The HRES is a complex (series with parallel) configuration system shown in Fig. 11. In series, the array of PV is coupled to the DC bus through the charge controller and the DC-DC converter to regulate the voltage to the requisite level. In parallel, WT is connected to the battery, and the inverter is connected in parallel to the alternating current load. The required additional power is balanced by the battery. This case

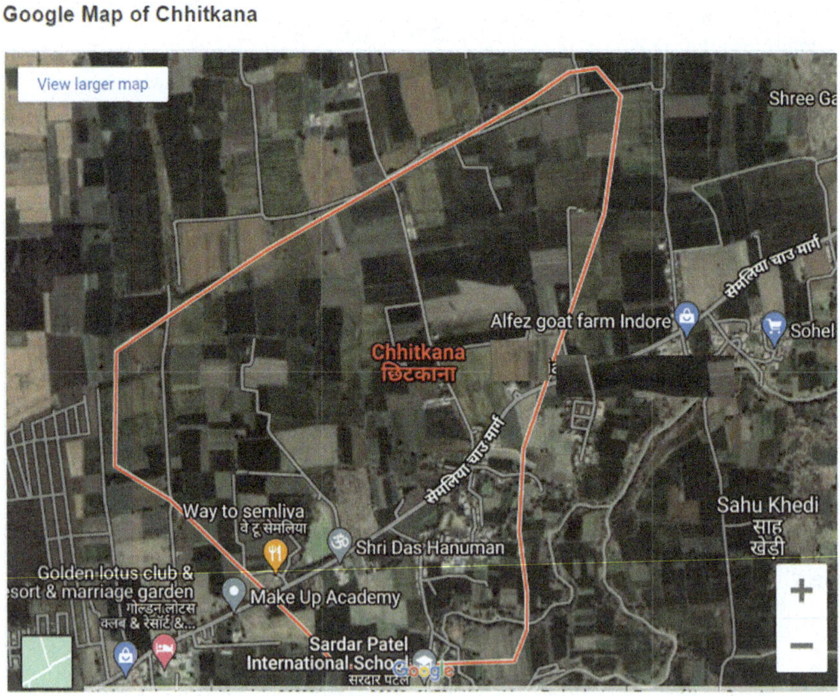

Fig. 10 Google map of chhitkana village [30]

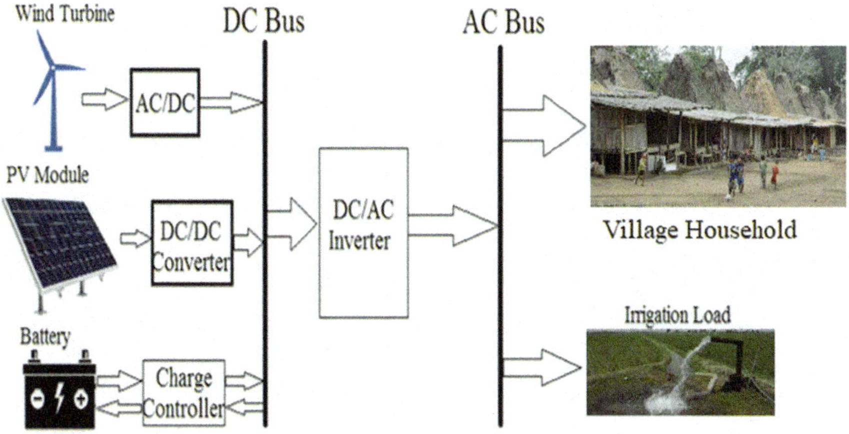

Fig. 11 Proposed HRES

requires less rating of the inverter while the efficiency obtained is appreciable. The appropriate harmonization is mandatory amid output and energy resources. This research work proposes HRES to accomplish the daily load demand of farmhouses as shown in Fig. 11.

PV and WT systems produce electricity. The battery bank stores surplus energy and retrieves energy when PV and WT show deficit energy production.

7.2 Load Profile

The electricity demand of the study area is evaluated based on the requirement of the village people. The electricity demand for Farmhouse is segregated into two sections—Farmhouse residence and irrigation. Figure 12 displays the mean electrical demand throughout the year.

- The demand for electricity in residence increases in the months of summer season (March–June) while the demand for power decreases in the irrigation section in these months. This demand is due to the rise in surrounding temperature in this season.
- The demand for electricity in the irrigation section from July to October is at its least value due to the rainy season. While the demand in farmhouse is almost constant.
- In India, crops such as wheat, grams, and potatoes are typically planted in October and harvested by February, depending on weather conditions [31]. This process necessitates a significant amount of water, resulting in increased electricity usage

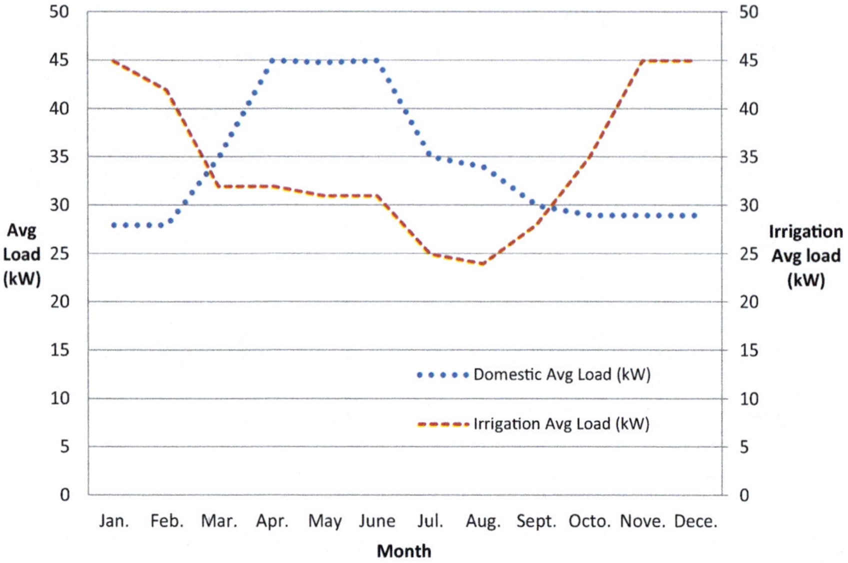

Fig. 12 Average electricity demand of load for farmhouse and irrigation

in the irrigation sector during this period. Conversely, electricity demand in farmhouse residences decreases due to colder temperatures. Despite these fluctuations, the overall average demand for electricity remains relatively constant.

The people residing in farmhouse generally utilize fluorescent lamps, CFLs, vanes, TV sets, grinders, and other loads. The individual electricity load is about 1.2 kW. In the village, only twelve 5hp submersible pumps have been used. The usual demand for power for total village houses is approximately 80.0 kW while the total electricity consumed is approximately 180.0 kWh each day as displayed in Table 1.

8 Results and Discussion

8.1 Technical Specifications

The technical parameters of HRES components are tabularized in Tables 2 and 3. Figure 13 displays the local area's typical annual weather conditions, including wind speed and solar radiation strength.

Table 1 Electrical demand for Farmhouse

Load sector	Appliances utilizing electricity	Rating of appliances (W)	Qt	Total power (W)	Running hrs (hrs./day)	Consumed energy (Wh/day)
Residential load	Fluorescent lamp	40	2	80	5	400
	LED lamp	9	3	27	7	189
	Vane	75	4	300	8	2400
	TV set	150	1	150	5	750
	Mixer	500	1	500	1	500
	Miscellaneous	100	1	100	1	100
	Total			1157		4339
	For total village (30 houses) load			34,710		130,170
Irrigation load	Water pump	3750	12	45,000	1.1	49,500
	Total			79,710		179,670

Table 2 Solar panel (ALPEX-250W) specification [32]

S. N	Nomenclature of technical specification	Capacity
1	Open circuit voltage (Voc)	37.201 V
2	Short circuit current (Isc)	8.751 Amp
3	Max. value of current (Impp)	8.35 Amp
4	Max. value of voltage (Vmpp)	29.951 V
5	Max. peak power (Pmpp)	250W (−+ 3.0%)
6	No. of cells in series	60
7	No. of modules	20
8	Insolation	1×10^3 W/m^2
9	Range of temperature	From 40 °C below zero to 85 °C above zero
10	NOCT*	46 °C (−+ 2 °C)

* NOCT Irradiance 1000 W/m, Ambient Temperature 25 °C, wind speed 1 m/sec

Table 3 WT specification (Whisper-200) [33]

Configuration	Rating
Number of blades	3.0
Speed of wind as cut-in/cut-off	3.11 m/s
Power	1×10^3 W (peak)
Voltage	120 V/240 V DC
Speed of wind	11.6–13.1 m/s
Swept area	5.81 m^2
Life span	25 years

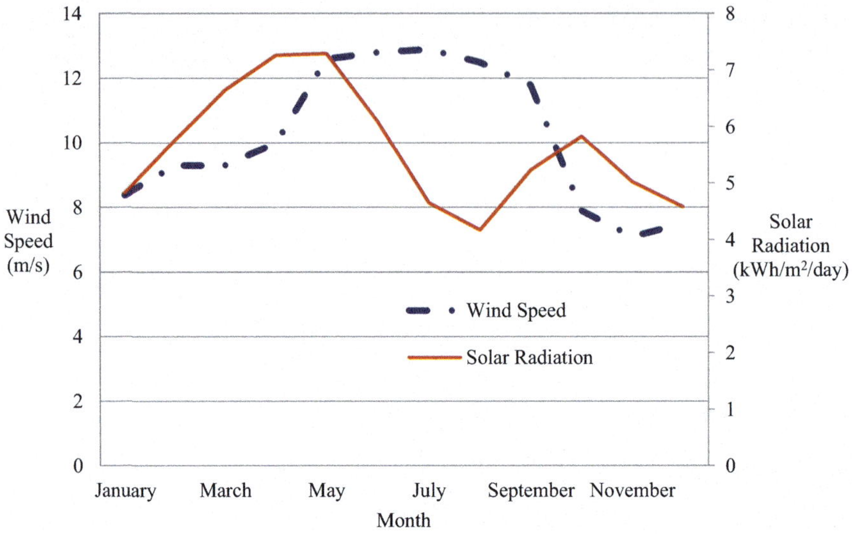

Fig. 13 Solar radiation and wind speed of located area [34, 35]

In this research work ALPEX-250W model solar panel is used. Table 2 lists the electrical parameters of the chosen PV panel. The nominal operating cell temperature (NOCT) gives the actual temperature of the cell of the module w.r.t. the operational surrounding circumstances. The solar generator is planned for 5.0 kilo watt considered.

This research work used the Whisper-200 model of WT. The WT rotation axis is horizontal and the rotation direction is clockwise. The material of the wind turbine is polypropylene and the blade material is glass reinforced. The average power is about 570W at 11 m/s and the rotor swept area is 5.8 m^2. The detailed specification of the Whisper-200 model WT is tabulated in Table 3.

8.2 Optimization Results

Using HOMER, PSO, GA, and GOA, the proposed HRES is technologically and economically optimized. A distant location's HRES is sized using a variety of PSO, GA, and GOA factors. These algorithms are utilized to compute optimized combinations of resources in remote areas with variable climate data, i.e., temperature, speed of wind, and sun radiation. Economic optimization with reliability-based performance indicators is computed using four methods. Table 4 depicts the various specifications of the used optimization algorithm.

Table 5 displays the comprehensive performance analyses of the optimization techniques GA, PSO, GOA, and HOMER software. All algorithms are limited to a

Table 4 Lists several PSO, GA, and GOA parameters

Algorithm	Parameter	Value
PSO	Swarm population	15
	Number of iterations	100
	Inertia weight	1
	Coefficient	$C_1 = 0.27$ and $C_2 = 1.55$
GA	Population	15
	Number of iterations	100
	Mutation frequency	0.2
	Crossover rate	0.8
GOA	Size of population	15
	Number of iterations	100
	Social constant	1.85
	Reduction factor	$C_{min} = 0.1$ and $C_{max} = 0.9$

maximum of 100 iterations. Since they are not workable systems, PV-only and WT energy systems are not taken into consideration. This paper takes into account the four integrated configurations PV_battery, WT_battery, PV_WT, and HRES.

The ideal PV, battery, inverter, and converter sizes according to GOA are 240, 160, 81, and 96.2, respectively. Furthermore, compared to other optimization approaches, COE and LPSRI have reduced costs. However, the load need cannot be satisfied by the PV_battery energy system. GOA-based size was determined to be the optimum option after an evaluation of the WT_battery energy system's ideal sizing. The LPSRI (0.0914%) and COE of the WT_battery energy system are also lower than those of other optimization methods. The WT_battery system is unable to fully provide the load demand, nevertheless.

PV_WT system sizing based on GOA is calculated, i.e., $N_{pv} = 280, N_{wind} = 60, N_{inv} = 81.4\, and\, N_{con} = 96.92$ with 15.55 ₹/kWh and 0.0021 LPSRI. PV_WT system fails to meet the required load. Configuration of the PV_WT_battery system, with optimal size determined by GOA based on parameters as $N_{pv} = 240, N_{wind} = 40, N_{bat} = 160, N_{inv} = 81\, and\, N_{con} = 96$ with 5.92 ₹/kWh and 0.0001 LPSRI.

PV_battery: HOMER software results shows system sizing $N_{pv} = 755, N_{bat} = 264, N_{inv} = 96\, and\, N_{con} = 24$. GOA-based results show that the lowest number of components, i.e., $N_{pv} = 645, N_{bat} = 224, N_{inv} = 87\, and\, N_{con} = 24$. GA and PSO-based sizing are higher than the GOA-based system.

WT-battery: HOMER software result shows system sizing $N_{wind} = 100, N_{bat} = 256, N_{inv} = 92\, and\, N_{con} = 24$. GOA-based results show that the lowest number of components, i.e., $N_{wind} = 84, N_{bat} = 200, N_{inv} = 83\, and\, N_{con} = 24$. WT_battery system does not satisfy the load requirement.

The number of components and LPSRI (%) calculated by HOMER, PSO, GA, and GOA for PV_WT_battery is shown in Figs. 14 and 15.

Table 5 Comprehensive comparisons of the optimization methods' performance

System	Algorithm	Npv	Nwind	Nbat	Ninv (kW)	Ninv @ 4 kW/inv	Ncon (kW)	Ncon @1 kW/con	NPC (₹)	COE (₹/kWh)	LPSRI (%)
PV_battery	Software										
	HOMER	755	0	264	96	24	104	104	10,737,200	7.53	0.1494
	PSO	687	0	248	90.08	24	100.68	104	11,881,696	7.99	0.0974
	GA	655	0	232	87.52	24	101.52	104	11,414,860	7.88	0.0454
	GOA	645	0	224	87	24	100.48	104	10,722,140	7.42	0.0344
WT_battery	HOMER	0	100	256	92	24	116	116	34,013,038	22.96	0.1412
	PSO	0	92	224	86.48	24	103.04	104	35,066,536	23.53	0.1314
	GA	0	88	216	83.12	24	99.96	100	34,314,036	22.88	0.1214
	GOA	0	84	200	82.2	24	99.32	100	33,862,532	22.15	0.0984
PV_WT	HOMER	348	80	0	88	24	108	108	23,190,708	15.95	0.0594
	PSO	320	68	0	86.08	24	102.68	104	25,135,728	16.28	0.0094
	GA	300	64	0	82.32	24	99.92	100	23,789,172	15.69	0.0082
	GOA	280	60	0	81.4	24	96.92	100	23,115,904	15.55	0.0021
PV_WT_battery	HOMER	308	72	192	88	24	104	104	8,951,511	5.99	0.0011
	PSO	264	48	176	84.48	24	100.68	104	9,011,684	6.02	0.0084
	GA	248	44	168	81.92	24	99.12	100	8,981,596	5.95	0.0054
	GOA	240	40	160	81	24	96.52	100	8,921,416	5.92	0.0001

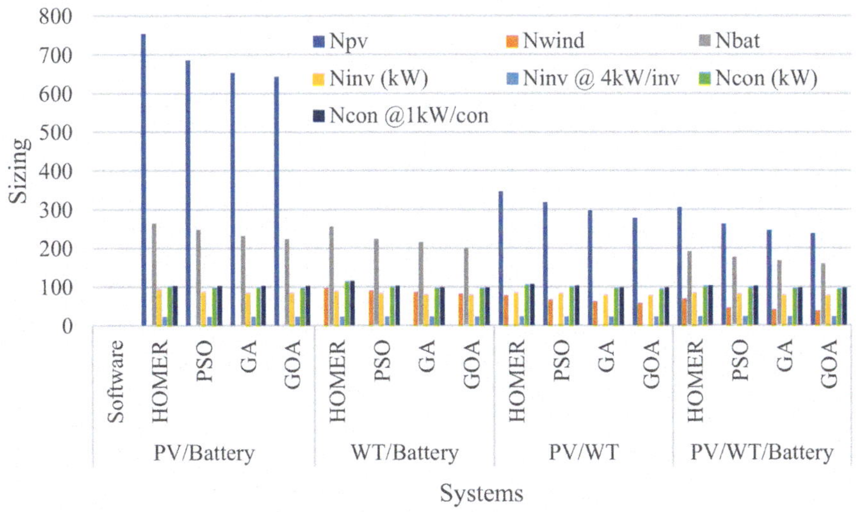

Fig. 14 Number of components calculated by PSO, HOMER, GA, and GOA

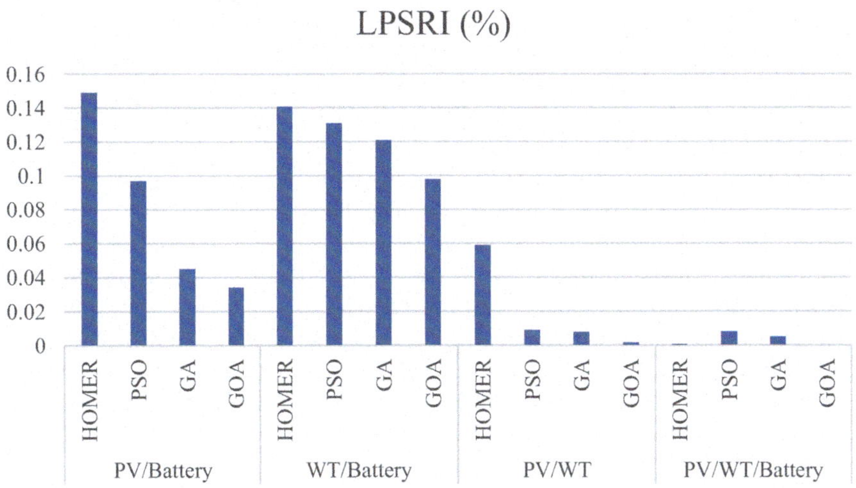

Fig. 15 LPSRI values calculated by HOMER, PSO, GA, and GOA

PV_WT: HOMER software result shows system sizing $N_{wind} = 80, N_{PV} = 348, N_{inv} = 88\, and\, N_{con} = 24$. GOA-based results show that the lowest number of components, i.e., $N_{pv} = 280, N_{wind} = 60, N_{inv} = 82\, and\, N_{con} = 24$. PV_WT system does not satisfy load requirement.

PV_WT-battery: GOA found the, e.g., $N_{pv} = 240, N_{wind} = 40, N_{bat} = 160, N_{inv} = 81\, and\, N_{con} = 24$. This system meets the daily load demand.

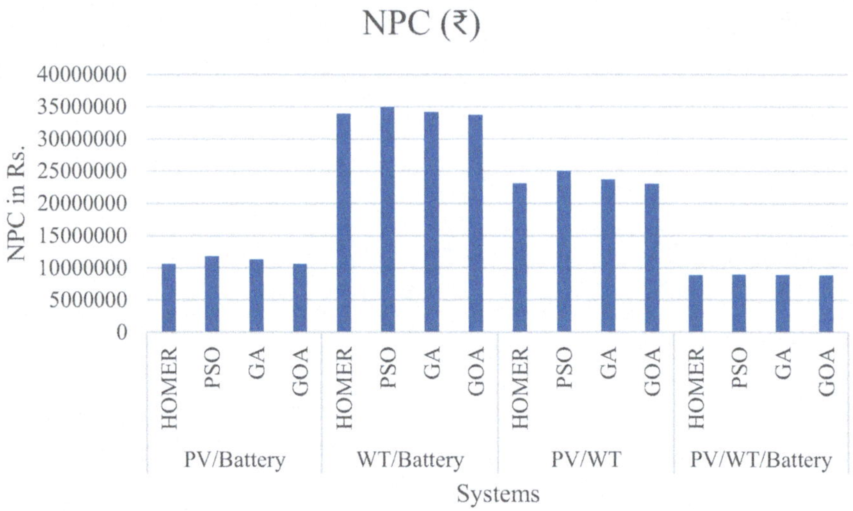

Fig. 16 NPC in all systems

The LPSRI of GOA results show WT_battery (0.0984%), PV_battery (0.0344%), PV_WT (0.0021%), and PV_WT_battery (0.0001%). The LPSRI of WT_battery is the highest and PV_WT_battery is the lowest.

Figure 16 demonstrates the high NPC of the WT_battery and PV_WT systems. PV_WT_battery and PV_battery have the lowest NPC.

Figure 17 shows the number of iterations required in computing COE.

The COE is highest in WT_battery system and lowest in hybrid PV_WT_battery system. The GOA algorithm shows the lowest COE compared to HOMER, PSO, and GA. Figure 17 shows the hybrid PV_WT with battery backup COE using PSO(6.02 ₹/kWh), GA(5.95 ₹/kWh), and GOA (5.92 ₹/kWh).

Figure 18 shows the COE comparison details computed by HOMER, PSO, GA, and GOA. The hybrid PV_WT_battery and PV_battery systems are cost-effective system in terms of COE. The WT_battery and PV_WT are the most expensive system.

9 Summary

The chapter presented various optimization techniques of RES for the electric necessity of farmhouses and irrigation load stipulation situated at remote zone. The optimum size of HRES has been procured to completely exploit the capacity of the complete unit. The economic scrutiny has been done for the diverse arrangement of HRES in the phrase of NPC, COE, and LPSRI. The four configurations of PV and bank of batteries, PV and WT integrated system, WT and bank of batteries and integrated PV, WT and bank of batteries energy system were considered in this chapter.

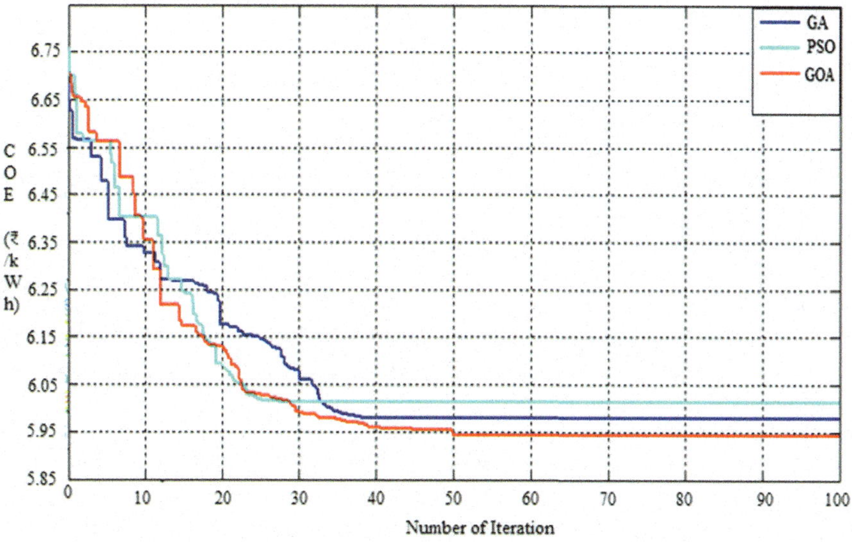

Fig. 17 Number of iterations required in computing COE

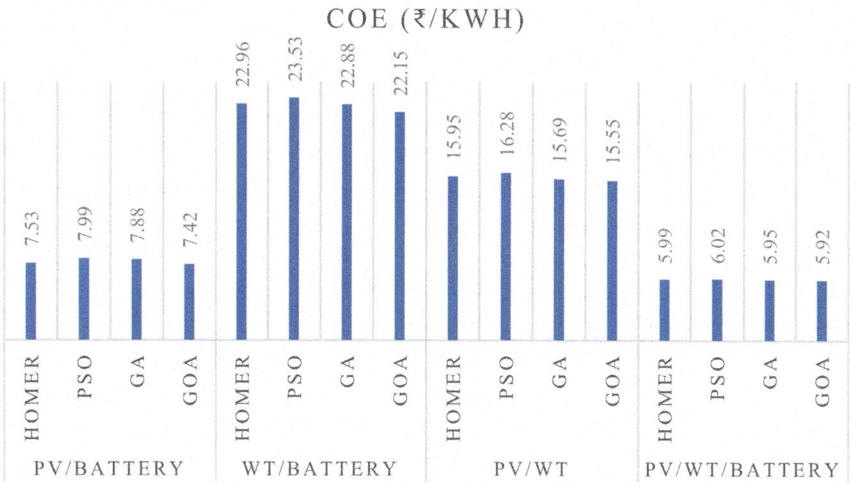

Fig. 18 COE values computed by HOMER, PSO, GA, and GOA

This chapter recommends a neoteric evolutionary method built on metaheuristic GOA to be utilized for choosing optimal HRES configuration.

Based on the results, it is concluded:

- The COE of PV_WT_battery is less by 16.71%, 73.64%, and 61.61% as compared to PV_battery, WT_battery, and PV_WT, respectively.

- The COE obtained for various systems is the least for PV_WT_battery system (5.9 ₹/kWh). So, the performance of this system presides over other systems.
- The COE value becomes constant after 25 iterations for PSO, 38 iterations for GA, and 50 iterations for GOA, respectively.
- The number of PV modules required for HPWBES calculated by HOMER, PSO, GA, and GOA are 308, 264, 248, and 240, respectively. The least number is obtained from GOA.
- The number of WT required for HPWBES calculated by HOMER, PSO, GA, and GOA are 72, 48, 44, and 40, respectively. The least number is obtained from GOA.
- The number of batteries required for HPWBES calculated by HOMER, PSO, GA, and GOA are 192, 176, 168 and 160, respectively. The least number is obtained from GOA.
- The number of inverters required are same when computed by HOMER, PSO, GA, and GOA but the amount of power in terms of kW for the HPWBES by GOA is the least.
- The number of converters required by HOMER and PSO is 104 while computed by GA and GOA is 100. Looking from the kW perspective, the least amount is computed by GOA.
- Among the LPSRI computed by HOMER, PSO, GA, and GOA, the least LPSRI is obtained through GOA.
- In comparison to GOA, PSO requires a greater number of iterations to achieve convergence.
- When dealing with a small population, GOA proves to be more effective in terms of reliability index and flexibility than both PSO and GA.

Considering the different parameter values obtained from HOMER, PSO, GA, and GOA, GOA computes the most optimum results. Based on the GOA result, HPWBES proves to be the most efficient for fulfilling load requirement. GOA also indicates that an integrated configuration of PV, WT, and battery is a cost-effective system as compared to other configurations.

References

1. Jahannoosh M, Nowdeh SA, Naderipour A, Kamyab H, Faraji Davoudkhani I, Jaromír Klemes J (2021) New hybrid meta-heuristic algorithm for reliable and cost-effective designing of photovoltaic/wind/fuel cell energy system considering load interruption probability. J Cleaner Prod 278:123406
2. Nasr R, Abou-Zalam B, Nabil E (2023) Metaheuristic optimization algorithm-based enhancement of photovoltaic energy system performance. Arab J Sci Eng.https://doi.org/10.1007/s13369-023-07847-0
3. Nassef AM, Abdelkareem MA, Maghrabie HM (2023) Review of metaheuristic optimization algorithms for power systems problems, 1–27

4. Kumar NM, Vishnupriyan J, Sundaramoorthi P (2019) Techno-economic optimization and real-time comparison of sun tracking photovoltaic system for rural healthcare building. J Renew Sustain Energy 11(1):015301
5. Oumaima M, Mouhaydine T, Janeiro FM, Abdeloawahed H, Khalid K (2019) "Estimation of photovoltaic panel parameters by a numerical heuristic searching algorithm. 8th International conference on renewable energy research and applications, Brasov, ROMANIA, Nov. 3–6, 2019
6. Hazra S, Pal T, Roy PK (2019) Renewable energy based economic emission load dispatch using grasshopper optimization algorithm. Inte J Swarm Intell Res 10(1)
7. Fathy (2018) Recent meta-heuristic grasshopper optimization algorithm for optimal reconfiguration of partially shaded PV array. Solar Energy 171:638–651
8. Hichem H, Elkamel M, Rafik M, Mesaaoud MT, Ouahiba C (2019) A new binary grasshopper optimization algorithm for feature selection problem. J King Saud Univ Comput Inf Sci. https://doi.org/10.1016/j.jksuci.2019.11.007
9. Raghuwanshi SS, Khare V (2018) Sizing and implementation of photovoltaic water pumping system for irrigation. IAES Int J Artif Intell (IJ-AI) 7(1):54–62. https://doi.org/10.11591/ijai.v7.i1.pp54-62
10. Zhou J, Xu Z (2023) Optimal sizing design and integrated cost-benefit assessment of stand-alone microgrid system with different energy storage employing chameleon swarm algorithm: a rural case in Northeast China. Renew Energy 202:1110–1137. https://doi.org/10.1016/j.renene.2022.12.005
11. Raghuwanshi SS, Arya R (2020) Design and economic analysis of a stand-alone hybrid photovoltaic energy system for remote healthcare center. Int J Sustain Eng (T&F) 13(5):360–372. https://doi.org/10.1080/19397038.2019.1629674
12. Avvari RK, DM VK (2023) A new hybrid evolutionary algorithm for multi-objective optimal power flow in an integrated WE, PV, and PEV power system. Electr Power Syst Res 214:108870
13. Rezk H, Olabi AG, Wilberforce T, Sayed ET (2023) A comprehensive review and application of metaheuristics in solving the optimal parameter identification problems. Sustainability 15:5732
14. Raghuwanshi SS, Arya R (2019) Economic and reliability evaluation of hybrid photovoltaic energy systems for rural electrification. Int J Renew Energy Res 9(1):515–524
15. Dulhare UN (2018) Prediction system for heart disease using Naive Bayes and particle swarm optimization. Biomed Res 29(12):2646–2649, ISSN 0970-938X
16. Sheng W, Li R, Yan T, Tseng M-L, Lou J, Li L (2023) A hybrid dynamic economics emissions dispatch model: distributed renewable power systems based on improved COOT optimization algorithm. Renew Energy 204:493–506
17. Sellami R, Sher F, Neji R (2022) An improved MOPSO algorithm for optimal sizing & placement of distributed generation: a case study of the Tunisian offshore distribution network (ASHTART). Energy Rep 8:6960–6975. https://doi.org/10.1016/j.egyr.2022.05.049
18. Mohammed AQ, Al-Anbarri KA, Hannun RM (2022) Using particle swarm optimization to find optimal sizing of Pv-Bs and diesel generator. J Eng Sustain Dev 25(3):51–59. https://doi.org/10.31272/jeasd.25.3.6
19. Liu H, Wu B, Maleki A, Pourfayaz F (2022) An improved particle swarm optimization for optimal configuration of standalone photovoltaic scheme components. Energy Sci Eng 10(3):772–789. https://doi.org/10.1002/ese3.1052
20. Hu K, Cao S, Li W, Zhu F (2019) An improved particle swarm optimization algorithm suitable for photovoltaic power tracking under partial shading conditions. IEEE Access 7:143217–143232. https://doi.org/10.1109/ACCESS.2019.2944964
21. Maharana HS, Dash SK (2023) Quantum behaved artificial bee colony based conventional controller for optimum dispatch. Int J Electr Comput Eng 13:1260
22. Liaquat S, Fakhar MS, Rahman Kashif SA, Rasool A, Saleem O, Padmanaban S (2020) Performance analysis of APSO and firefly algorithm for short term optimal scheduling of multi-generation hybrid energy system. IEEE Access 8:177549–177569. https://doi.org/10.1109/ACCESS.2020.3027436

23. Xu L, Tu Y, Zhang Y (2020) 'A grasshopper optimization-based approach for task assignment in cloud logistics. Math Problems Eng, Art. no. 3298460
24. Bhuyan M, Barik AK, Das DC (2020) 'GOA optimised frequency control of solar-thermal/sea-wave/biodiesel generator based interconnected hybrid microgrids with DC link.' Int J Sustain Energy 39(7):615–633
25. Raju VA, Srilatha P, Anumandla KK, Patnaik S, Prabu AV (2020) 'Grasshopper algorithm based channel assignment for cognitive radio networks. Mater Today Proc
26. Azizi M, Aickelin U, Khorshidi HA, Baghalzadeh Shishehgarkhaneh M (2023) Energy valley optimizer: a novel metaheuristic algorithm for global and engineering optimization. Sci Rep 13(1):1–23. https://doi.org/10.1038/s41598-022-27344-y
27. Raghuwanshi SS, Raghuwanshi P, Masih A, Singh P (2023) Modeling and optimization of hybrid renewable energy with storage system using flamingo swarm intelligence algorithms. Energy Storage.https://doi.org/10.1111/EST2.470
28. Yousaf S, Mughees A, Khan MG, Amin AA, Adnan M (2020) A comparative analysis of various controller techniques for optimal control of smart nano-grid using GA and PSO algorithms. IEEE Access 8:205696–205711. https://doi.org/10.1109/ACCESS.2020.3038021
29. Sharma AK et al (2023) Role of metaheuristic approaches for implementation of integrated MPPT-PV systems: a comprehensive study. Mathematics 11(2):269. https://doi.org/10.3390/math11020269
30. https://www.google.com/maps/place/Chhitkana,+Madhya+Pradesh+452016/@22.7561964,75.9869126,15z/data=!4m5!3m4!1s0x3962e1ff9e271add:0xb386d5580a19e8b!8m2!3d22.7561964!4d75.9956673
31. Dulhare UN, Gouse S (2022) Automation of rice cultivation from ploughing–harvesting with diseases, pests and weeds to increase the yield using AI" ICCCE 2021: Proceedings of the 4th international conference on communications and cyber physical engineering (pp 505–513). Singapore: Springer Nature Singapore
32. http://alpexsolar.com/solar-power-packs.html.
33. Masih A, Verma HK, Optimum sizing and simulation of hybrid renewable energy system for remote area. Energy Environ, 1–19. DOI: https://doi.org/10.1177/0958305X211030112
34. Weather online data, August 2022, https://justweather.org/India/Madhya-Pradesh
35. Raghuwanshi SS, Masih A, Raghuwanshi P (2022) Design and optimisation of stand-alone hybrid energy system for rural areas. Int J Ambient Energy.https://doi.org/10.1080/01430750.2022.2103185

Chapter 4
Stochastic Optimization of Renewable Energy Sources in Distribution Networks

Hussein Abdel-Mawgoud and Salah Kamel

Abstract Integration of natural resources in distribution systems is increasingly popular due to their sustainability and minimal environmental impact. This chapter introduces three metaheuristic algorithms—arithmetic optimization algorithm (AOA), slime mold algorithm (SMA), and beluga whale optimization (BWO)—to obtain the preferable sizing and placement of wind turbine (WT) and photovoltaic (PV) units in Radial Distribution System (RDS). In this chapter, the probabilistic of system load demand with PV and WT generation is considered to minimize the objective function of overall energy loss. The preferable allocation of PV and WT alone or together is determined to study the changes in system performance. The presented algorithms are applied to the IEEE 69-bus RDS. From simulation results, incorporating WT and PV by enabling their inverters achieves preferable results than installing WT and PV alone or together. Also, simultaneous integration of WT and PV achieves preferable results than installing WT and PV alone in RDS.

Keywords Optimization · Uncertainty · PV · WT · AOA · BWO · SMA · RDS

1 Introduction

Energy plays a crucial role in promoting social and economic development. Currently, the majority of worldwide energy is produced from fossil fuels such as oil, coal, and natural gas which are widely used in power stations worldwide. Fossil fuels-based energy accounts for approximately 75% of worldwide energy [1–3]. However, due to energy security and environmental concerns, countries worldwide are reducing their dependence on fossil fuels [4–7]. As a result, there has been a growing trend toward integrating renewable energy sources (RES) into distribution networks over the past few years. RES are considered cleaner and more sustainable than fossil fuels and

H. Abdel-Mawgoud · S. Kamel (✉)
Electrical Engineering Department, Faculty of Engineering, Aswan University, Aswan 81542, Egypt
e-mail: skamel@aswu.edu.eg

U. N. Dulhare and E. H. Houssein (eds.), *Machine Learning and Metaheuristics: Methods and Analysis*, Algorithms for Intelligent Systems,
https://doi.org/10.1007/978-981-99-6645-5_4

can help meet the increasing demand for electricity [8–11]. Biomass, wind turbines (WT), photovoltaic (PV), and hydropower are some of the RES commonly used in power systems [12–14].

PV devices generate electrical energy directly from solar energy, and their integration into the Radial Distribution System (RDS) has become increasingly popular because of its rapid development and decreasing costs over the last decade [15, 16]. Similarly, WT generates electrical energy from wind energy and is installed in RDS to improve system reliability and stability during transient conditions, thanks to its inertia [17]. WT operations are dependent on wind velocity, which must exceed cut-in and cut-out values for the WT to start and stop, respectively. Additionally, WT generates power when wind velocity exceeds its rated value. Reactive power injection/absorption using PV and WT inverters is employed for power loss minimization and voltage regulation. Since PV and WT power generation is subject to climate conditions, their probabilistic must be considered when obtaining their superior allocations in the electric grid. Probability distribution functions (PDFs) such as Beta and Weibull have been used to account for these uncertainties [18, 19]. Recently, several researchers have explored the superior sizing and placement of WT and PV into distribution networks from various angles, including optimization, energy management, design, and planning [20–22]. The maximum reduction in voltage deviation, power loss, and line loading by incorporating PV and WT with BES in RDS using ABC [23]. The energy loss is reduced by installing WT and PV with battery using the GWO algorithm [24]. In [25, 26], the minimization of system losses, voltage deviation, and cost of energy generated is obtained by incorporating PV and BES in RDS using the GA algorithm. GWO algorithm was utilized to obtain the superior sizing and placement of electric vehicle and inverter with WT by minimizing the energy loss [27]. In [28], the MMRFO algorithm has been applied to minimize the total energy loss by installing WT and inverter-based PV with battery in RDS. HHO-PSO is applied to maximize the techno-economic benefits by installing WT and PV in RDS [29]. This work aims to obtain the preferable sizing and placement of WT and PV by using up-to-date metaheuristic algorithms in RDS. The optimal allocation will then be applied to IEEE 69-bus RDS, with the objective function of total energy loss, taking into account the uncertainties for PV, WT, and system load. The study will also investigate the effect of incorporating WT and PV on the voltage profile and power loss of the system, as well as the change in system power loss by enabling the inverters of WT and PV.

2 Objective Function

Figure 1 represents two nodes of the main feeder in the distribution system. The system load flows are obtained using a forward-backward approach [30]. Equations (1, 2) are utilized to evaluate the reactive and active load flows in a backward direction through the branches.

$$P_3 = P_4 + P_{L,4} + r_{3,4}\left(\frac{(P_4 + P_{L,4})^2 + (Q_4 + Q_{L,4})^2}{|V_4|^2}\right) \tag{1}$$

$$Q_3 = Q_4 + Q_{L,4} + x_{3,4}\left(\frac{(P_4 + P_{L,4})^2 + (Q_4 + Q_{L,4})^2}{|V_4|^2}\right) \tag{2}$$

where $Q_{3,4}$ and $P_{L,4}$ are the reactive and active loads at bus 4, respectively. P_3, $r_{3,4}$, Q_3 and $x_{3,4}$ represent the real power, resistance, reactive power, and reactance among buses 3 and 4, respectively.

Equation (3) evaluates the voltage magnitude of bus 4.

$$V_4^2 = V_3^2 - 2(P_3 r_{3,4} + Q_3 x_{3,4}) + (r_{3,4}^2 + x_{3,4}^2)\left(\frac{P_3^2 + Q_3^2}{V_3^2}\right) \tag{3}$$

where the voltage value of buses 3 and 4 are represented by V_3 and V_4, respectively.

Incorporating PV and WT in RDS will change the active load flows that are evaluated by Eq. (4). Also, enabling WT and PV inverters will change the reactive load flows that are evaluated by Eq. (5).

$$P_3 = P_4 + P_{L,4} + r_{3,4}\left(\frac{(P_4 + P_{L,4})^2 + (Q_4 + Q_{L,4})^2}{|V_4|^2}\right) - P_{DG} \tag{4}$$

$$Q_3 = Q_4 + Q_{L,4} + x_{3,4}\left(\frac{(P_4 + P_{L,4})^2 + (Q_4 + Q_{L,4})^2}{|V_4|^2}\right) \pm Q_{DG} \tag{5}$$

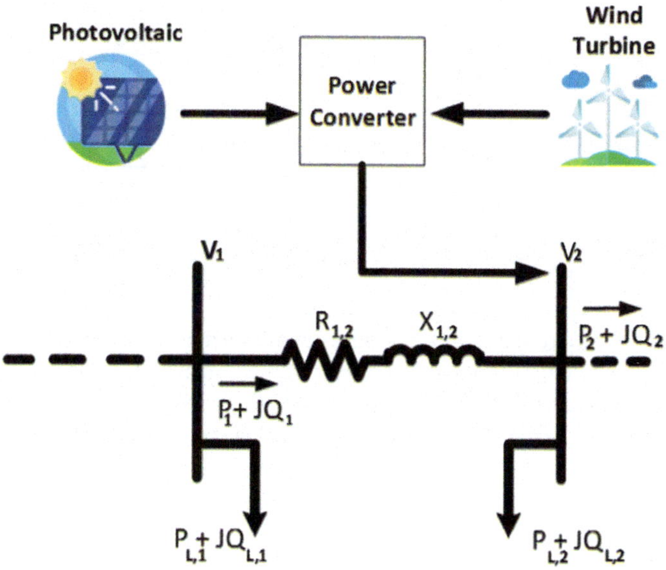

Fig. 1 Section of two buses in RDS

where

$$P_{DG} = P_{PV} + P_{WT} \tag{6}$$

$$Q_{DG} = Q_{PV} + Q_{WT} \tag{7}$$

P_{PV} and P_{WT} are the real outputs from PV and WT, respectively. Q_{PV} and Q_{WT} are the reactive outputs from PV and WT inverters, respectively.

Single-objective function is utilized by minimizing the total energy loss as follows:

$$F_{obj} = \left(\sum_{h=1}^{24} P_{loss}(h) \right) \tag{8}$$

where $P_{loss}(h)$ is the system loss in period h.

The used constraints can be presented as shown next [31–33].

A. Equality constraints

The power balance and power flow equations are the equality constraints. Equations (9, 10) represents the power flow equations. Equations (11, 12) represents the power balance between the power generation from slack bus, WT, and PV and the consumption power to the system loss and load demand.

$$P_3 = P_4 + P_{L,4} + r_{3,4}\left(\frac{(P_4 + P_{L,4})^2 + (Q_4 + Q_{L,4})^2}{|V_4|^2} \right) \tag{9}$$

$$Q_3 = Q_4 + Q_{L,4} + x_{3,4}\left(\frac{(P_4 + P_{L,4})^2 + (Q_4 + Q_{L,4})^2}{|V_4|^2} \right) \tag{10}$$

$$P_s + \sum_{m=1}^{p_{PV}} P_{PV}(m) + \sum_{l=1}^{p_{WT}} P_{WT}(l) = \sum_{e=1}^{p_B} P_L(e) + \sum_{r=1}^{p_{Br}} P_{Loss}(r) \tag{11}$$

$$Q_s + \sum_{m=1}^{p_{PV}} Q_{PV}(m) + \sum_{l=1}^{p_{WT}} Q_{Loss}(l) = \sum_{e=1}^{p_B} Q_L(e) + \sum_{r=1}^{p_{Br}} Q_{Loss}(r) \tag{12}$$

where P_B and P_{Br} are the number of buses and branches, respectively. $Q_{Loss}(r)$ and $P_{Loss}(r)$ are the reactive and real power loss of the branch (r), respectively. p_{WT} and p_{PV} are the number of WT and PV, respectively. P_s and Q_s are the active and reactive power supplied from the grid, respectively. $P_L(e)$ and $Q_L(e)$ are the real and reactive load at bus (r).

B. Inequality constraints

(1) System voltage constraints

$$V_{Do} \leq V_r \leq V_{Up} \tag{13}$$

where V_r is the voltage of bus (r). The maximum and minimum value of bus voltage can be represented by V_{Up} and V_{Do}, respectively.

(2) WT and PV sizing constraints

$$\left(\sum_{m=1}^{p_{PV}} P_{PV}(m) + \sum_{l=1}^{p_{WT}} P_{WT}(l)\right) \leq \left(\sum_{e=1}^{p_B} P_L(e) + \sum_{r=1}^{p_{Br}} P_{Loss}(r)\right) \tag{14}$$

$$\left(\sum_{m=1}^{p_{PV}} Q_{PV}(m) + \sum_{l=1}^{p_{WT}} Q_{WT}(l)\right) \leq \left(\sum_{r=1}^{p_{Br}} Q_{Loss}(r) + \sum_{e=1}^{p_B} Q_L(e)\right) \tag{15}$$

$$P_{PV,L} \leq P_{PV,r} \leq P_{PV,H} \tag{16}$$

$$P_{WT,L} \leq P_{WT,r} \leq P_{WT,H} \tag{17}$$

where the upper and lower of PV output are represented by $P_{PV,H}$ and $P_{PV,L}$, respectively. The upper and lower of WT output are represented by $P_{WT,H}$ and $P_{WT,L}$, respectively.

(3) Branches current constraints

The maximum current and the current of the branch (r) are represented by ($I_{\max,Br}$) and (I_r), respectively [34].

$$I_r \leq I_{\max,Br} \quad r = 1, 2, \ldots\ldots, P_{Br} \tag{18}$$

3 Load Modeling

Figure 2 represents the modeling uncertainty for commercial load during the day [35]. This model depends on voltage-dependent load and time-varying during 24-h [36]. Equations (19, 20) are utilized to obtain the commercial load during the day.

$$P_r(T) = P_{L,r}(T) \times V_r^{n_a} \tag{19}$$

$$Q_r(T) = Q_{L,r}(T) \times V_r^{n_r} \tag{20}$$

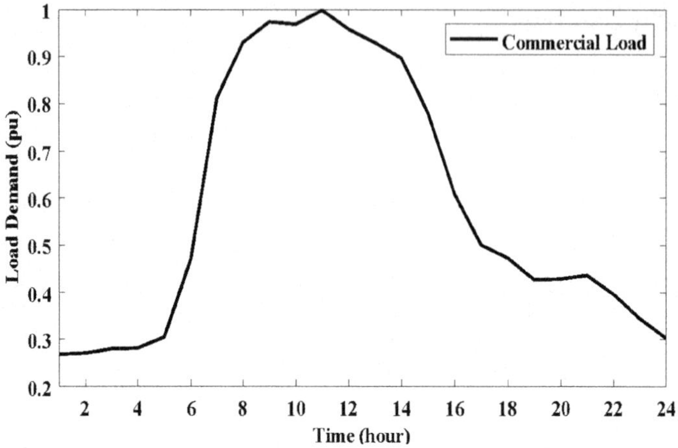

Fig. 2 The uncertainty for system load during the day

where n_a and n_r represent the active and reactive load modeling voltage indexes (3.4 and 1.51, respectively [36]). Q_r and P_r are the reactive and real power injections at bus (r). $Q_{L,r}$ and $P_{L,r}$ represent the reactive and active loads at bus (r).

4 WT Modeling

Figure 3 represents the modeling uncertainty for WT generation that based on wind velocity. The wind velocity (s) is represented by the Weibull probability density function $F_k(s)$ as shown in Eq. (21).

$$F_k(s) = \frac{n}{y}\left(\frac{s}{y}\right)^{n-1} \exp\left[-\left(\frac{s}{y}\right)^{n}\right] \tag{21}$$

where$n$ and y are the shape and scale parameters that can be evaluated using (22, 23) [19, 27, 37]. These parameters are obtained by the mean (μ_s) and standard deviation σ_s of the wind speed.

$$n = \left(\frac{\sigma_s}{\mu_s}\right)^{-1.086} \tag{22}$$

$$y = \frac{\mu_s}{\Gamma(1/k + 1)} \tag{23}$$

The output of WT is calculated using the following equation.

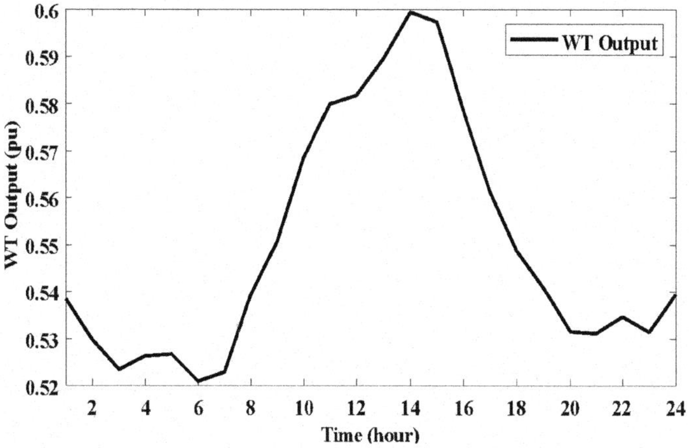

Fig. 3 The uncertainty for the output of WT during the day

$$P_k(s_m) = \begin{cases} 0, & 0 \le s_m < s_{ci} \\ P_{rated} \times \frac{(s_m - s_{ci})}{(s_r - s_{ci})}, & s_{ci} \le s_m < s_r \\ P_{rated}, & s_r \le s_m < s_{co} \\ 0, & s_m \ge s_{co} \end{cases} \tag{24}$$

where P_{rated} and s_m are the rated power of WT and the average wind speed. $P_k(s_m)$ is WT output. s_{ci}, s_r and s_{co} are the cut-in speed, rated speed of WT, and cut-out speed. Equation (25) describes the average WT output in a period (t).

$$P_{WT,\,a} = \int_0^t F_k(s) P_k(s_m)_{dx} \tag{25}$$

5 PV Modeling

Figure 4 describes the output of PV considering the uncertainty of solar irradiance [18]. The solar irradiance (c) can be represented using the Beta function $F_k(c)$ as shown in Eq. (26). φ and ρ are Beta function parameters that can be calculated by the standard deviation θ and mean γ of solar irradiance as given in (27).

$$F_k(c) = \begin{cases} \frac{\Gamma(\rho+\varphi)}{\Gamma(\rho)\Gamma(\varphi)} c^{(\varphi-1)}(1-c)^{(\rho-1)} & 0 \le c \le 1,\ \rho, \varphi \ge 0 \\ 0 & otherwise \end{cases} \tag{26}$$

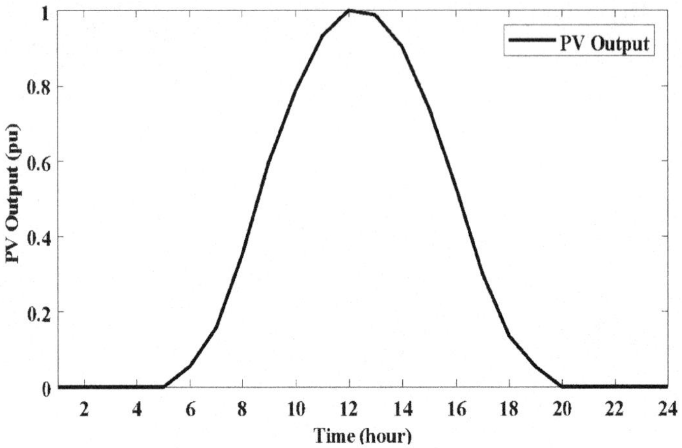

Fig. 4 The uncertainty of PV output during the day

$$\rho = (1-\gamma)\left(\frac{\gamma(1+\gamma)}{\theta^2} - 1\right), \varphi = \frac{\gamma \times \rho}{1-\gamma} \tag{27}$$

The ambient temperature and solar irradiance are the parameters that affect the PV output. Equation (28) evaluates the maximum power of PV output (P_k). The utilized parameters of the PV module are given in [38].

$$P_k(c) = d \times k \times r \times n \tag{28}$$

$$t_e = t_m + c\left(\frac{n_n - 20}{0.8}\right) \tag{29}$$

$$d = c[I_s + k_u \times (t_c - 25)] \tag{30}$$

$$k = v_o - k_o \times t_c \tag{31}$$

$$r = \frac{v_{MP} \times I_{MP}}{v_o \times I_s} \tag{32}$$

where n represents the number of PV modules. n_n, t_e and t_m are the temperature of nominal, cell, and ambient, respectively. I_s and v_o are the short-circuit current and open-circuit voltage. k_o and k_u are the temperature coefficient of voltage and current, respectively.

Equation (33) represents the average power of PV($P_{PV,\, av}$).

$$P_{PV,\, a} = \int_0^t f_k(c) P_b(c)_{dc} \tag{33}$$

6 Sizing of WT Inverter and PV Inverter

The inverter of WT and PV is used to absorb or inject reactive power from or to the system. The inverter rated power of PV and WT are equal to the rated power of PV and WT as shown in Eqs. (32, 33), respectively [39, 40]. Consequently, the maximum reactive power drawn from the inverter of PV and WT can be obtained when the power generation from WT and PV is equal to zero as shown in Eqs. (34, 35), respectively. Also, the reactive power drawn from the inverter of WT and PV is equal to zero when the maximum power generation from WT and PV are obtained, respectively.

$$S_{inverter,\,PV} = Max(P_{PV}) \tag{34}$$

$$S_{inverter,\,WT} = Max(P_{WT}) \tag{35}$$

$$Q^{max}_{inverter,PV} = \mp\sqrt{S^2_{inverter} - P^2_{PV}} \tag{36}$$

$$Q^{max}_{inverter,WT} = \mp\sqrt{S^2_{inverter} - P^2_{WT}} \tag{37}$$

where PV and WT inject or absorb reactive power to RDS with boundaries of $Q^{max}_{inverter,PV}$ and $Q^{max}_{inverter,WT}$, respectively.P_{WT} and P_{PV}, respectively, are the WT and PV outputs. $S_{inverter,WT}$ and $S_{inverter,PV}$ are the inverter rated power of PV and WT, respectively.

7 Metaheuristic Optimization Algorithms

The following recent metaheuristic algorithms are used to determine the superior sizes and locations of PV and WT in RDS.

7.1 Slime Mold Algorithm (SMA)

SMA is inspired by slime molds behaviors in nature [41]. SMA has superior exploitation and exploration phases to obtain the global solution with minimum populations. When slime molds search for food, they use positive and negative feedback behaviors to go to the global results obtained so far and avoid the local solution. Slime molds depend on the odor in the air to search for the quality food location. When the vein reaches the best location of quality food, the vein width will be increased. The SMA algorithm flowchart is illustrated in Fig. 5. The mathematical expression steps for SMA are discussed as shown next.

Step 1: Read maximum iteration (E), number of search agents (w), and system data.

Step 2: Produce a random number of search agents among the upper (U_v) and lower (L_v) values of dimensions by Eq. (36).

$$J(r, w) = L_v(r, w) + rand(U_v(r, w) - L_v(r, w)) \quad (38)$$

where *rand* is a random value among the value of 1 and 0. *w* represents the number of dimensions (control variables).

Fig. 5 The flow chart of the SMA algorithm

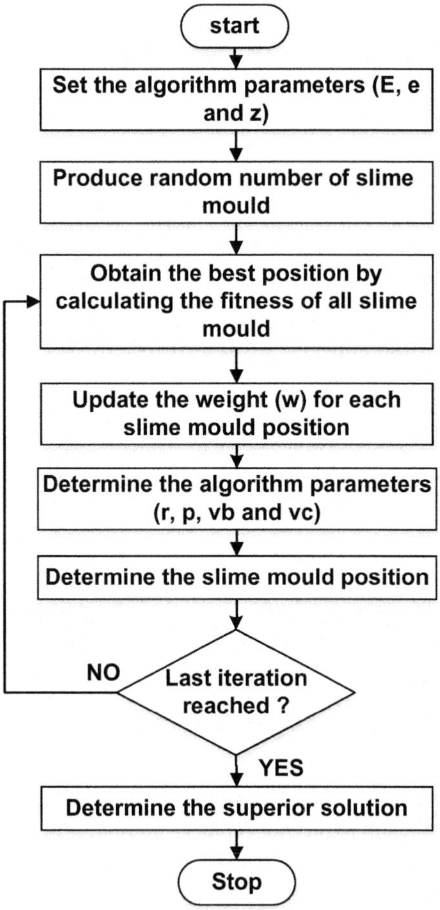

Step 3: The generated solutions are the position of slime molds that are determined by Eq. (39).

$$L = \begin{bmatrix} L_{1,1} & \cdots & L_{1,w-1} & L_{1,w} \\ L_{2,1} & \cdots & L_{2,w-1} & L_{2,w} \\ \vdots & \ddots & \vdots & \vdots \\ L_{N,1} & \cdots & L_{N,w-1} & L_{N,w} \end{bmatrix} \quad (39)$$

where L represents the slim mold position.

Step 4: Obtain the worst and best fitness by evaluating the fitness of slime mold, then obtain the best value of slime mold position.

Step 5: Update the weight (w) of slime mold position by Eq. (40).

$$w = \begin{cases} 1 + r \times \log\left(\frac{B_F - S_F(i)}{B_F - W_F} + 1\right), \textit{Condition} \\ 1 - r \times \log\left(\frac{B_F - S_F(i)}{B_F - W_F} + 1\right), \textit{others} \end{cases} \quad (40)$$

where the first half of slime mold fitness can be represented as Condition. $SF(i)$ is the fitness of slime mold.

Step 6: Produce a new value of slime mold position.

$$\mathrm{L}_{\mathrm{new}} = \begin{cases} vc \times L(t), & r \geq p \\ L_b(t) + vb \times (w \times L_A(e) - L_B(t)), & r < p \\ rand \times (U_v - L_v) + L_v, & rand < z \end{cases} \quad (41)$$

$$a = \arctan h(-(\frac{e}{E}) + 1), \quad (42)$$

$$vb = [-a, a], \quad (43)$$

$$p = \tanh|SF(i) - df|, \quad (44)$$

$$r = rand, \quad (45)$$

$$vc = [-1, 1], \quad (46)$$

where L_A and L_B are the randomly selected value of slime mold position.z is a parameter that equal to 0.03 and df is the best fitness. vb is a value among a and -a and amounting to (0) at the last iteration. vc is a value among 1 and -1 that amounting to (0) at last iteration. E and e are the maximum and the current iteration, respectively.

Step 7: When the last iteration is not obtained, go to Step 4.

Step 8: Obtain the superior position of slime mold.

7.2 Arithmetic Optimization Algorithm (AOA)

The arithmetic operators of mathematics that are called subtraction, addition, multiplication, and division are modeled to create a recent metaheuristic algorithm called the AOA algorithm [42]. In the AOA algorithm, addition and subtraction operators are used in the exploitation phase while the division and multiplication operators are utilized in the exploration phase. Search agents move in the search space to find the promising area during the exploration phase, they move in the promising area to find the global solution during the exploitation phase [43, 44]. The exploitation and exploration phases are determined by a coefficient function called the math optimizer accelerator (MOA). Another coefficient function called math optimizer probability (MOP) is updated during the simulation algorithm. Equations (48, 47, 46, 45) are the mathematical models for addition, subtraction, division, and multiplication operators that are used to determine the new solution position. Figure 6 demonstrates the mathematical expression steps for the AOA algorithm.

$$X_{i,j}(e+1) = best\left(X_j\right) \times MOP \times \left(\left(UB_j\_LB_j\right) \times \mu + LB_j\right) \tag{47}$$

$$X_{i,j}(e+1) = best\left(X_j\right) \div (MOP + \varepsilon) \times \left(\left(UB_j\_LB_j\right) \times \mu + LB_j\right) \tag{48}$$

$$X_{i,j}(e+1) = best\left(X_j\right) - MOP \times \left(\left(UB_j\_LB_j\right) \times \mu + LB_j\right) \tag{49}$$

$$X_{i,j}(e+1) = best\left(X_j\right) + MOP \times \left(\left(UB_j\_LB_j\right) \times \mu + LB_j\right) \tag{50}$$

$$MOP(e) = 1 - \frac{e^{\frac{1}{\alpha}}}{E^{\frac{1}{\alpha}}}. \tag{51}$$

$$MOA(e) = min + e \times \left(\frac{max - min}{E}\right) \tag{52}$$

The mathematical expression steps for MAOA can be explained as follows:

Step 1: Read maximum iteration (E), number of search agents (w), and system data.

Step 2: Determine the AOA parameters such asmin,max, α, and μ to 1, 0.2, 5, and 0.499, respectively.

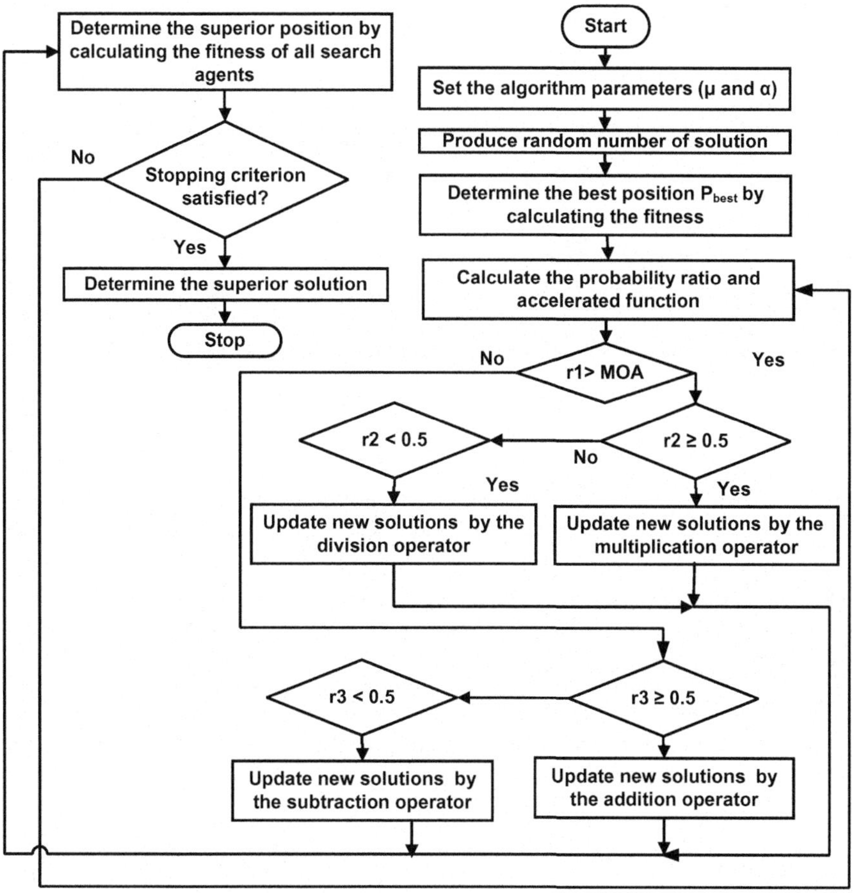

Fig. 6 The flow chart of AOA algorithm

Step 3: Produce a random number of search agents (solutions) among the upper (U_v) and lower (L_v) values of dimensions by Eq. (51).

$$J(r, w) = L_v(r, w) + rand(U_v(r, w) - L_v(r, w)) \tag{53}$$

where *rand* is a random value among the value of 1 and 0.w represents the number of dimensions (control variables).

Step 4: The generated solutions of AOA can be formulated by Eq. (52).

$$L = \begin{bmatrix} L_{1,1} & \cdots & L_{1,D-1} & L_{1,D} \\ L_{2,1} & \cdots & L_{2,D-1} & L_{2,D} \\ \vdots & \ddots & \vdots & \vdots \\ L_{N,1} & \cdots & L_{N,D-1} & L_{N,D} \end{bmatrix} \tag{54}$$

where L represents the search agents position.

Step 5: Obtain the best fitness by evaluating the search agent fitness.

Step 6: Update the accelerated function and the probability ratio.

$$MOP(e) = 1 - \frac{e^{\frac{1}{\alpha}}}{E^{\frac{1}{\alpha}}} \tag{55}$$

$$MOA(e) = min + e \times \left(\frac{max - min}{E} \right) \tag{56}$$

where *max minmax* and *min* are the maximum and minimum magnitudes of the accelerated function that equal to 0.2 and 1, respectively. E and e, respectively are the maximum and current iterations.

Step 7: If r1 > MOA, the exploration phase will operate.

If r2 < 0.5, Eq. (46) will be used as a division operator to determine the updating positions of search agents $X_{i,j}(C_{Iter} + 1) = best(X_j) \div (MOP + \varepsilon) \times ((UB_j\_LB_j) \times \mu + LB_j)$.

Else (r2 $\geq$ 0.5), Eq. (45) will be used as a multiplication operator to determine the updating positions of search agents $X_{i,j}(C_{Iter} + 1) = best(X_j) \times MOP \times ((UB_j\_LB_j) \times \mu + LB_j)$.

Else (r1 $\leq$ MOA), the exploitation phase will operate:

If r3 < 0.5, Eq. (47) will be used as a subtraction operator to determine the updating positions of search agents.

Else (r3 $\geq$ 0.5), Eq. (48) will be used as an addition operator to determine the updating positions of search agents $X_{i,j}(C_{Iter} + 1) = best(X_j) + MOP \times ((UB_j\_LB_j) \times \mu + LB_j)$.

Step 8: Calculate the fitness of the updating positions and determine the superior function.

Step 9: When the last iteration is not obtained, go to Step 4.

Step 10: Obtain the superior solution position.

7.3 Beluga Whale Optimization (BWO) Algorithm

BWO algorithm is inspired by beluga whale behaviors in nature [45, 46]. Beluga whales are living in the sea and have a round and stocky body with pure white color of adults. The beluga whale behaviors including whale fall, preying, and swimming are represented mathematically corresponding to whale fall, exploitation, and exploration phases. In the BWO algorithm, a balance factor (β_f) is utilized to allow the algorithm to move between exploration and exploitation smoothly. Also, the convergence characteristic of the BWO algorithm is enhanced by presenting a levy flight strategy in its exploitation phase. Figure 7 presents the flowchart of the BWO algorithm. The mathematical expression can be formulated as shown next.

$$\beta_f = \beta_0(1 - \frac{e}{2E}) \quad (57)$$

where E and e are the maximum and current iteration number. β_0 is a random magnitude among (1 and 0).

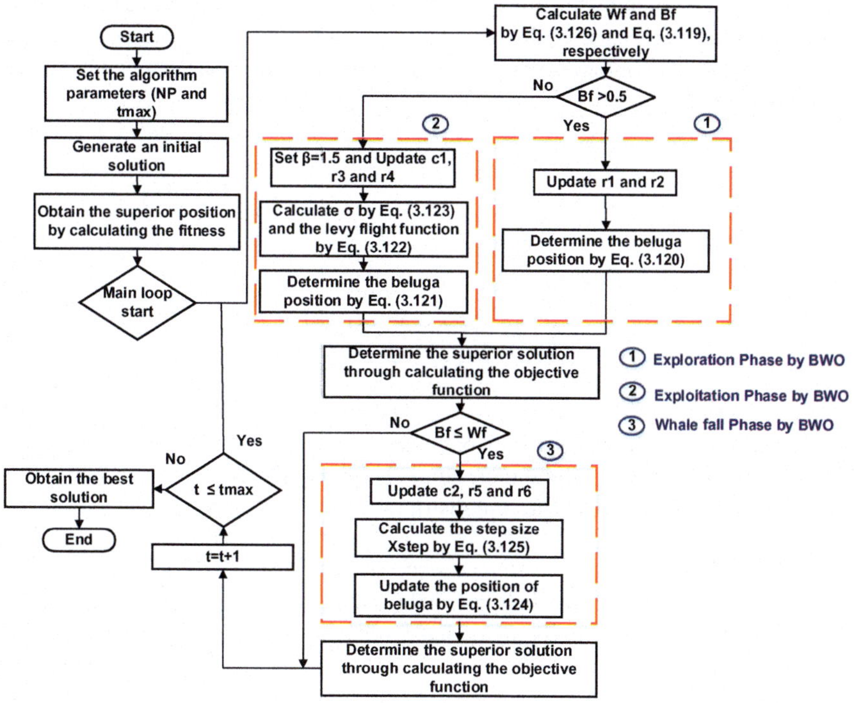

Fig. 7 The flow chart of the BWO algorithm

(1) Exploration phase

This phase describes the swimming behavior of beluga whales. They have social behaviors under different postures such as the pair swim of two beluga whales closely together in a synchronized or mirrored manner. Therefore, the updating position of beluga whales can be represented by Eq. (56).

$$X_{i,j}^{e+1} = \{X_{i,pj}^{e} + \left(X_{r,p1}^{e} - X_{i,pj}^{e}\right)(1 + r_1) sin\, sin(2\pi r_2),$$
$$j = even X_{i,pj}^{e} + \left(X_{r,p1}^{e} - X_{i,pj}^{e}\right)(1 + r_1) cos cos(2\pi r_2), j = odd \tag{58}$$

where $X_{i,j}^{e+1}$ and $X_{i,pj}^{e}$ are the updating position of i-th beluga whale in j-th and pj-th dimensions, respectively. $X_{r,p1}^{e}$ is the updating position of r-th beluga whale in p1-th dimension. r_1 and r_2 are random numbers among [0,1].

(2) Exploitation phase

This phase describes beluga whale behavior in preying. They move near to the neighboring whales, so they are sharing the information with each other to obtain the superior positions for catching the prey. Levy strategy plays a major role in enhancing the convergence characteristic of the exploitation phase. Therefore, beluga whales can reach their prey with a levy strategy and their updating positions can be formulated as follows:

$$x_i^{e+1} = r_3 x_{best}^{e} - r_4 x_i^{e} + C_1 L_F (x_r^{e} - x_i^{e}) \tag{59}$$

$$L_F = 0.05 \times \frac{\mu \times \sigma}{|v|^{1/\beta}} \tag{60}$$

$$\sigma = \left(\frac{(1 + \beta) \times sin(\pi\beta/2)}{((1 + \beta)/2) \times \beta \times 2^{(\beta-1)/2}}\right)^{1/\beta} \tag{61}$$

where x_i^e and x_r^e are the current position and random position of beluga whales, respectively. x_{best}^e is the best position of beluga whale in the swarm and L_F is the levy flight function. r_3 and r_4 are random numbers among [1, 0]. β is a constant with a value equal to 1.5. μ and v represent distributed random numbers.

(3) Whale Fall

During preying behavior, the beluga whale is threatened by humans, polar bears, and killer whales. However, beluga whales are sharing the information with each other to escape from threats, and a small number of them are fallen into the deep sea. This behavior is defined as whale fall behavior. Also, the step size of whale fall is calculated to ensure the number of beluga whales is constant or decreased by fallen them into deep sea. The probability of beluga whale fall can be modeled by decreasing the danger of beluga whale as long as they are close to the prey. Therefore the mathematical expression of this behavior can be formulated as shown next.

$$X_i^{e+1} = r_5 X_i^e - r_6 X_r^e + r_7 X_{step} \tag{62}$$

$$X_{step} = (up - lo)exp(-C_2 e/E) \tag{63}$$

$$w_f = 0.1 - 0.05e/E \tag{64}$$

where X_{step} is the step size of the beluga whale fall. r_5, r_6 and r_7 are random values among [1, 0].w_f is the probability of whale fall that is decreased from 0.1 to 0.05 through the simulation optimization algorithm. *lo* and *up* are the lower and upper limits of variables.

BWO algorithm can be expressed in the pseudo-code as shown next.

```
The pseudo code of BWO algorithm
Input  : system data, system constraints, E and N
Output: best solution (PV location and PV size)
        Produce random value of beluga whales
        Obtain the best fitness by calculating the fitness of all beluga whales
        While e ≤ E Do
          Determine B_f by equation (2.55) and w_f by equation (2.62)
          For i = 1 to N Do
            If B_f(i) > 0.5
           Update r_1, r_2                      // Apply the exploration phase
            Update the beluga whale position by equation (2.56)
             Else If B_f(i) ≤ 0.5
             Update C_1, r_3 and r_4            // Apply the exploitation phase
              calculate σ by equation (2.59)
             Set β=1.5 and evaluate the levy flight function by equation (2.58)
             Calculate the beluga whale position by equation (2.57)
             End If
             Determine the superior position by calculating the whale fitness
           End For
          For i = 1 to N Do
                                       // Apply the beluga whale fall phase
           If B_f(i) ≤ w_f
            Update C_2, r_5 and r_6 ,and calculate the step size X_step
             Update the beluga whale position by equation (2.60)
             Obtain the best fitness by calculating the fitness of all beluga whales
            End If
          End For
            e=e+1
          End While
```

8 Simulation Results

IEEE 69-bus RDS includes a load of 3801.5 kW and 2694.6 kVAr with sixty-eight branches and sixty-nine buses as shown in Fig. 8 [47]. Also, 100 MVA and 12.66 kV are used as base values. The studied system is used with variable commercial load during the day considering the system operating constraints. Without installing PV and WT in RDS, the minimum voltage at bus 65 is 0.9093pu with a power loss of 2173.851 kW. Table 1 presents the system constraints and the algorithm parameters. The simulation results can be presented and discussed as follows:

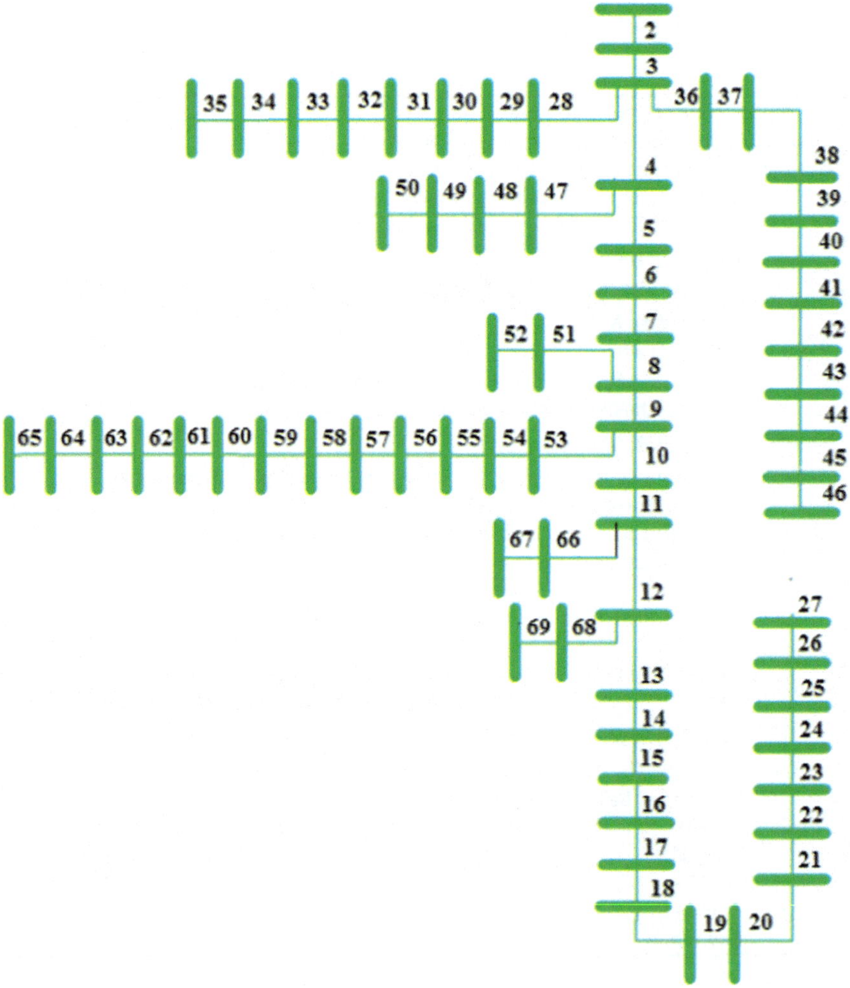

Fig. 8 IEEE 69-bus RDS

Table 1 The algorithm parameters and system constraints

Item	Value
P_{PV} sizing limits	$0\ \text{kW} \leq P_{PV} \leq 4000\ \text{kW}$
P_{WT} sizing limits	$0\ \text{kW} \leq P_{WT} \leq 4000\ \text{kW}$
Voltage limits	$0.90 \leq V_j \leq 1.05$
Maximum iteration	500
Agent number	50

8.1 Installing PV and WT Alone Using SMA in RDS

This studied case illustrates the superior allocation of PV and WT alone using the SMA algorithm in RDS. This case can be discussed into two subcases as shown next.

8.1.1 Integrating PV Alone in RDS

This studied case discusses PV integration in RDS to supply real power only during the day. Table 2 illustrates the superior allocation of three, two, and one PV units in RDS. The superior size of one PV at bus 61 is 2168.238 kW. Also, the superior sizes of the two PVs at buses 17 and 61 are 611.57 kW and 2063.052 kW, respectively. Finally, the superior sizes of three PVs at buses 11, 17, and 61 are 599.51 kW, 439.53 kW, and 1991.45 kW, respectively. Integrating three, two, and one PV units in RDS decreases the system loss to 1021.694 kW, 1038.229 kW, and 1124.54 kW, respectively. Figures 9, 10 and 11 illustrate PVs during the day. The energy of two PVs on buses 17 and 61 are 4603 kWh and 15,527 kWh, respectively, while the energy of one PV on bus 61 is 16319 kWh. The energy for three PVs at buses 17, 11, and 61 is 3308.1 kWh, 4512.1 kWh, and 14,989 kWh, respectively. From Table 2, the least bus voltage is enhanced to 0.9407pu at bus 65 by incorporating three, two, and one PVs in RDS.

8.1.2 Installation of WT Alone in RDS

This studied case discusses WT integration in RDS to supply real power only during the day. Table 3 illustrates the superior allocation of three, two, and one WT units in RDS. The best size of one WT at bus 61 is 2051.9 kW. Also, the best sizes of two WTs at buses 17 and 61 are 573.8 kW and 1954.23 kW, respectively. Finally, the best sizes of three WTs at buses 17, 11, and 61 are 415.747 kW, 552.113 kW, and 1888.498 kW, respectively. Integrating three, two, and one WT units in RDS decreases the power loss to 902.379 kW, 920.147 kW, and 1015.892 kW, respectively. Figures 12, 13 and 14 illustrate of WTs during the day. The energy of two WTs at buses 17 and 61 are 7570.8 kWh and 25,784 kWh, respectively while the energy of one WT at bus 61 is 27073 kWh. The energy for three WTs at buses 17, 11, and 61 are 5485.4 kWh,

Table 2 Simulation results using the SMA algorithm for installing PV alone in RDS

Item	Bus (PV size (kW))	Bus (WT size (kW))	P_{loss} (kW)	Bus (maximum voltage (pu))	Bus (minimum voltage (pu))	Maximum voltage of bus $_{(PV)}$
Without PV and WT	–	–	2173.8506	2 (1.0000)	65 (0.9093)	–
1-PV alone	61 (2168.238)	–	1124.54	2 (1.0000)	65 (0.9407)	61 (0.9967)
2-PV alone	61 (2063.052) 17 (611.57)	–	1038.229	2 (1.0000)	65 (0.9407)	61 (0.9968) 17 (0.9982)
3-PV alone	61 (1991.45) 17 (439.53) 11 (599.51)	–	1021.694	2 (1.0000)	65 (0.9407)	61 (0.9968) 17 (0.9983) 11 (0.9981)

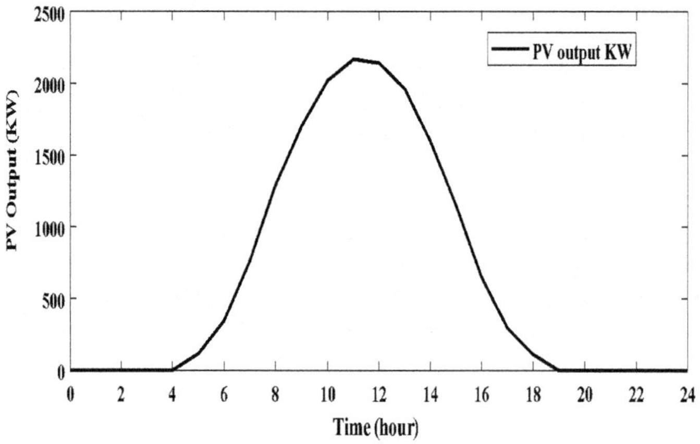

Fig. 9 PV output during 24 h by integrating 1-PV in RDS

7284.6 kWh, and 24,917 kWh, respectively. From Table 2, the least bus voltage is enhanced to 0.9549pu at bus 65 by incorporating three, two, and one PVs in RDS.

8.2 *Simultaneous Allocation of PV and WT Using BWO Algorithm in RDS*

This studied case discusses the simultaneous installation of PV and WT in RDS to supply real power only during the day. Table 4 illustrates the superior allocation of three, two, and one PV and WT units in RDS.

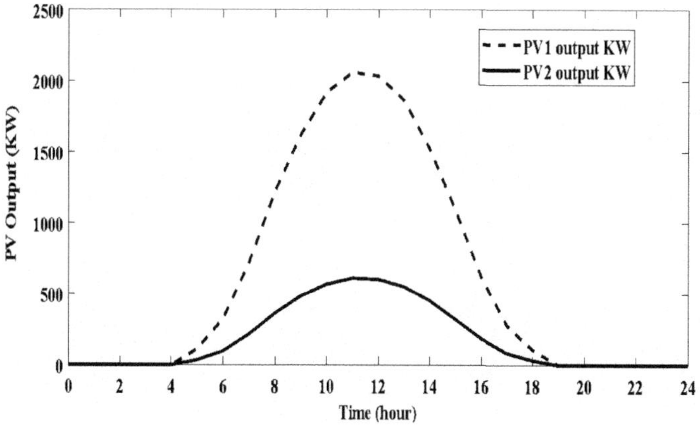

Fig. 10 PV output during 24 h by integrating 2-PV in RDS

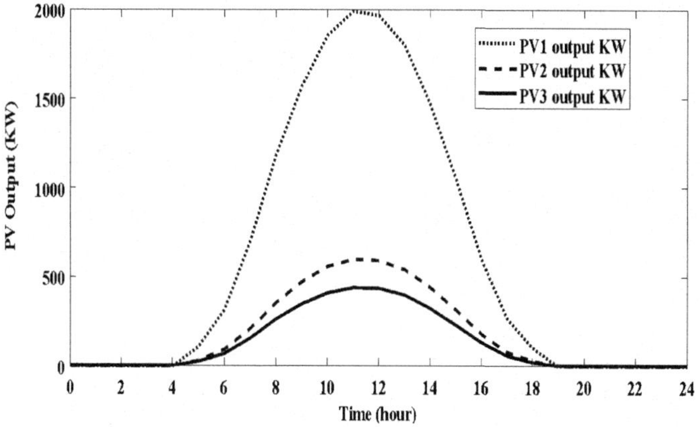

Fig. 11 PV output during 24-h by integrating 3-PV alone in RDS

Table 3 Simulation results using the SMA algorithm for integrating WT alone in RDS

Item	Bus (PV size (kW))	Bus (WT size (kW))	Bus (minimum voltage)	P_{loss} (kW)	Maximum voltage of bus $_{(WT)}$	Bus (maximum voltage)
Without PV and WT	–	–	65 (0.9093)	2173.8506	–	2 (1.0000)
1WT alone	–	61 (2051.9)	65 (0.9549)	1015.892	61 (1.0154)	61 (1.0154)
2WT alone	–	61 (1954.23) 17 (573.8)	65 (0.9549)	920.147	61 (1.0154)	61 (1.0154)
3WT alone	–	61 (1888.5) 11 (552.2) 17 (415.6)	65 (0.9549)	902.37	61 (1.0155)	61 (1.0155)

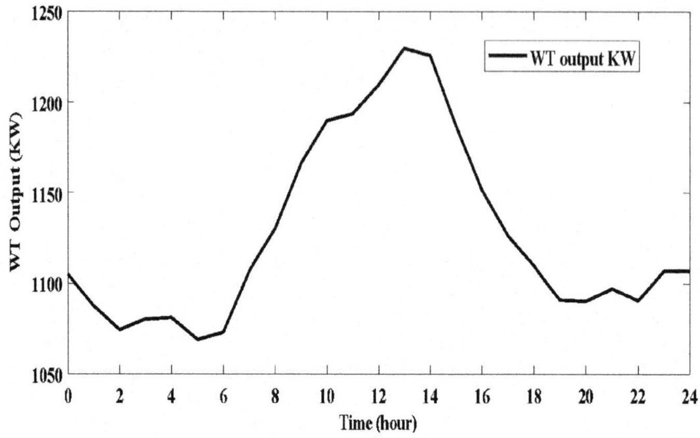

Fig. 12 WT output during 24 h by integrating 1 WT in RDS

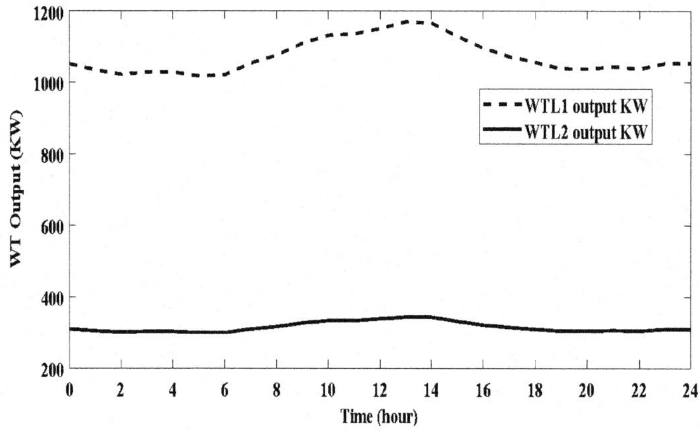

Fig. 13 WT output during 24 h by integrating 2WT in RDS

The superior sizes of one PV and WT at buses 62 and 61 are 1118.95 kW and 1345.58 kW, respectively. The best sizes of two PVs and WTs are 1003.655 kW and 702.779 kW at buses 62 and 11 for PV and 1267.06 kW and 450.948 at buses 61 and 17 for WT, respectively. Also, the best sizes of three PVs and WTs are 1063.331 kW, 908.577 kW, and 330.418 kW at buses 62, 49, and 17 for PV and 1214.872 kW, 209.2 kW, and 560.32 kW at buses 61, 21, and 11 for WT, respectively. Simultaneous installation of three, two, and one units of PV and WT improve the minimum voltage to 0.9589pu at bus 65 and minimize the power loss to 735.342 kW, 751.632 kW, and 875.069 kW, respectively. From Figs. 15 and 16, the total energies of one unit of PV and WT at buses 62 and 61 are 8421.8 kWh and 17,754 kWh, respectively. From Figs. 17 and 18, the total energies of two units of PV and WT are 7554 kWh and 5289.4 kWh at buses 62 and 11 for PV and 16,718 kWh and 5949.8 kWh at

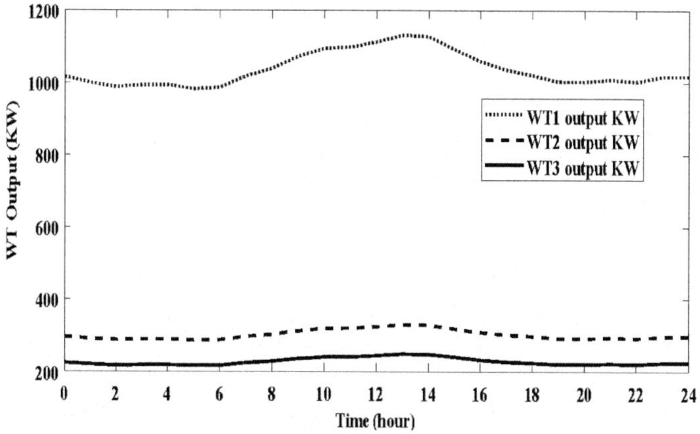

Fig. 14 WT output during 24 hours by integrating 3WT in RDS

Table 4 Simulation results for simultaneous allocation of WT and PV using BWO algorithm in RDS

Item	Bus (WT size (kW))	Bus (PV size (kW))	Maximum voltage of bus (WT)	Maximum voltage of bus (PV)	Bus (maximum voltage)	Bus (minimum voltage)	P_{loss} (kW)
Without PV and WT	–	–	–	–	2 (1.0000)	65 (0.9093)	2173.8506
1 (WT + PV)	61 (1345.58)	62 (1118.95)	61 (1.0029)	62 (1.0028)	61 (1.0029)	65 (0.9589)	875.069
2 (WT + PV)	61 (1267.06) 17 (450.948)	62 (1003.655) 11 (702.779)	61 (1.0029) 17 (1.0028)	62 (1.0028) 11 (0.9996)	61 (1.0029)	65 (0.9589)	751.632
3 (WT + PV)	61 (1214.872) 21 (209.2) 11 (560.32)	62 (1063.331) 49 (908.577) 17 (330.418)	61 (1.0029) 21 (1.0014) 11 (1.0015)	62 (1.0028) 49 (0.9986) 17 (1.0011)	61 (1.0029)	65 (0.9589)	735.342

buses 61 and 17, respectively. Finally, the total energies of three units of PV and WT are 7392.9 kWh, 2760.2 kWh, and 16,029 kWh at buses 11, 21, and 61 for WT and 2486.9 kWh, 8003.1 kWh and 6838.4 kWh at buses 17, 62, and 49 for PV as shown in Figs. 19 and 20, respectively.

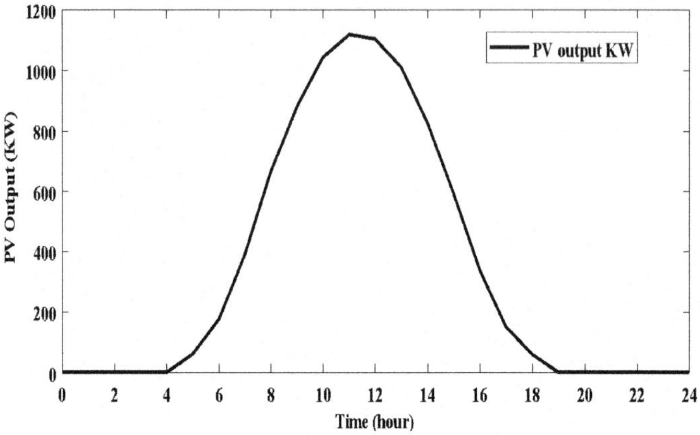

Fig. 15 PV output during 24 h by integrating 1 (PV + WT) in RDS

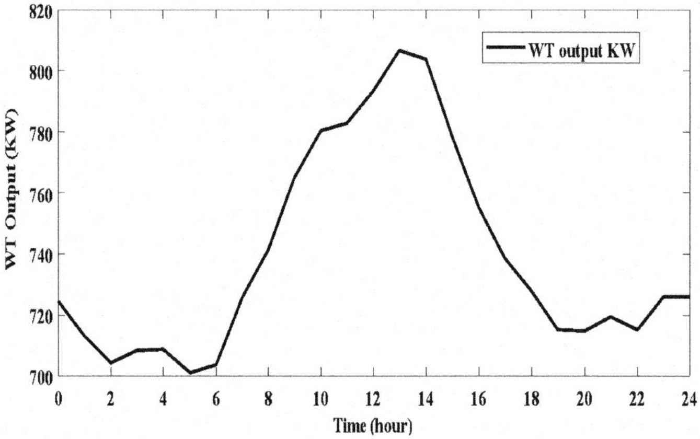

Fig. 16 WT output during 24 h by integrating 1 (PV + WT) in RDS

8.3 Installation of Inverter-Based WT and Inverter-Based PV Using AOA in RDS

In this case, the superior allocation of inverter-based WT and inverter-based PV is obtained using the AOA algorithm in RDS. This case can be divided into two subcases as follows.

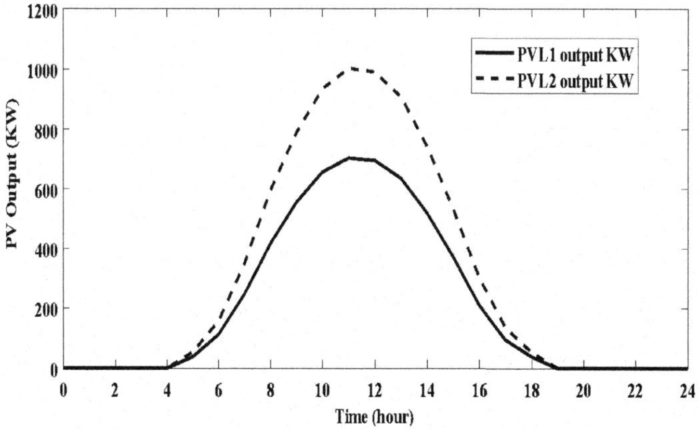

Fig. 17 PV output during 24 h by integrating 2 (PV + WT) in RDS

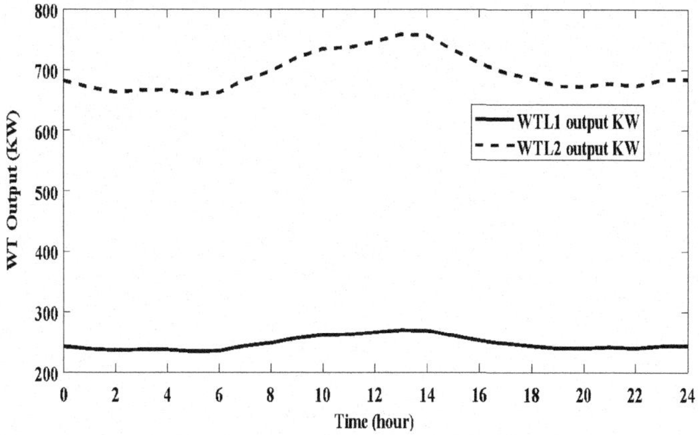

Fig. 18 WT output during 24 h by integrating 2 (PV + WT) in RDS

8.3.1 Installation of Inverter-Based WT in RDS

Incorporating inverter-based WT gives a preferable solutions than installing WT alone in RDS. In this thesis, integrating WT at the same size and location of WT alone is utilized to determine the inverter sizing in RDS. Incorporating one inverter-based WT in RDS improves the minimum voltage to 0.9683pu at bus 27 and minimizes the power loss to 421.855 kW. From Fig. 21, integrating one inverter-based WT in RDS with WT sizing of 2051.9 kW injects reactive energy of 17,497 kVArh by the inverter. Incorporating two inverter-based WT units in RDS enhances the least voltage at bus 65 to 0.9704pu and decreases the power loss to 286.0918 kW, respectively.

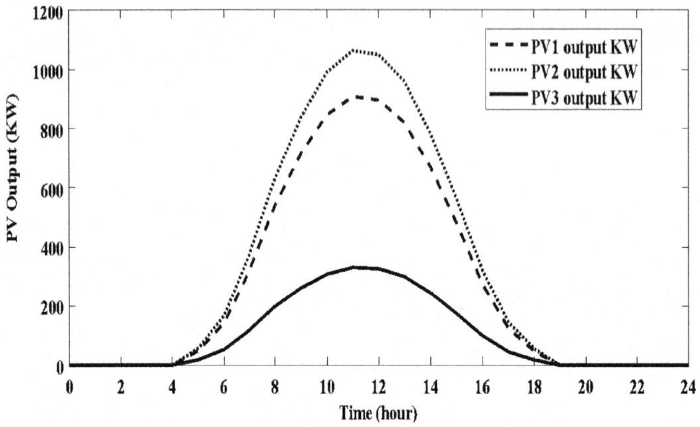

Fig. 19 PV output during 24 h by integrating 3 (PV + WT) in RDS

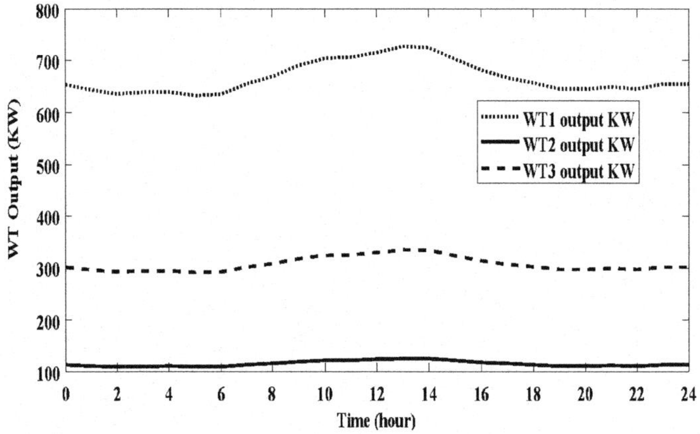

Fig. 20 WT output during 24 h by integrating 3 (PV + WT) in RDS

From Table 5, the maximum reduction in power loss by integrating one, two, and three inverter-based WT units in RDS is 80.6%, 86.8%, and 87.95%, respectively. By incorporating two inverter-based WT units, the WT sizing of 573.8 kW and 1954.2 kW at buses 17 and 61 with inverter energies sizing are 5108.6 kVArh and 16,959 kVArh, respectively as shown in Fig. 22. From Fig. 23, incorporating three inverter-based WT minimize the power loss to 261.884 kW and supply reactive energy of 15,909 kVArh, 5276.7 kVArh and 3864.1 kVArh at buses 61, 11, and 18 by the inverters, respectively.

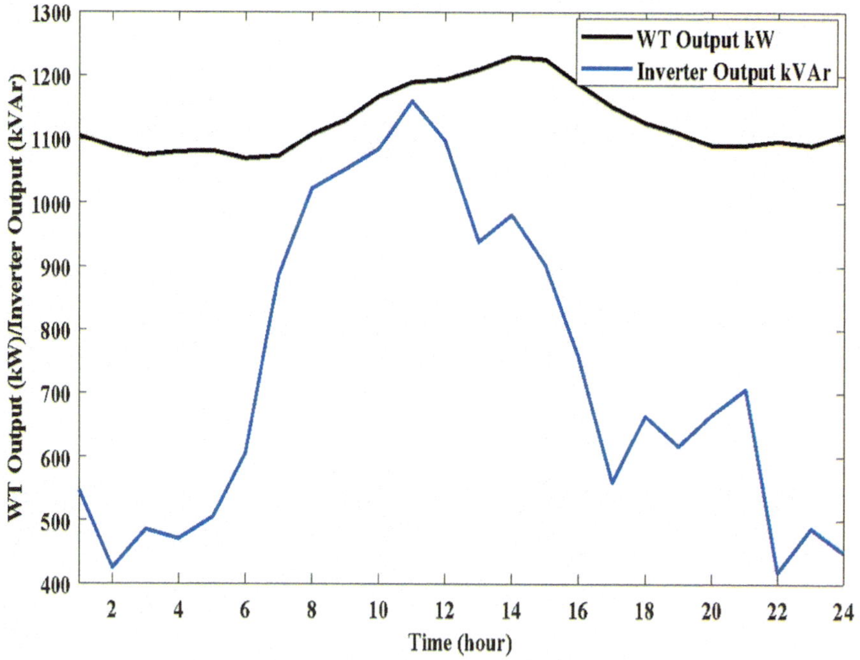

Fig. 21 The output of inverter and WT by integrating one inverter-based WT in RDS

Table 5 Simulation results using the AOA algorithm by integrating inverter-based WT in RDS

Item	Without WT	3-(WT-inverter)	2-(WT-inverter)	1-(WT-inverter)
Location (WT (kW))	–	61 (1888.5) 18 (415.6) 11 (552.2)	61 (1954.2) 17 (573.8)	61 (2051.9)
Location (inverter (kVAr))	–	61 (15,909) 18 (3864.1) 11 (5276.7)	61 (16,959) 17 (5108.6)	61 (17,497)
P_{loss} (kW)	2173.9	261.884	286.0918	421.855
Bus (minimum voltage (pu))	65 (0.9093)	65 (0.9713)	65 (0.9704)	27 (0.9683)
Maximum reduction (%)	–	87.95	86.8	80.6

8.3.2 Installation of Inverter-Based PV in RDS

The simulation results cleared that the installation of inverter-based PV obtains superior solutions than PV integration without enabling its inverter in RDS. Also, PV generation with enabling its inverter is equal to the same value as PV generation without enabling its inverter in RDS. Integrating one, two, and three inverter-based-PV units in RDS reduces the power loss to 631.442 kW, 525.553 kW, and 479.777 kW with power loss reduction of 70.95%, 75.82%, and 77.93% as shown in Table 6. Also,

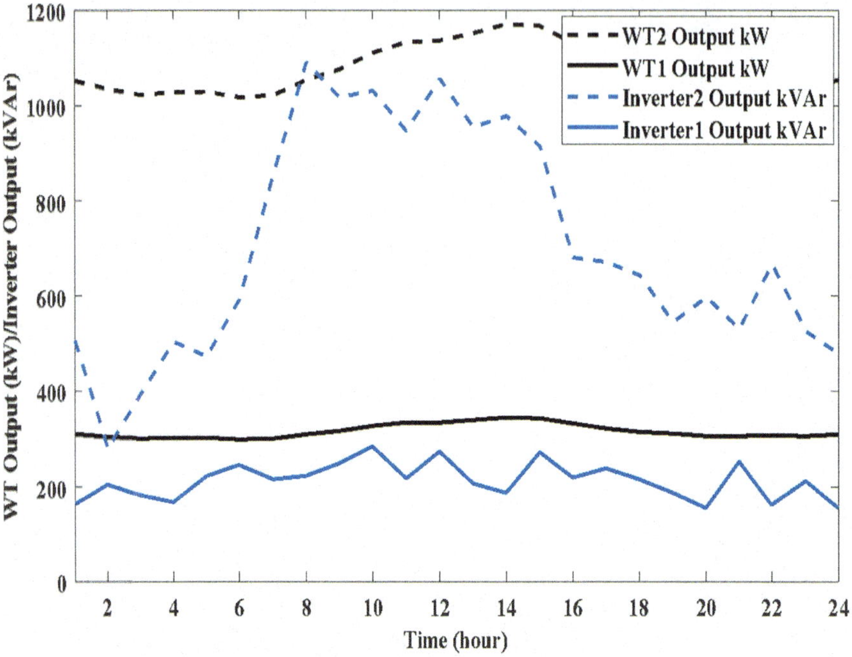

Fig. 22 The output of inverter and WT by integrating two inverter-based WT in RDS

the minimum bus system voltages are enhanced to 0.9559pu, 0.9564, and 0.9566pu at bus 65 by installing one, two, and three of inverter-based-PV units in RDS, respectively. From Fig. 24, the reactive power energy from one PV inverter is 15055 kVArh at bus 61 with a PV output of 2168.238 kW. Also, the injection energy from two PV inverter units is 13078 kVArh at bus 61 and 5108 kVArh at bus 17 as shown in Fig. 25. From Fig. 26, installing three inverter-based-PV units inject reactive power energies of 13,355 kVArh, 3469 kVArh, and 5336 kVArh to RDS at buses 61, 18 and 11, respectively.

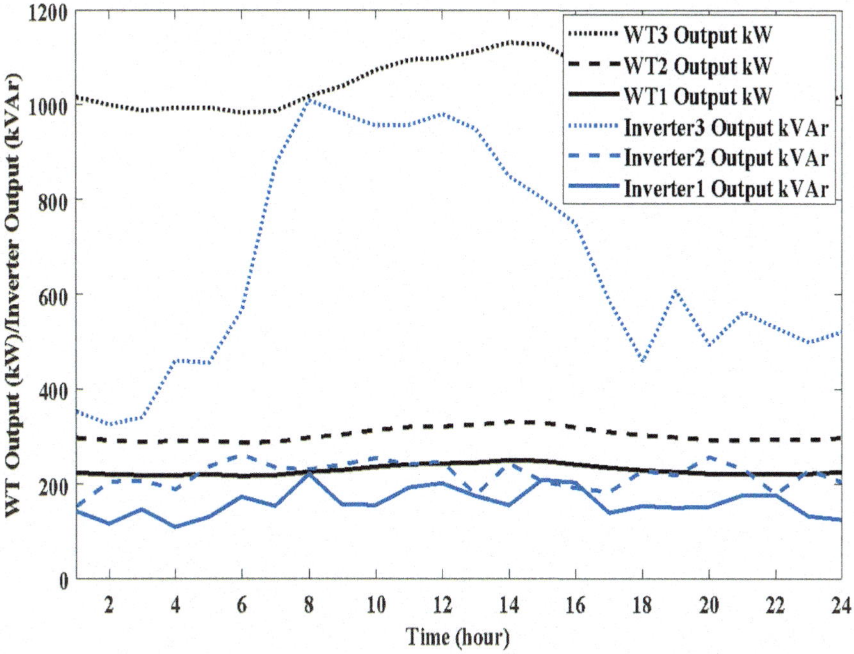

Fig. 23 The output of inverter and WT by integrating three inverter-based WT in RDS

Table 6 Simulation results by installing inverter-based PV using the AOA algorithm in RDS

Item	Bus (PV size (KW))	Bus (inverter energy (kVArh))	Bus (minimum voltage (pu))	P_{loss} (KW)
Without PV	–	–	65 (0.9093)	2173.8506
1-PV	61 (2168.238)	61 (15,055)	65 (0.9559)	631.442
2-PV	61 (2063.052) 17 (611.57)	61 (13,078) 17 (5108)	65 (0.9564)	525.553
3-PV	61 (1991.45) 18 (439.36) 11 (599.68)	61 (13,355) 18 (3469) 11 (5336)	65 (0.9566)	479.777

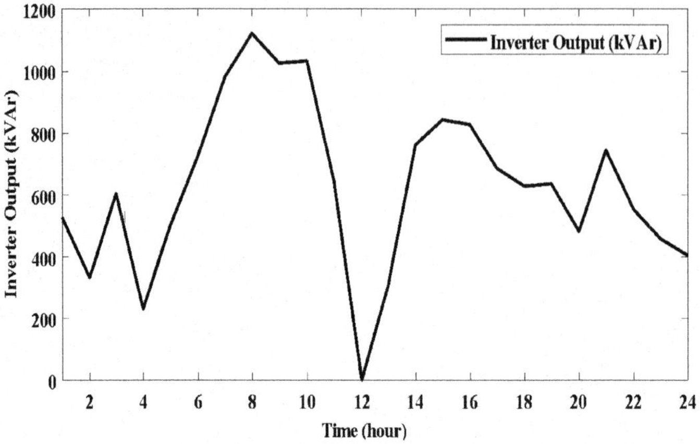

Fig. 24 The output of the inverter during 24 hours by installing one inverter-based PV in RDS

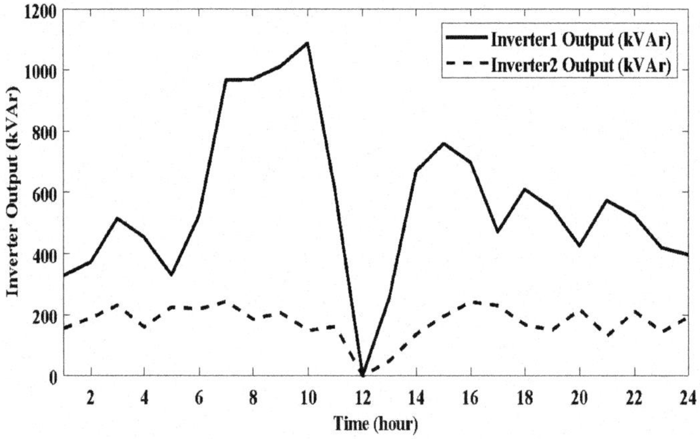

Fig. 25 The output of the inverter during 24 hours by integrating two inverter-based PV in RDS

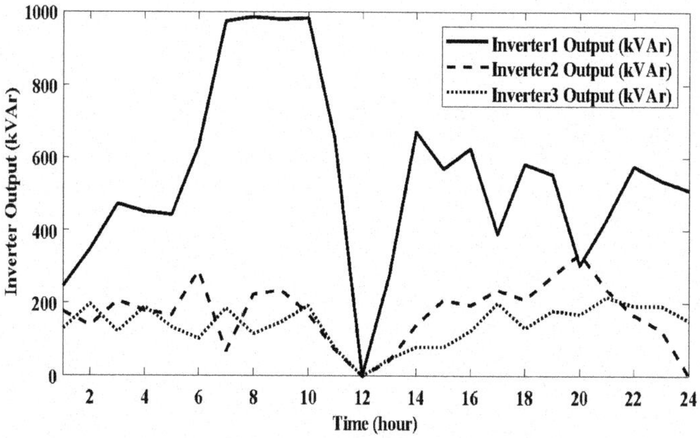

Fig. 26 The output of inverter during 24 hours by installing three inverter-based PV in RDS

9 Conclusion

The aim of this chapter was to examine the integration of multiple and single WT and PV in RDS under different scenarios, using various recent metaheuristic algorithms such as AOA, SMA, and BWO. The objective function used in this study was to decrease total energy loss while taking into account the uncertainties for variable load demand with WT and PV outputs. The findings showed that integrating multiple PV and WT into RDS was more effective than integrating single WT and PV. Furthermore, the results indicated that integrating both WT and PV simultaneously was more advantageous than integrating them individually. In addition, the results revealed that integrating WT with an enabled inverter produced better outcomes than integrating PV with an enabled inverter in RDS.

References

1. Ramli MA, Bouchekara H, Alghamdi AS (2018) Optimal sizing of PV/wind/diesel hybrid microgrid system using multi-objective self-adaptive differential evolution algorithm. Renew Energy 121:400–411
2. Rehman S, Al-Hadhrami LM (2010) Study of a solar PV–diesel–battery hybrid power system for a remotely located population near Rafha, Saudi Arabia. Energy 35(12):4986–4995
3. Ashari M, Nayar C (1999) An optimum dispatch strategy using set points for a photovoltaic (PV)–diesel–battery hybrid power system. Sol Energy 66(1):1–9
4. Trop P, Goricanec D (2016) Comparisons between energy carriers' productions for exploiting renewable energy sources. Energy 108:155–161
5. Lucas H, Pinnington S, Cabeza LF (2018) Education and training gaps in the renewable energy sector. Sol Energy 173:449–455
6. Momete DC (2018) Analysis of the potential of clean energy deployment in the European Union. IEEE Access 6:54811–54822

7. Zubo RH, Mokryani G, Rajamani H-S, Aghaei J, Niknam T, Pillai P (2017) Operation and planning of distribution networks with integration of renewable distributed generators considering uncertainties: a review. Renew Sustain Energy Rev 72:1177–1198
8. Paska J, Biczel P, Kłos M (2009) Hybrid power systems–An effective way of utilising primary energy sources. Renew Energy 34(11):2414–2421
9. Rizzi F, van Eck NJ, Frey M (2014) The production of scientific knowledge on renewable energies: worldwide trends, dynamics and challenges and implications for management. Renew Energy 62:657–671
10. Mekhilef S, Saidur R, Kamalisarvestani M (2012) Effect of dust, humidity and air velocity on efficiency of photovoltaic cells. Renew Sustain Energy Rev 16(5):2920–2925
11. Abdmouleh Z, Gastli A, Ben-Brahim L, Haouari M, Al-Emadi NA (2017) Review of optimization techniques applied for the integration of distributed generation from renewable energy sources. Renew Energy 113:266–280
12. Ludin NA et al (2018) Prospects of life cycle assessment of renewable energy from solar photovoltaic technologies: a review. Renew Sustain Energy Rev 96:11–28
13. Ashfaq A, Ianakiev A (2018) Features of fully integrated renewable energy atlas for Pakistan; wind, solar and cooling. Renew Sustain Energy Rev 97:14–27
14. Raheem A et al (2016) Renewable energy deployment to combat energy crisis in Pakistan. Energy Sustain Soc 6(1):1–13
15. Liu F, Li R, Li Y, Yan R, Saha T (2017) Takagi–Sugeno fuzzy model-based approach considering multiple weather factors for the photovoltaic power short-term forecasting. IET Renew Power Gener 11(10):1281–1287
16. Dincer F (2011) The analysis on photovoltaic electricity generation status, potential and policies of the leading countries in solar energy. Renew Sustain Energy Rev 15(1):713–720
17. Bevrani H, Ghosh A, Ledwich GJIRPG (2010) Renewable energy sources and frequency regulation: survey and new perspectives, vol 4, no 5, pp 438–457
18. Hung DQ, Mithulananthan N, Bansal R (2014) Integration of PV and BES units in commercial distribution systems considering energy loss and voltage stability. Appl Energy 113:1162–1170
19. Abul'Wafa AR (2013) Optimization of economic/emission load dispatch for hybrid generating systems using controlled Elitist NSGA-II. Electric Power Syst Res 105:142–151
20. Siddaiah R, Saini R (2016) A review on planning, configurations, modeling and optimization techniques of hybrid renewable energy systems for off grid applications. Renew Sustain Energy Rev 58:376–396
21. Dawoud SM, Lin X, Okba MI (2018) Hybrid renewable microgrid optimization techniques: a review. Renew Sustain Energy Rev 82:2039–2052
22. Zia MF, Elbouchikhi E, Benbouzid M (2018) Microgrids energy management systems: a critical review on methods, solutions, and prospects. Appl Energy 222:1033–1055
23. Das CK, Bass O, Kothapalli G, Mahmoud TS, Habibi D (2018) Optimal placement of distributed energy storage systems in distribution networks using artificial bee colony algorithm. Appl Energy 232:212–228
24. Kalkhambkar V, Kumar R, Bhakar R (2017) Joint optimal sizing and placement of renewable distributed generation and energy storage for energy loss minimization. in 2017 4th international conference on advanced computing and communication systems (ICACCS), pp 1–9
25. Chedid R, Sawwas A (2019) Optimal placement and sizing of photovoltaics and battery storage in distribution networks. Energy Storage 1(4):46
26. Teng J-H, Luan S-W, Lee D-J, Huang Y-Q (2012) Optimal charging/discharging scheduling of battery storage systems for distribution systems interconnected with sizeable PV generation systems. IEEE Trans Power Syst 28:1425–1433
27. Ali A, Mahmoud K, Raisz D, Lehtonen M (2020) Optimal allocation of inverter-based WTGS complying WITH THEIR DSTATCOM functionality and PEV eequirements. IEEE Trans Veh Technol 69(5):4763–4772
28. Abdel-Mawgoud H et al (2021) A modified manta ray foraging optimizer for planning inverter-based photovoltaic with battery energy storage system and wind turbine in distribution networks. IEEE Access 9:91062–91079

29. Elkadeem M, Abd Elaziz M, Ullah Z, Wang S, Sharshir SW (2019) Optimal planning of renewable energy-integrated distribution system considering uncertainties. IEEE Access 7:164887–164907
30. Eminoglu U, Hocaoglu MH (2008) Distribution systems forward/backward sweep-based power flow algorithms: a review and comparison study. Electric Power Components Syst 37(1):91–110
31. Ali E, Elazim SA, Abdelaziz A (2016) Ant lion optimization algorithm for renewable distributed generations. Energy 116:445–458
32. Abdel-mawgoud H, Kamel S, Ebeed M, Youssef A-R (2017) Optimal allocation of renewable dg sources in distribution networks considering load growth. in 2017 Nineteenth international middle east power systems conference (MEPCON), IEEE, pp 1236–1241
33. El-Fergany A (2015) Optimal allocation of multi-type distributed generators using backtracking search optimization algorithm. Int J Electr Power Energy Syst 64:1197–1205
34. Aman M, Jasmon G, Bakar A, Mokhlis H (2014) A new approach for optimum simultaneous multi-DG distributed generation Units placement and sizing based on maximization of system loadability using HPSO (hybrid particle swarm optimization) algorithm. Energy 66:202–215
35. Lopez E, Opazo H, Garcia L, Bastard P (2004) Online reconfiguration considering variability demand: applications to real networks. IEEE Trans Power Syst 19(1):549–553
36. Price W et al (1995) Bibliography on load models for power flow and dynamic performance simulation. IEEE Power Eng Rev 15(2):70
37. Jangamshetti SH, Rau VG (1999) Site matching of wind turbine generators: a case study. IEEE Trans Energy Convers 14(4):1537–1543
38. Teng J-H, Luan S-W, Lee D-J, Huang Y-Q (2012) Optimal charging/discharging scheduling of battery storage systems for distribution systems interconnected with sizeable PV generation systems. IEEE Trans Power Syst 28(2):1425–1433
39. Abdel-mawgoud H, Kamel S, Shahinzadeh H (2022) Strategic planning for incorporating inverter-based wind turbine in distribution networks using arithmetic optimization algorithm. in 2022 9th Iranian conference on renewable energy & distributed generation (ICREDG), pp 1–7
40. Ali A, Mahmoud K, Raisz D, Lehtonen M (2020) Probabilistic approach for hosting high PV penetration in distribution systems via optimal oversized inverter with watt-var functions. IEEE Syst J 15:684–693
41. Li S, Chen H, Wang M, Heidari AA, Mirjalili S (2020) Slime mould algorithm: a new method for stochastic optimization. Futur Gener Comput Syst 111:300–323
42. Abualigah L, Diabat A, Mirjalili S, Abd Elaziz M, Gandomi AH (2021) The arithmetic optimization algorithm. Computer methods in applied mechanics and engineering, vol 376, p 113609
43. Cheng S, Lu H, Lei X, Shi Y (2018) A quarter century of particle swarm optimization. Complex Intell Syst 4:227–239
44. Kaur S, Awasthi LK, Sangal A, Dhiman G (2020) Tunicate Swarm Algorithm: a new bio-inspired based metaheuristic paradigm for global optimization. Eng Appl Artif Intell 90:103541
45. Zhong C, Li G, Meng Z (2022) Beluga whale optimization: a novel nature-inspired metaheuristic algorithm. Knowl-Based Syst 251:109215
46. Houssein EH, Sayed A (2023) Dynamic candidate solution boosted beluga whale optimization algorithm for biomedical classification. Mathematics 11(3):707
47. Sahoo NC, Prasad K (2006) A fuzzy genetic approach for network reconfiguration to enhance voltage stability in radial distribution systems. Energy Convers Manage 47(18–19):3288–3306. https://doi.org/10.1016/j.enconman.2006.01.004

Chapter 5
Metaheuristic Algorithms for the Classification and Prediction of Skin Lesions: A Comprehensive Review

Azmath Mubeen and Uma N. Dulhare

Abstract One of the most challenging tasks in diagnosing skin lesions is classifying their types and predicting their severity. In many of the cases, classification of skin lesions and extraction of features becomes a more important issue. Accuracy in classification is one of the major drawbacks in many of the systems found in the literature. Many of the researchers have suggested binary classification which may not be sufficient for diagnosing various skin lesions. Metaheuristic algorithms have played a major role in solving sophisticated problems. Metaheuristic methods have shown efficient results in predicting and classifying different diseases. With the use of metaheuristic techniques, the computing accuracy of the models can be improvised. To diagnose skin lesions at an early stage, classification of different types using metaheuristic techniques is a good idea. As the patient's health care costs and slow course of diagnosis, new techniques that can diagnose the skin disease at early stages have become particularly important for society. This work reviews and analyses various metaheuristic techniques like Hybrid Whale Optimization (HWO), Bat Algorithm (BA), Apiary the ABC (Artificial Bee Colony), Ant Colony Optimization (ACO), and Firefly optimization for skin lesions categorization and the accuracy of methods in prediction and detection of various skin cancers. The work analyses the working and performances of the different metaheuristic algorithms.

Keywords Metaheuristic techniques · Skin lesions · BA · HWO · ABC · Firefly · ACO · Classification

---

A. Mubeen (✉)
Department of Computer Science and Engineering, University College of Engineering, Osmania University, Hyderabad, Telangana, India
e-mail: azmathresearch2018@gmail.com

U. N. Dulhare
Department of AI & Computer Science and Engineering, Muffakham Jah College of Engineering & Technology, Hyderabad, Telangana, India

U. N. Dulhare and E. H. Houssein (eds.), *Machine Learning and Metaheuristics: Methods and Analysis*, Algorithms for Intelligent Systems,
https://doi.org/10.1007/978-981-99-6645-5_5

1 Introduction

Medical experts ought to demand a basic comprehension of the skin in order to comprehend skin cancer properly. The epidermis, dermis, and subcutaneous fat are the three major layers [1]. Skin cancer develops when an abnormal skin cell formation goes unchecked [2]. Fresh derma cells replace the old derma cells regularly as they die. However, if this process goes in the wrong direction, it can lead to a situation in which the former units are no longer in the process of fading, but instead pass away and new cells develop where none are required. A mass of tissue is formed by the excess skin cells, which develop into a tumour [3, 4]. Dermoscopy has been implemented to improve the diagnosis procedure; it is a non-surgical method that is ignited and imageries of skin specks improved. The dermatologist uses this method to detect skin cancer instead of the more time-consuming and inaccurate visual inspection and manual screening that was previously used [5]. But thanks to the most recent machine learning methods for timely detecting of lethal malignant diseases, improvements have been made. If found early enough, certain skin tumours are benign and treatable, seldom progressing to malignancy. Skin tumours in the melanoma group are amongst the most dangerous and devastating. There are countless further subtypes of skin tumours, and each cancer kind is based on how the aberrant cells behave. Melanoma [6], squamous, and basal cell carcinoma (together known as nonmelanocytic skin cancers) are the three kinds of skin cancer. Merkel cell skin cancer [7]. The human body undergoes many alterations, just as skin cancer has a variety of symptoms. The skin's colour fading, the mole's size growing, and ulceration are the changes that take place. These signs might, however, differ from one type to another. Basal cell carcinoma (BCC) typically manifests as a tumour with blood vessels that are visible and a smooth, swelling surface on the neck and shoulder. These tumours form as a result of bleeding from the tumour's central location, yet they are entirely curable. Cancer manifests in the shape of nodules which appear firm when present in squamous cell carcinoma (SCC). Ulceration and tumour-related haemorrhage are potential outcomes. If this kind of cancer is not treated right away, it spreads rapidly throughout the skin. Merkel skin carcinoma (MCC) cancer is mistaken for a cyst because it appears as a painless red patch [8]. Ultraviolet (UV) is a dangerous, reversible damaging cause for skin tumour and other skin conditions that are influenced by the environment. UV has a conflicting impact on individuals well-being because it is moreover regarded as a significant source of vitamin D. UV exposure in excess can lead to health problems like cancer, altered pigmentation, wrinkles, and atrophy. According to epidemiological and molecular research, various types of skin cancer are linked to the UV. Another factor that may contribute to skin diseases is the genetic component [9]. Since it is well known that the prevalence of skin illnesses is rising annually, it is crucial to take preventative steps to avoid these conditions. Skin protection from UV radiation is essential because they are a common environmental cause of skin cancer. Avoiding direct midday sun exposure is the best way to protect yourself from UV radiation, along with using sunscreen and wearing suitable clothing. To prevent such dangerous diseases, it is crucial to perform self-examinations and schedule routine check-ups with a dermatologist [10, 11].

1.1 Motivation and Impact

The work aims to cover as many facets and components of the problem domain as possible. This review's objective is to present the skin cancer computer-aided diagnosis techniques that are now accessible. Methods currently in use are appraised in accordance with predetermined evaluation standards after systematic reviews have been gathered. The information is all gathered from major bases, and following study, it provides details regarding the techniques used, the outcomes, and any problems the researchers ran across. Skin tumour exhibits an exceptionally high mortality rate today and is expected to continue to do so in the future. These facts encourage us to participate by attempting to review as much material as possible in order to better comprehend the problem domain and the suggested remedies.

1.2 Purpose and Goals

In an early stage, the detecting tumour and its types with the use of technology can start and open a new avenue in study and minimise manual labour. The serious situation brought on by the increase in skin cancer instances makes early detection of the disease crucial in saving the patient's life. This study focuses on classifying skin tumour by assessing several detection methods. Image processing, computer vision, and machine learning techniques handle all machine-based processing.

2 Related Work

A computer-aided diagnostic system has several stages, which are described below with different metaheuristic algorithms. Various algorithms have been proposed by different researchers to provide accurate classification and prediction of skin lesions. Different researchers have suggested various algorithms and a sequence of phases for the categorization and prediction. Some of the works are discussed below [12–24].

Essam et al. [25] suggested the opposition-based golden jackal optimizer (IGJO), an effective variant of the newly created golden jackal optimisation (GJO) method. With Otsu's approach as the actual task, the multilevel thresholding problem is solved using the IGJO algorithm. Four common implementation procedures—peak signal-to-noise ratio, structure similarity index, feature similarity index, and mean square error—are used to evaluate the functioning of the algorithm. According to experimental findings, the method proposed by Essam et al. successfully solves the segmentation problem and outperforms competing algorithms in terms of PSNR, SSIM, FSIM, and MSE segmentation metrics [25].

In [26] Mahum et al. proposed a model in which a GF filter was used in the preprocessing stage to eliminate noise from the imageries. After that using Local Binary Patterns (LBP) and Inception V3 for automated attribute extraction, the features were manually extracted. In addition, the learning rate modifications were made using an Adam optimizer. LSTM network was deployed on fused characteristics to classify skin tumour into malignant and benign forms. The advantages of both ML- and DL-based algorithms [27–31] were used in the suggested system. The researchers used the skin lesion DermIS dataset, which can be found on the Kaggle website and contains 1000 photos, 500 of which are benign and 500 of which are malignant. 98.7% precision, 98.66% recall, and 99.4% accuracy were obtained for the model [26].

In [32] the study presents a rapid identification method that, with the aid of computer vision architectures, can quickly identify the monkeypox virus. This may be taken into account for a preliminary assessment of skin abrasions and assist the sufferer in isolating self as they may be careful and help halt the spread of the disease. The authors used deep learning models; much research has been done to identify the monkeypox. In this study [32], the researchers evaluated the test results from deep learning CNN models AlexNet and GoogLeNet using a transfer learning strategy to select the most effective model. When the algorithms were tested the model was able to achieve the maximum accurateness of 83.61% for AlexNet and 82.64% for Google Neural Network by varying the batch sizes and number of epochs [32].

In our present work some other contributions of various researchers have been divided and explained in various steps of how a model works and which researcher has employed which algorithm or technique in each of the phases. The various phases and the different methods employed by various researchers are given below.

2.1 *Image Acquisition*

Datasets

Afza et al. proposed [33] multiclass skin lesion classification method. The method recommended by the authors is composed of a sequence of phases. The primary phase is acquiring of images. The authors have used two datasets in their work, they are—HAM-10000 and ISIC-2018. The two datasets' details are provided below.

Dataset HAM1000: One of the largest datasets is the HAM-10000 "Human Against Machine with 10,000 Training Images" dataset, that comprises 10,015 dermoscopy photos in total that are used to recognise pigmented skin lesions. The seven classifications that make up this dataset are melanocytic nevus (nv = 6705), actinic keratosis (akiec = 327), dermatofibroma (df = 115), basal cell carcinoma (bcc = 514), vascular lesion (vacs = 115), benign keratosis (bkl = 1099), and melanoma

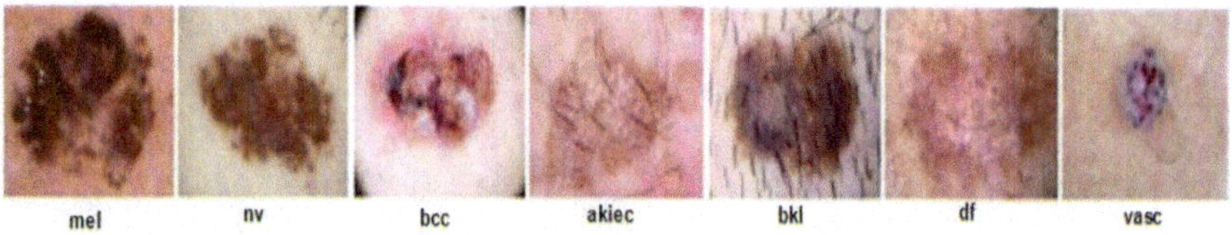

Fig. 1 Sample images of HAM1000 dataset. melanoma (mel), melanocytic nevus (nv), basal cell carcinoma
(bcc), actinic keratosis intraepithelial carcinoma (akiec), benign keratosis (bkl), dermatofibroma (df), and vascular lesion (vasc) [33]

(mel = 1113). There are 54% male and 45% female skin abrasion photos in the collection. It is a large, complicated dataset in which the abrasion imageries have low inter-class and high intra-class variance concerns, making it complicated to classify these skin classes and increasing the likelihood of misdiagnosis. Figure 1 displays a number of imageries of various skin abrasions from the HAM dataset.

ISIC Dataset 2018: The International Skin Imaging Collaboration (ISIC) released the ISIC 2018 dataset, a sizable dataset of dermoscopy images that had more than 12,500 images. Lesion segmentation, trait discovery, and categorization of disease are the three tasks that the dataset completes [34]. Additional 10,000 imageries from 7 different categories make up this dataset for the classification challenge [35]. Figure 2 presents the sample images. There are two main issues with the ISIC 2018 challenge: first, the scanty quantity of imageries in classes and secondarily, the unequal number of imageries in the various categories makes it challenging for the classifier to correctly classify some classes.

Khan et al. [36] in their work have used the results of 4 datasets, comprising ISBI 2016, ISBI 2017, ISIC 2018, PH2, and HAM-10000, were developed to verify their frameworks. The above section already discusses about ISIC 2018 dataset and the HAM-10000 dataset the datasets ISBI 2016, ISBI 2017, ISIC 2018 and PH2 datasets will be discussed.

ISBI 2016 Dataset: This dataset [24] has 379 testing images and 900 training images altogether. The segmentation task can be validated using the openly accessible ground truth images. Melanoma and benign classes are represented in the training images. While the remaining test images were expended to assess the newly developed CNN prototype for the segmentation job, these imageries were used to train a CNN model for the lesion segmentation task. Figure 3 includes a reproduction of a few representative images.

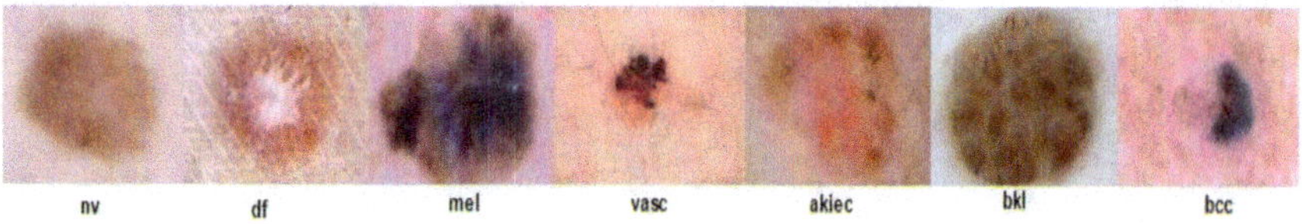

Fig. 2 Sample images of ISIC 2018 dataset [33]

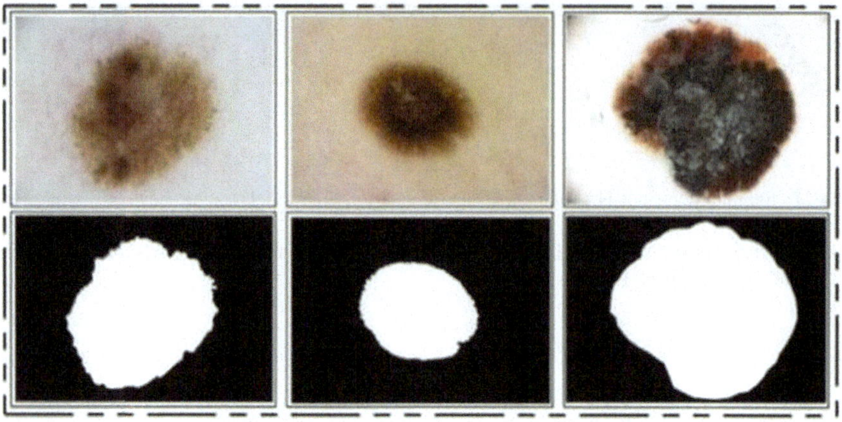

Fig. 3 Sample images of ISBI 2016 for skin lesion segmentation [33]

ISIC 2017 collection: The collection includes 2750 dermoscopy images altogether. Two thousand of those are expended for training, one hundred for validation, and six hundred for testing [33]. Also available to the public are the reference data samples of this dataset, that were expended to check the segmentation algorithm's accuracy. In Fig. 4, a few examples images are displayed. Images were divided into three groups for the classification task: melanoma (374), seborrheic keratosis (254), and nevi (1372). Due to the unbalanced nature of these classes, the authors augmented the data [36] (Fig. 5).

The ISBI 2018 data source is split into 3 sections: lesion segmentation, attribute detection, and lesion type classification, an aggregate of 2594 training imageries and

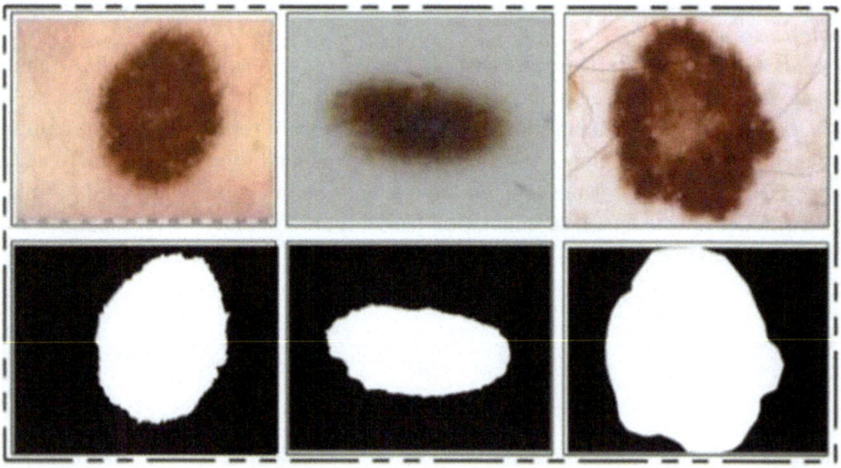

Fig. 4 Sample images of the ISIC 2017 dataset [36]

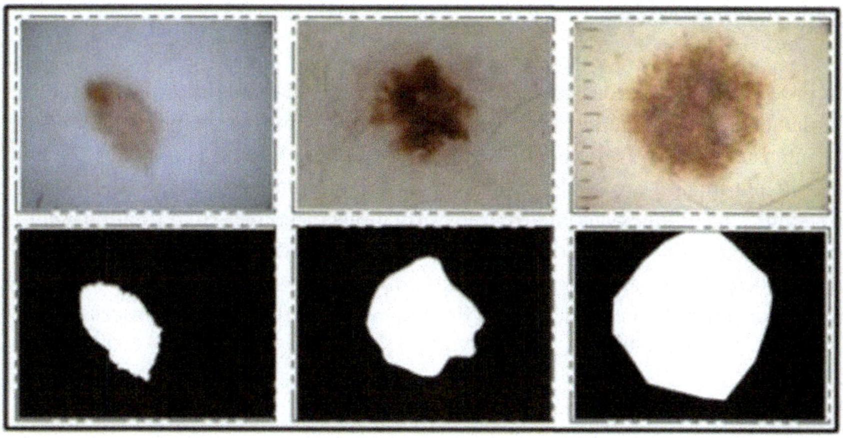

Fig. 5 Sample images of the ISBI 2018 dataset [36]

real-world images are specified for segmentation, whereas 1000 and 100 imageries are provided for testing and validation, correspondingly. Figure 3 displays a few sample imageries coupled with experimental imageries.

Malik et al. [37] have used three openly available data sources- ISIC-2016,2017 and 2018. The authors have implemented their model using a hybrid metaheuristic pre-processor called BA-ABC to boost images' contrast while maintaining their brightness. Their model performed well, which was confirmed by the performance matrix and the visual results of the boundary estimate techniques. In the final results, authors obtained the dice coefficient to 94.6%.

In [38] Garg et al. used the dermatological facility of Hospital Pedro Hispano (PH2) dataset and the International Skin Imaging Collaboration (ISIC) records dataset, the sources of the input sample images.

The PH2 dataset was created for benchmarking and investigation aims to promote relative analysis on dermoscopic image segmentation and categorization methods. The PH2 database contains dermoscopic images that were gathered at the Hospital Pedro Hispano in Matosinhos, Portugal's dermatology department.200 dermoscopy imageries of melanocytic lesions, including 80 common nevi, 80 atypical nevi, and 40 melanomas, are incorporated in this image archive. All of the images in the PH2 database have medical annotation, such as lesion segmentation, clinical and histological diagnosis, and evaluation of a number of dermatoscopic conditions (colours, pigment network, dots/globules, streaks, regression regions, blue-whitish veil, etc.).

2.2 Preprocessing

Preprocessing is the word for operations on images that are performed at the most fundamental stage; together the keyed in and resultant images are intensity imageries. An intensity imagery is frequently denoted by a matrix of illumination rates, and these identifiable images are of the same type as the original sensor data. Because they use similar techniques, geometric transformations of images (like rotation, scaling, and translation) are classified here as preprocessing methods. However, the objective of preprocessing is an improvement of the imagery facts that dominate unintended deformities or enhance some imagery features essential for further processing.

2.2.1 Enhancing the Contrast

Enhancing the local contrast of a picture using contrast is a crucial stage in the subject of medical imagery. Contrast stretching is typically used in medical image processing to identify skin lesions accurately; in this step, contrast stretching is used to extract more valuable features. There are several methods for improving global contrast that are discussed in the literature, but there aren't many that concentrate on local contrast stretching.

In the work [33] Afza et al. proposed a hybrid technique for contrast stretching that relies on both a skewness function and absolute mean deviation is proposed. This technique makes the image more vivid by increasing the contrast of the skin abrasion spot. This increases the image's value for subsequent processing.

Consider Dermoscopic database Δ, and input picture X having a dimension of $N \times M$. Let n stand for the overall quantity of picture elements in the imagery and *xi* for each individual pixel. Equations (1) and (2) are used to formulate the Absolute Mean Deviation (AMD) and skewness [33].

$$\text{AMD} = \frac{1}{n}\sum_{i=1}^{n} |x_i - m(X)| \tag{1}$$

$$\text{Skewness (SK)} = \frac{\sum_i^N \left(X_i - \overline{X}\right)^3}{(N-1) * \sigma^3} \tag{2}$$

$$I_1 = |\text{MD}^{\sim}(i) + X| \tag{3}$$

$$I_F = |I_1 - \text{SK}^{\sim}(i)| \tag{4}$$

where I stands for the image's pixels and I_F represents the final converted image. Prior to deep learning model training, this procedure is used on all of the chosen datasets. Figure 6 presents a few representative outcomes. This graphic makes it

(a)

(b)

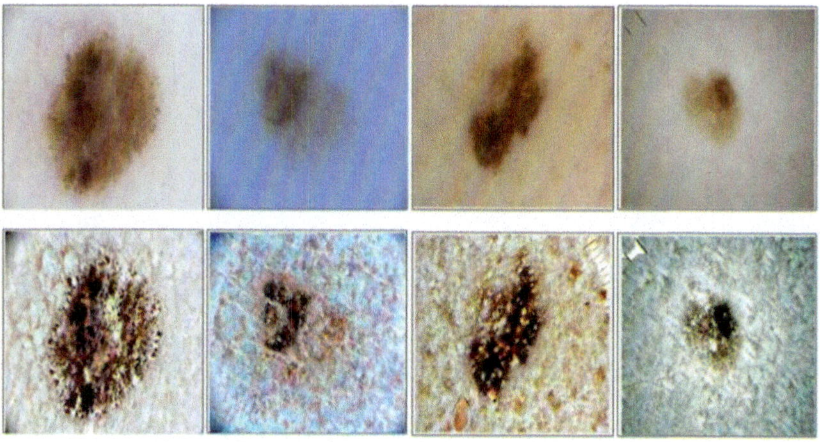

Fig. 6 Contrast enhancement results. **a** Original dermoscopic images; **b** Suggested contrast boost [33]

crystal evident how much simpler it is to locate the lesion site using the suggested change.

2.2.2 Lesion Contrast Stretching

Contrast boosting is one of the fundamental parameters for evaluating image quality. Image enhancement is the technique of enhancing a picture's quality by enhancing a select number of features or reducing blur between specific image pixels. The improvement of image quality over the original image is the main goal of this step. Our main goal was to make the lesion region more contrasty so that the region of interest (ROI) could be extracted with ease. Histogram equalisation (HE), one of the most well-known strategies, is one of many that are discussed in the literature. In HE, all picture elements are enhanced, but the lowlier and top boundaries are adjusted to solely improve the lesions specific area [36] (Fig. 7).

The authors in [38] have performed preprocessing of the imageries of the ISIC dataset. The presence of hairs makes it more challenging to segment lesion images because hair pixels affect crucial properties like texture and border. The hairs also make the process of extracting features more challenging. In [38] the authors combined the threshold and morphological operations and proposed an improved technique that has been used to remove the hair and other noise artefacts in their research [38].

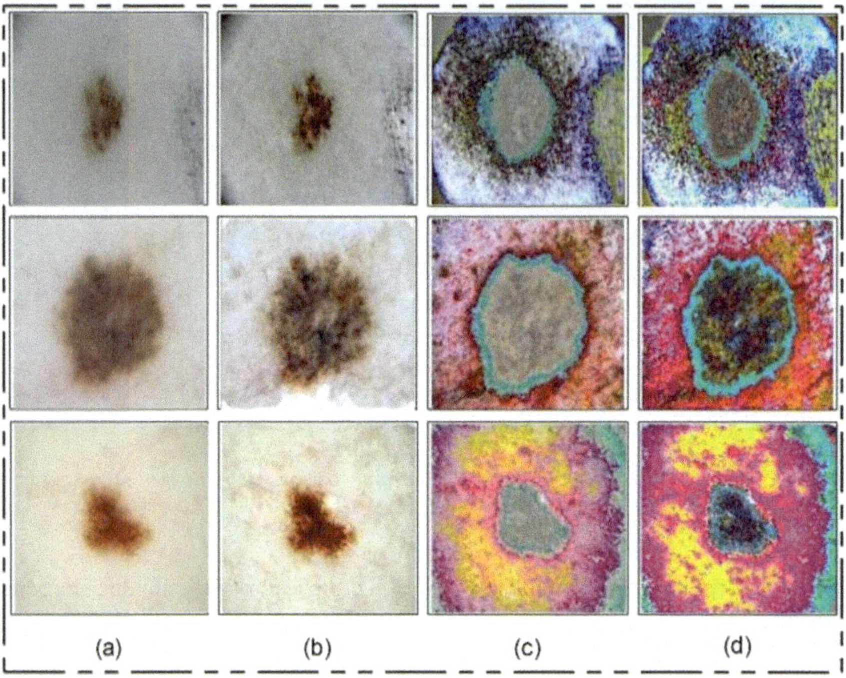

Fig. 7 Lesion contrast stretching results. **a** Original Image; **b** fused image; **c** output of first fitness function; **d** output of second fitness function [36]

2.2.3 Lesion Segmentation

Lesion Segmentation may also be considered as one of the preprocessing steps. An important step in locating diseased areas is the segmentation of skin lesions [39]. Saliency approaches and convolutional neural network-based approaches are just two of the many techniques for lesion segmentation that are given in the major works. Although saliency-based strategies are simpler than CNN-based algorithms, saliency-based strategies are not much accurate. A high number of ground truth images were needed by the CNN-based algorithms in order to train a model, which was then used to the detection process.

According to the method described by the authors in [36], (i) a regular CNN model with ten layers is built; (ii) characteristics from the final convolutional layer are visualised and combined into a single imagery; (iii) super pixels from the combined imagery are calculated; (iv) a threshold is employed for the concluding segmentation; and (v) borders are sketched on segmented areas employing an effective delineation method to localise skin lesions.

In [37] the authors performed preprocessing of skin lesion images, they have recommended a new variance growth (Differential Evolution) bat algorithm (DE-BA) employing a hybrid metaheuristic, an additional hybrid technique, differential evolution-artificial bee colony (DE-ABC). The authors suggested unique techniques for the algorithms to approximate the parameter set. Effective boundary estimation is further aided by the efficient preprocessing.

The fundamental operation of the glare-conserving, contrast stretching technique is shown in the diagram below. The procedures demonstrate how an input image is altered to help segmentation paradigms. Moreover, the authors cross-compared the boundary estimation results of the primary lesions with the BA-ABC contrast-adjusted models to validate the effectiveness of the benchmark presented.

In addition to obtaining the source image and the output image, authors in [40] propose a system that goes through two primary stages. Preprocessing and edge detection are the two main steps. Preprocessing is carriedout in 3 phases: colour-space conversion, contrast enhancement, and noise removal via filtering. The test image's edge map is then obtained using the well-known Sobel, Canny, and Prewitt edge operators. Utilising ACO further improves the edges produced by this method. The effectiveness of the improved ACO-based edge detection is then evaluated by contrasting the entropy of existing edge detection methods with that of the [40] approach.

In [38] the authors proposed a model in which preprocessing and segmentation are two crucial processes, unwanted artefacts like hair, light, and many other artefacts are eliminated in a preprocessing scenario by an improved technique using threshold and morphological operations. And the second phase involves segmenting the skin lesion with high accuracy utilising the k-means segmentation algorithm and an improved firefly algorithm (FFA) approach.

In [38] the authors recommended skin abrasion segmentation method comprising 2 phases: K-means clustering to pinpoint the precise abrasion area and FFA algorithm optimisation of the clustering to obtain high accuracy. The K-mean clustering approach is a popular tool used by researchers to solve a variety of practical clustering problems. The same approach, however, is ineffective for resolving non-globular clustering issues. A recently developed metaheuristic algorithm called the Firefly algorithm, which further imitates the flash behaviour of fireflies, is developed to tackle this problem.

K-mean and the Firefly algorithm have been combined to achieve greater accuracy and segmentation in the final product. K-mean and the Firefly algorithm have been combined to provide results with improved segmentation and accuracy. The firefly method aids in escaping local optima by employing the global search capability to address the K-mean technique constraint. Effective cluster analysis has been performed using the combined recommended approach. The outcomes of the suggested method are also contrasted with those of other benchmark techniques. FFA frequently eliminates unnecessary extract characteristics. FFA is reliable and has a comparably greater convergence rate [38].

2.2.4 Clustering via K-means

This method is employed to group the pixels in order to generate the RGB foreground and background region. With this technique, the first cluster midpoint is automatically selected from the image pixels. The remaining image input pixels are now chosen as the next cluster midpoint based on the probability of their square stretch from the imagery picture elements closest to the group core. The lesion's interior, the lesion's backdrop, the background picture, and the lesion boundary are all represented by pixels in the image. The produced clusters using this method retain the cluster's pixel colour, which is the same as the colour of the skin lesion. This technique automatically looks for pixel-level information to identify a mass as a skin lesion location. Figure 8 shows the original and pre-processed images. Figure 9 displays the outcomes of the K-mean algorithm applied to pre-processed images [38].

2.2.5 Colour Space Conversion

The colour space of the keyed skin lesion imagery is converted into a grayscale value in the system given by authors in [40] utilising the weighted sum of the R, G, and B elements as provided by Eq. (5).

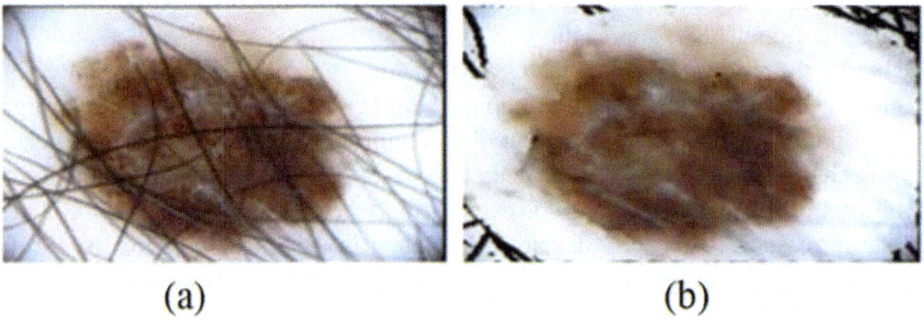

Fig. 8 Primary and pre-processed image **a** Primary image taken from ISIC dataset **b** Pre-processed image [38]

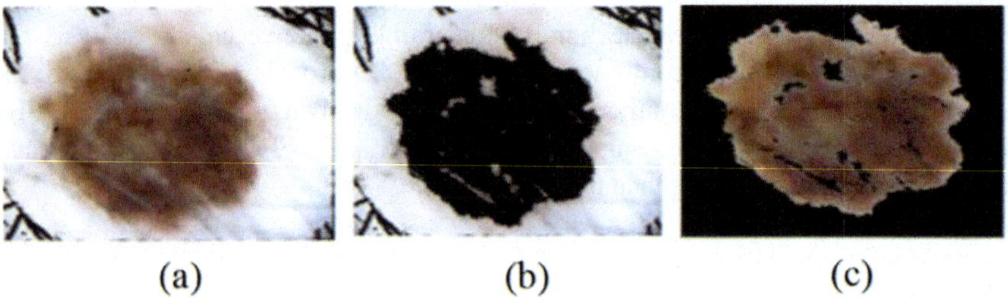

Fig. 9 Results using K-Mean clustering segmentation for input image taken from ISIC dataset **a** pre-processed image **b** surrounding area following segmentation **c** front region after segmentation [38]

$$\mathrm{R} \times 0.2989 + \mathrm{G} \times 0.5870 + \mathrm{B} \times 0.1140 \tag{5}$$

The GNU Octave image processing function is used to carry out this process. The RGB image will be converted to grayscale in this step by eliminating the unnecessary saturation and hue information by preserving brightness.

2.2.6 Enhancing the Contrast

In order to make the data in the image more comprehensible, contrast enhancement techniques are crucial in image processing. As a result, numerous features are easier to see. In [40] focuses on enhancing the disparities in an image. The most widely used methods for enhancing contrast include histogram equalisation, morphological enhancement, grayscale adjustment, thresholding, and interpolation. By transferring the intensity values from the keyed in skin abrasion imagery to a latest value, the method in [40] increases the contrast. 1% of the image's upper and lowest pixel values are saturated by this new value. By adjusting the contrast of the skin lesion image, this technique ensures that 1% of the data is inundated at both elevated and lowly concentrations [40].

The authors in [40] used median filtering in their work. This filter typically decreases noise in an image while maintaining all of the important information. In midpoint sifting, every pixel in the image is contrasted to its neighbouring communities to see whether it accurately depicts the area. After that, the pixel value is restored using the neighbourhood's median value. The surrounding pixel values are first sorted into numerical order in order to determine the median. After that, the pixel value in question is swapped with the middle pixel [40] (Figs. 10, 11 and 12).

3 Selecting Features

A crucial stage is choosing the optimal features, which has the benefit of increasing classification accurateness and decreasing calculation time. Two strategies are used in the work [33] to choose the finest features. A hybrid whale optimisation (HWO) algorithm is suggested in the first technique. The HWO algorithm was first presented by Mirjalili [41]. The two stages of the algorithm's execution are (i) surrounding the target and (ii) searching for the target. The two steps in simple words may be defined as the utilisation phase and the investigation phase are carried out. The stages are expressed mathematically as

$$\psi = |V_1 * B *^{\sim} (t) - B^{\sim}(t)| \tag{6}$$

$$B^{\sim}(t+1) = B *^{\sim} (t) - V_2 \cdot \psi \tag{7}$$

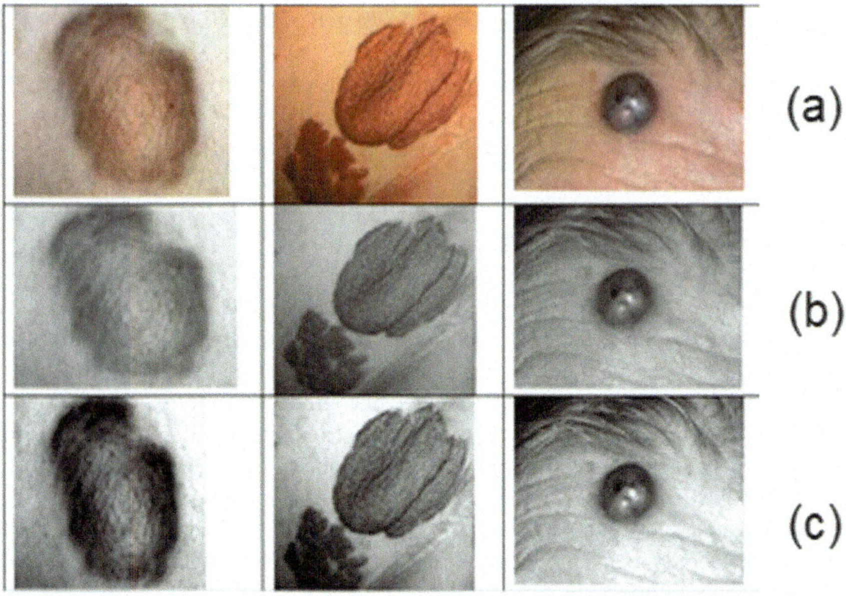

Fig. 10 Realisation of preprocessing: **a** primary image, **b** outcome of colour-space conversion, **c** outcome of disparity enhancement [40]

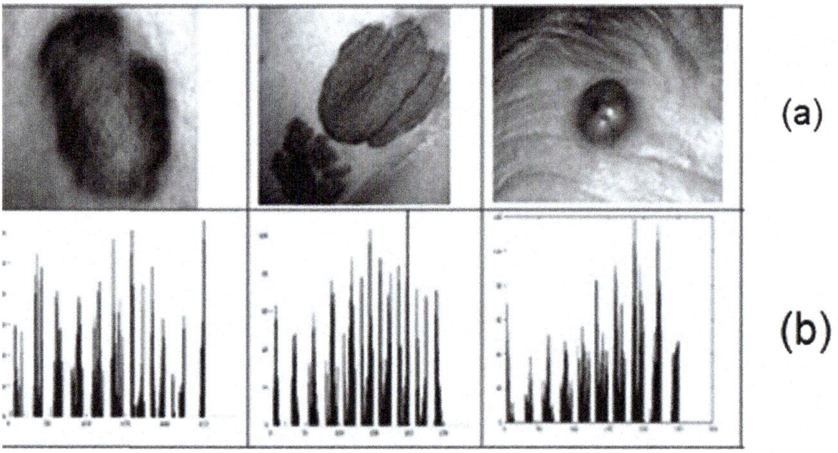

Fig. 11 **a** Skin lesion images following the preprocessing step **b** histogram of subsequent images[40]

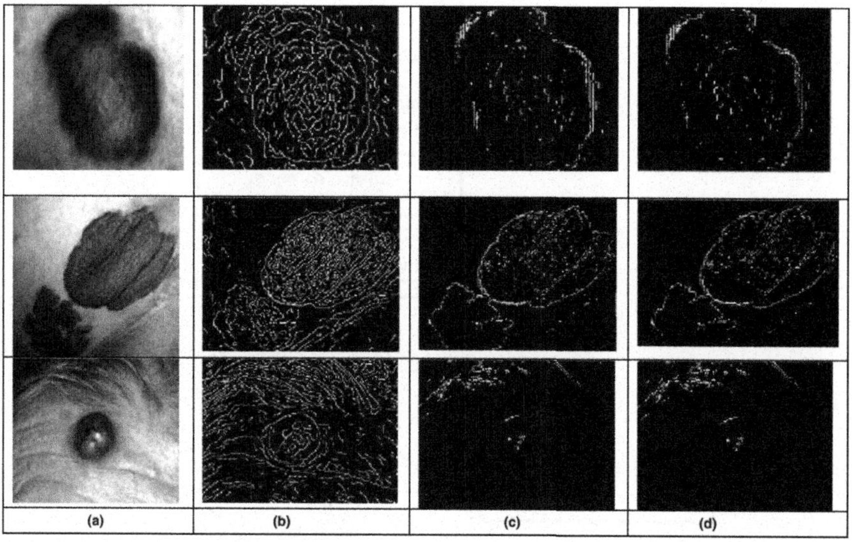

Fig. 12 Skin abrasion(lesion) image: **a** primary, **b** with ACO—Canny edge detection, **c** with ACO—Sobel edge detection, and **d** with ACO—Prewitt edge detection

$$V_1 = 2.r \tag{8}$$

$$V_2 = 2\alpha.r - \alpha \tag{9}$$

From the above equation we define r as a random vector with values [0, 1], B is the locus vector, B is the finest solution, V_1 and V_2 are the constant vectors, ψ is a decay linearly from 2 to 0 throughout the repetitions, and $|\alpha|$ is the definite value, correspondingly. According to the best solutions B, the $B\,(t+1)$ revises the locations. Corresponding to their finest results, the whale contains the costs V_1 and V_2. The following equation updates the shrinking behaviour. Figure 13 shows the flow chart for Hybrid whale optimization.

4 Data Augmentation

In [36], the authors first equalised the data utilising a data expansion phase for the multiclass classification. To achieve this, they carried out the following operations on the initial and reversed images: right-flip, left-flip, and swap. After balancing the skin classes, we used ResNet101 and DenseNet201, two pre-trained deep learning paradigms. By using inductive transfer for extracting features, retrained both models. Below is a description in more detail. In [36], the direct links amongst the layers are

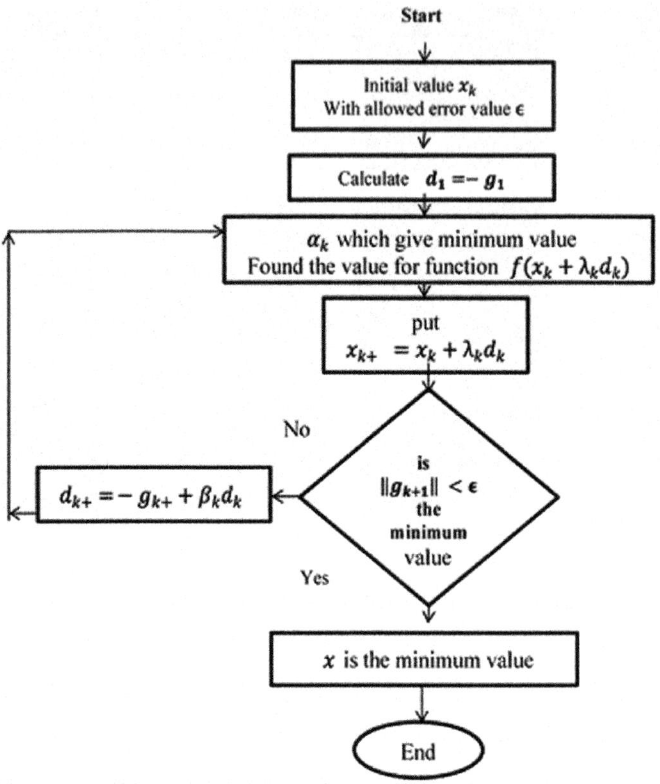

Fig. 13 Flow chart for hybrid whale optimization

formed while a few connections are merely bypassed. The "bottleneck" construction components in ResNet101 are utilised to minimise the parameters.

5 Metaheuristic Algorithms for Optimization

5.1 Feature Optimization Using Moth Flame

The process of feature selection involves removing features from a feature collection that are unnecessary, annoying, and redundant. The choice of the finest characteristics reduces overfitting and complexity, speeds up training, and makes machine learning algorithms simpler to understand [36].

In this study, Improved Moth Flame Optimisation (IM FO) was utilised to choose the finest features within the highly structural data. The moths' natural navigational mechanism is imitated by MFO. In order to assign plausible solutions (i = 1, 2,

3....... N) with characteristic measurement ij, ($j = 1, 2, 3, \ldots$ Dim), this approach first allocates a population, denoted by N.

$$\Psi\text{mfo} = (\Psi_{\text{IP}}, \Psi_{\text{UP}}, \Psi_{\text{SP}}) \tag{10}$$

where UP stands for an updated phase, IP stands for an initial phase, and SP stands for the stopping condition. The population is produced at random during the first phase, with N equal to 20. Applying the fitness function to assess the

$$\text{Fitness (Vi)} = r \times \text{ Error (Vi)} + (1 - r) \times (|\Psi\text{i}|\text{Dim}) \tag{11}$$

where the fitness function is represented by Error(Vi). Considering the cubic SVM, the preferred kernel method is cubic, and the procedure is solitary versus the whole thing; r is an arbitrary number lying in [0, 1] expended to equalise the classifier accuracy; Vi signifies keyed characteristics, and i represents nominated characteristics in the i-th repetition. We initially use an initiation mechanism for the first-stage characteristic range, and then the moths are renewed depending on the blazes. The definition of the activation function is

$$\Psi i = \begin{cases} 1, \text{ for } 11 + e - \Psi i > V \\ 0, \text{ Otherwise} \end{cases} \tag{12}$$

The features are updated using.

$$\Psi i = \text{dielbcos}\,(2\pi \text{l}) + \text{Flame}\,(u) \times w + (1 - w) \times u \tag{13}$$

where $l = [1, -1]$ and i, $u = 1,2,\ldots N$, di $=$ Flame(u)i. This expression is used to update the moths' positions in relation to the flames. In this case, the distance from the flame is represented by $l = -1$ and the new closest position is represented by $l = 1$. After sorting the updated flames according to Up(i), which determines their descending order, the randomness cost H(Up(i)) is calculated.

The algorithm for the moth flame optimization is given below:

Algorithm

Step 1: The initial population's random position is determined.

Step 2: The positions are updated.

Step 3: Each population's fitness is calculated.

Step 4: The algorithm searches through all of the potential solutions to find the best one.

5.2 Bat Algorithm for Finding Optimal Value

In [37] authors have used Bat Algorithm for optimization. A population is created by combining different parameter sets (Fig. 14).

Therefore, the search operation must begin with a population that has been randomly initialised. Following that, the BA begins calculating the parameter set's optimal value using the cost function [37].

$$Xj = \psi 1 + \text{rand}(0, 1) \times (\psi_u - \psi_1) \tag{14}$$

Yang examined bat behaviour and food-finding methods more than ten years ago, developing these methods in accordance with crucial bat features such rapidity, location, pulsation discharge rate, and wavelength. The phrase "echo location" signifies the process of locating and predicting prey. The primary characteristic of bats is the use of different ultrasonic pulses to avoid the predetermined environment. Using the aforementioned approach, recurrence is calculated as

$$f_{\tau j} = f_{\min} + \phi(f_{\max} - f_{\min}) \tag{15}$$

The parameters that the authors utilised in the are still in effect; in this case, is an arbitrary value between (0, 1). The following equations also include the numbers for velocity Vj and location *Xj*.

$$V_{\tau j} = V\tau_{-1j} + fj\left(X_{\tau j} - X_*\right) \tag{16}$$

$$X_{\tau j} = X_{\tau} -_{1j} + V_{\tau j} \tag{17}$$

Fig. 14 Encholocation behaviour of bats

Built on the prior speed, spot, current frequency, and global best, updated velocity values are calculated. Next, the current position is updated as well. The algorithm then chooses the population's top parameter set following these adjustments. Then, using the method, it begins an arbitrary walk like to a bat to estimate a personal best.

$$X\text{new} = X\text{old} + \theta L\tau, \quad \theta \in [-1, 1] \tag{18}$$

In the general neighbourhood of the previously established ideal spot, bats move at random. The average loudness L and the global optimum X are multiplied by an arbitrary factor to arrive at this calculation. Every bat in the search area has a pulse rate and loudness that are approximated using random integers. The bat's loudness reduces and its pulse rate rises as it gets closer to its objective, just as the temperature cooling process in simulated annealing. The following equations, where € and are n constants in this case, can be used to calculate the revised pulse rate j and loudness Li at time $+1$.

$$L\tau + 1j = \varepsilon L\tau j, \quad \varepsilon \in [0, 1] \tag{19}$$

$$\phi\tau + 1j = \phi 0j(1 - e - \eta\tau), \eta > 0 \tag{20}$$

The cost function is used to estimate population fitness, and if a lower cost can be found, both values are updated. The contrast modification function receives the top factor series determined by the primary metaheuristic process and improves the contrast of the intensity channel. The last contrast-enhanced model is created by merging the intensity channel, which has improved contrast, with the other two modified channels. To produce a concluding result with increased contrast and brightness, the shaded region is converted backwards to RGB. The results demonstrate the system's implementation done with experiments and contrasted it with border estimate systems [37].

5.3 *Artificial Bee Colony*

Apiary, ABC (Artificial Bee Colony) which is inspired by the living world, is applied to the parameter sets after a number of repetitions. With this method, the parameters are further optimised based on distinctive qualities. Self-association functions in the context of local hunt info instead of globalised data, taking into account adjustments in search responses, advice, and interactions with other employee bees. Employee bee category also pertains to the simultaneous completion of particular tasks. In ABC, unlike BA, the actual operation expense is not determined using a fitness formula; rather, the fitness cost is determined [37].

5.3.1 Phase of Employed Bees

Employee bees look for new food sources (m) that are closer to the food source (xm) and contain more nectar. They discover a neighbouring food resource and evaluate its capability (fitness). For instance, the formula provided by Eq. (21) can be used to determine a neighbouring food source, m:

$$v_{\mathrm{mi}} = x_{\mathrm{mi}} + \phi \mathrm{mi}(x_{\mathrm{mi}} - x_{\mathrm{ki}}) \tag{21}$$

in the above equation x_k represents an arbitrarily selected foodstuff resource, i represents an arbitrarily nominated factor key, and mi represents a randomly generated integer between $[a, a]$. Once the new food source m has been created, its condition is determined and a greedy selection is made between m and x_m.

5.3.2 The Observer Bees Phase

There are two types of unemployed bees: observers and scouts. Spectator bees that are waiting in the beehive obtain communication from working bees about their staple sources, and they use this information to probabilistically select their food sources. According to the probability values estimated using the fitness values provided by worker bees, an observer bee in ABC picks a food source.

Following the probabilistically selected food supply xm for an observer bee, a neighbourhood Eq. (21) is used to identify source m, and its fitness value is calculated. A greedy selection is used between m and xm, much like in the employed bee's phase. As a result, more observers are drawn to richer sources, and favourable feedback behaviour manifests.

Self-association functions in the context of local search info instead of the global data, taking into account adjustments in exploration responses, advice, and interactions with other employee bees.

The parameter set with the most trials is swapped out for the arbitrary set in the last stage of the ABC algorithm. Only when the trials exceed the previously indicated threshold, the random set is formed in the similar to that as the recent populace existed. At this point, there is no strict selection because the member whose trials have reached the threshold is eliminated. The array is updated to include the fitness cost of the recently inserted participant, which is assessed with the identical formulas. Its trial value is also set to zero concurrently. The output of the work [37] hybrid metaheuristic algorithm is the global best solution after a predetermined number of iterations or if the termination requirement is satisfied. The image transforming function is used to contrast stretch the power frequency. To create an HSI image, additional stages comprise combining the other 2 modified paths with the contrast stretched channel. The last stage is changing the shaded space from HSI to RGB to create lesions that maintain brightness and stretch contrast. The improved images can then be used to estimate lesion masks using the boundary estimation approach

Table 1 IoU and F1 score with and without preprocessing [37]

Algorithm	Dataset	IoU (%)	F1 (%)
Without preprocessing BAT	ISIC-2016	85.2	92.0
	ISIC-2017	86.2	92.6
	ISIC-2018	86.3	92.4
Prototype pre-processed with BA-ABC	ISIC-2016	86.9	93.1
	ISIC-2017	87.0	93.0
	ISIC-2018	87.2	93.2

Table 2 Entropy of ACO-generated images [40]

Image type	Canny edge detector	Enhanced Canny edge detector using ACO	Sobel edge detector	Enhanced Sobel edge detector using ACO	Prewitt edge detector	Enhanced Prewitt edge detector using ACO
Trial image 1	1.4693	1.94997	0.5894	0.8059	0.6996	0.7528
Trial image 2	1.15293	3.3643	0.9442	0.62694	0.2874	0.4633
Trial image 3	1.9369	1.9604	0.9779	1.04774	1.4444	1.0278

Table 3 Classification accuracy on the HAM-10000 dataset using several metaheuristic feature selection methods

Classifier type	Optimization algorithm used	Accuracy (%)	PPV (%)	FDR (%)	Time (s)
ELM	HWOA	85.09	84.86	15.14	147.6329
KELM	IMFO	90.668	90.20	9.33	133.4406
KELM	MFO	86.273	86.199	13.694	–

[37]. Table 1 shows the IoU and F1 score for the BA without preprocessing and with preprocessing using ABC (Tables 2, 3 and 4).

5.4 *Ant Colony Optimization*

In the optimisation of various techniques, ant colony optimisation (ACO) has been gaining popularity. An approach for issue optimisation that draws inspiration from nature is ant colony optimisation. The actions of actual ants that forage foodstuff from a resource and bring it to the ant colony are the basis for this algorithm. In this method, agents move around a region that symbolises all potential solutions like ants to find the best ones. In order for further iterations to provide better answers, these agents keep track of the position and quality of their solutions.

Table 4 EMI-based feature selection for ISIC dataset

Dataset	Classifier type	Optimization algorithm used	Accuracy (%)	PPV(%)
ISIC-2018	ELM	HWOA	85.96	85.52
ISIC-2016	KELM without preprocessing	BAT	85.2	92.0
ISIC-2017	KELM without preprocessing	BAT	86.2	92.6
ISIC-2018	KELM without preprocessing	BAT	86.3	92.4
ISIC-2016	KELM with preprocessing	BAT	86.9	93.1
ISIC-2017	KELM with preprocessing	BAT	87.0	93.0
ISIC-2018	KELM with preprocessing	BAT	87.2	93.2
ISIC-2017	K-means	FFA	99.4	95.7

In [40] the authors have used the ACO for optimization in their work for skin lesion edge detection. The ACO technique is a well-known optimisation technique that is used to provide solutions in a variety of areas, including image processing, assignment problems, set problems, and work scheduling. ACO is an effective optimisation tool.

To handle the complex nature of dermoscopic images, enhanced edge operator-based skin lesion identification approaches with ACO support are proposed in this work. For initialising the ants, the edge plots are produced with the well-known edge operatives viz Sobel, Prewitt, and Canny. Local variations in pixel intensity that are produced reflect a shift in the ant's position. As a result, the ACO method is added to the conventional edge detection operator process. The three upgraded edge detection approaches' output images are quantitatively evaluated to identify the most effective improved approach using the ACO algorithm [40].

An all-purpose escalation technique inspired on ant social behaviour is called ant colony optimisation. In 1997, Dorigo and Gambardella created this method. The behaviour of ants seeking for food is taken advantage of by ACO. An ant species called Somers leaves a chemical residue on the ground called "pheromone" to mark a prime route taken by another Somers ant from the same colony. The primary driving force behind ACO is the deposition, updating, and use of pheromone for the selection of the best path and subsequent use of the path indicated by pheromone [40].

5.4.1 The Metaheuristic for ACO

Step 1: Initialise the pheromone matrix, that is represented by 0 and K ant positions.

Step 2: In the subsequent phase of the construction method, where $n = 1{:}\ N$
The ant key is $k = 1{:}\ K$.

According to the probabilistic transition (Pn) matrix with regard to the M1 M2 M1 M2 image, shift the *k*th ant for L moves.

Step 3: Modify the pheromone matrix.

Step 4: Employ the stop criterion that is choice based on n in Step four.

The updating of the probability matrix (n) and construction of the probabilistic transition matrix (Pn) are the core components of the ACO algorithm. The *k*th ant is advanced from point (node) i to point (node) j for the generation of the transition matrix by adhering to the probabilistic approach.

In ACO, the pheromone is updated twice. Local update refers to the initial update. In this instance, the kth ant is moved inside the nth building phase before the pheromone matrix is updated. The ideal tour depends on the traveller. The finest exploration is provided by the existing building stage or the finest result identified with the inception of the algorithm. Sometimes a tour that combines the two is the best option.

After all k ants have moved through each construction phase, the pheromone matrix undergoes a second global update [40].

5.4.2 Firefly Method with K-means

Yang introduced the Firefly Algorithm (FFA), a swarm-based metaheuristic algorithm. It draws inspiration from stochastic, metaheuristic, and natural algorithms that can be utilised to solve the most difficult optimisation problems. The attraction and flashing characteristics of fireflies served as the basis for this algorithm, which was also motivated by the social behaviour of these insects. The application mimics how fireflies exchange messages by flashing their lights. A firefly's attraction is directly proportionate to its glare, that is determined by the objective function, hence the algorithm takes the assumption as considering all the fireflies are unisex and that every firefly can be drawn to any another firefly (Fig. 15).

In order to attain the optimum result, K-mean clustering is optimised in this study using an FFA method, and to implement this method, the authors used MATLAB

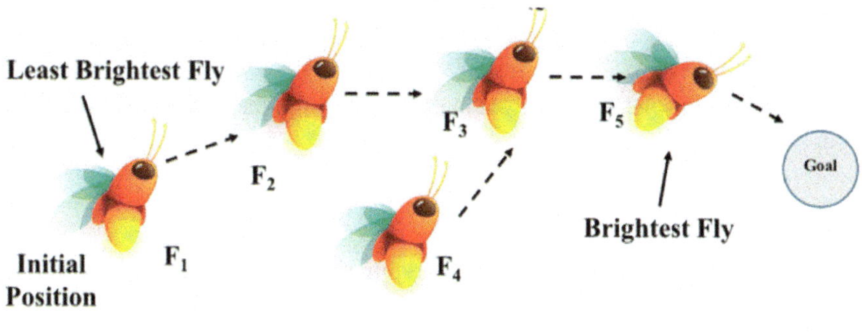

Fig. 15 Working of firefly algorithm

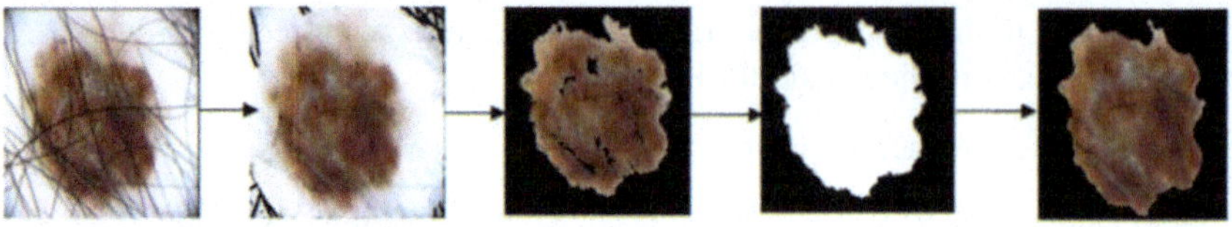

Fig. 16 Different images acquired in various phases of segmentation of skin abrasion areas using the K-means method **a** Given keyed representation, **b** Pre-processed imagery **c** Front obtained image using K-mean **d** Segmented Binary mask using K-mean with FFA **e** Segmented abrasion area using the K-mean and firefly method

2014 or a later version. The [38] suggested enhanced technique, which is based on threshold and morphological operations, pre-processes input sample images. The enhanced image is then subjected to K-mean clustering. The FFA technique was optimised to produce results with a high degree of precision.

The K-Mean method's segmented image is additionally dealt by applying the firefly algorithm. This is employed to discover the finest result with the least amount of population. When K-mean and FFA are combined, segmentation is done using K-mean first, and FFA is then optimised using the K-mean results as an input. At this point, FFA chooses an improved threshold estimate that can substitute the incorrect segment data in the K-means outturn with the chosen threshold value. With the help of a preprocessing phase and the extrication of the front and back areas using the K-Means and K-Means with FFA techniques, the procedures for segmenting the skin abrasion area with the suggested scheme are shown in Fig. 16e.

A firefly moves towards the firefly which is more noticeable. Additionally, as shown in Eq. (22), the brightness diminishes with distance according to the inverse square law.

$$I \prec \frac{1}{r_2} \tag{22}$$

The concentration of light at an expanse of r from the origin can be calculated using Eq. (23) if the light is travelling across a channel with a light intentness coefficient

$$I = I_0 e^{-\gamma r 2} \tag{23}$$

where I_0 represents the source's light intensity. The brightness, β, can also be stated as in Eq. (24).

$$\beta = \beta_0 e^{-\gamma r^2} \tag{24}$$

Equation (25) provides a generalised brightness function. In reality, we can use any monotonically declining function.

$$\beta = \beta e^{-\gamma r^{\omega}} \quad (25)$$

Established on what way the firefly performs in the goal task, which is generated arbitrarily, the approach assigns them a light intensity. This power will be expended to calculate the firefly's illumination, which is directly proportionate to its brightness intensity. The solution with the lowest functional value in minimization problems will be given the most attention. Once the solutions' brightness or intensity has been assigned, each firefly will then follow fireflies that are emitting more light. The firefly that is most visible will leisurely walk across its neighbourhood as part of a local hunt.

6 Current Literature and Challenges

The task of processing skin-infected images is particularly challenging for a variety of reasons. Low contrast is the key factor, that makes it hard to tell the difference between diseased and healthy skin. Additional aspects comprise the existence of objects made of ink, bubble, hair, and ruler. As a result, it is essential to consider these artefacts when creating algorithms. Numerous training sessions, photos of people with light skin in typical datasets, low inter-class variance, photographs of various sizes and shapes, and unbalanced dataset classes are further challenges.

7 Reference Datasets

Undermentioned is a listing of the commonly open datasets:

HAM-10000: The sample of skin lesions falls into seven different categories. The dermoscopic images are stored using a variety of techniques.

ISIC: The ISIC archive provides access to 10,015 samples in total, the dataset has 170 illustrations.

PH2: It has 200 dermatoscopic imageries and 3 categories of normal, abnormal, and melanoma.

DermoFit: DermoFit is composed of macroscopical imageries, and the source has one thousand three hundred (1300) imageries organised into ten categories and simultaneously segmented using binary masks.

ISBI 2016: The dataset is derived from the ISIC collection. A total of 1279 illustrations make up the dataset, out of that 900 serve as training images, and 379 serve as testing imageries.

ISBI 2017: It contains a total of 2750 photos, of which 150 are utilised for validation, 2000 for training, and 600 for testing.

ISIC 2018: The ISIC is a huge databank in the world with 12,500 photos and three jobs. The first step is lesion segmentation, followed by attribute identification and lesion disease categorization. The dataset is made up of 7 categories.

ISIC 2019: Information for ISIC 2019 was gathered from the HAM-10000, BCN_20000, ViDIR Group, and unnamed sources. 25,331 dermoscopic samples are present in the collection, which is divided into eight categories.

ISIC 2020: The collection contains 33,126 samples. The pixels in the file are JPEG-encoded, and it displays two different DICOM formats.

8 Conclusion

In this work an extensive study has been conducted on HWOA, Moth Flame, BA-ABC, Firefly with K-Means, and ACO. In this study, statistical markers like the regional area, standard deviation, and regional average were also observed to improvise the contrast of skin abrasions. The statistical transformation function was dependent on few key variables that could only be determined using the hybrid metaheuristic technique devised. The study delves into the methodology to the test using a few cutting-edge boundary estimation models in order to confirm its effectiveness. A thorough analysis of various metaheuristic methods used for an automated skin cancer detection is presented. The leading studies and frequently used datasets for skin cancer are compared. It has been determined that there is room to improve automated skin lesion diagnosis and new methods can provide better classification and prediction provided proper preprocessing is done with better methods. The study also covers different preprocessing techniques and segmentation techniques which give better results. Its objective is to use approaches from several studies to make this work a standard for future works in the area of skin malignancy detection. Additionally, in the field of dermatology, there is still much space for growth and improvement in terms of the recall, True Negative Rate, and accuracy of identifying skin abrasions.

Furthermore, skin specialists need to be the first to accept the new methods as a supplementary tool to support their diagnoses. There are many opportunities, promises, and obstacles when it comes to implementing end-to-end solutions in the healthcare industry. New methods and techniques which can help the medical experts to easily discover and classify or categorise the various classes of skin lesions have to be adopted by the dermatologists (Figs. 17, 18, 19 and 20).

9 Future Prospects

Researchers may examine skin cancer detecting systems in the future with more sectional categorization. Region-based, clustering-based, threshold-based, and other categories of traditional segmentation are also available. Similar to this, the extracted features should be divided into colour, shape, and texture categories. Despite the impressiveness of the outcomes produced by metaheuristic algorithms, several problems with optimisation and generalisation still exist and can be resolved in the future.

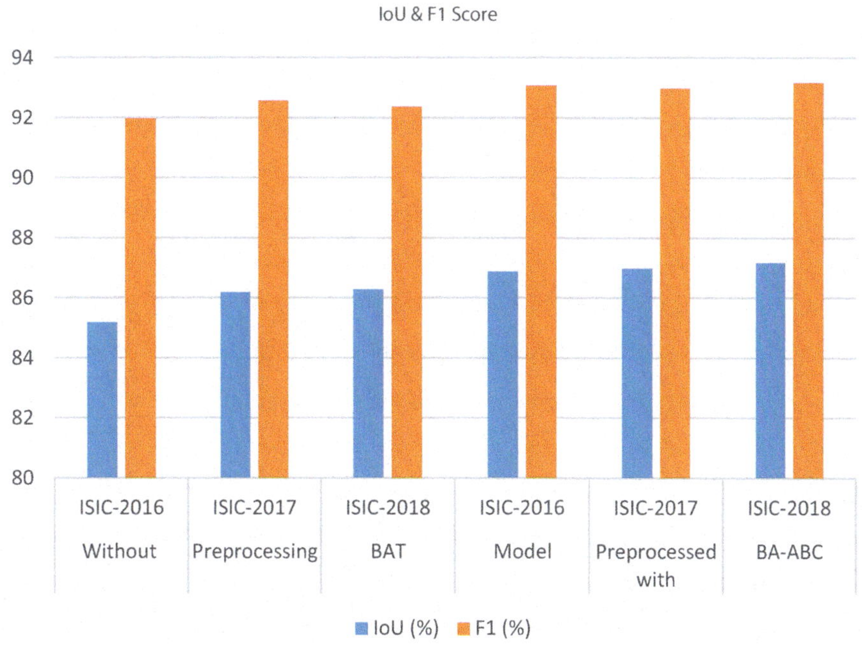

Fig. 17 IoU and F1 score using BAT and ABC with and without preprocessing

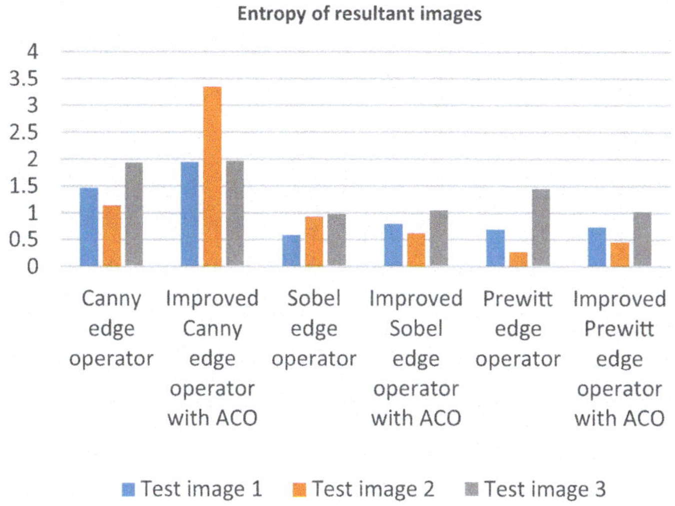

Fig. 18 Entropy values of images using ACO

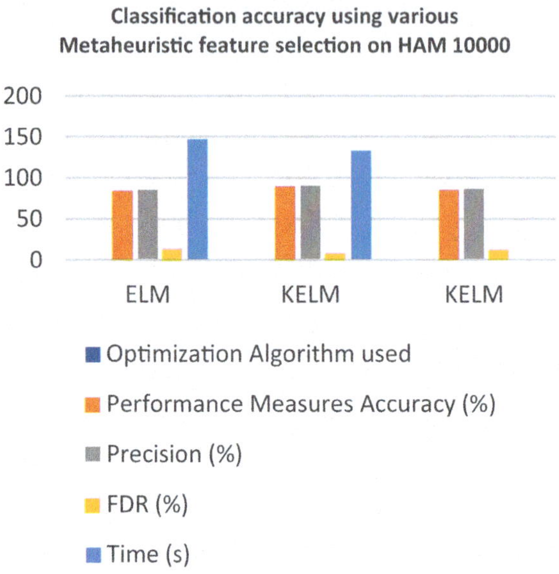

Fig. 19 Classification accuracy of various optimization algorithms on HAM-10000

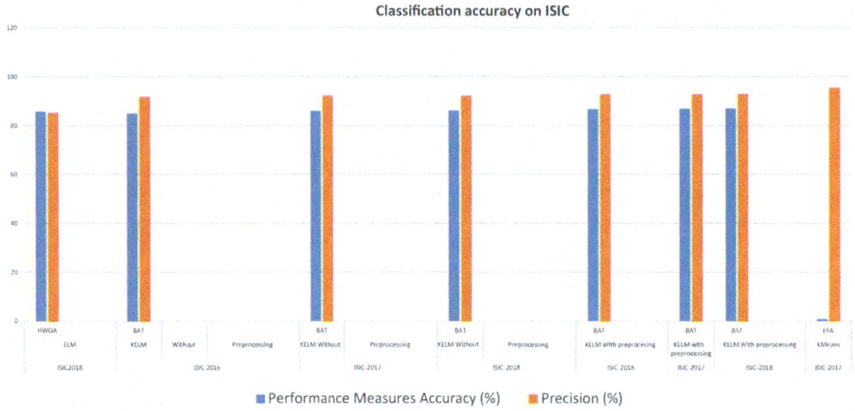

Fig. 20 Classification accuracy of various optimization algorithms on ISIC dataset

Funding The work carried out did not receive any monetary support.

Disputes of Concern By no means disputes of concern of interest are disclosed by the authors.

References

1. Gordon R (2013) Skin cancer: an overview of epidemiology and risk factors. Semin Oncol Nurs 29:160–169
2. Javed R, Rahim MSM, Saba T, Rehman A (2020) A comparative study of features selection for skin lesion detection from dermoscopic images. Netw Model Anal Health Inform Bioinform 9:4
3. Zhang N, Cai Y-X, Wang Y-Y, Tian Y-T, Wang X-L, Badami B (2020) Skin cancer diagnosis based on optimized convolutional neural network. Artif Intell Med 102:101756
4. Khan MA, Sharif MI, Raza M, Anjum A, Saba T, Shad SA (2022) Skin lesion segmentation and classification: A unified framework of deep neural network features fusion and selection. Expert Syst 39:e12497
5. Yu L, Chen H, Dou Q, Qin J, Heng PA (2016) Automated melanoma recognition in dermoscopy images via very deep residual networks. IEEE Trans Med Imaging 36:994–1004
6. Rezvantalab A, Safigholi H, Karimijeshni S (2018) Dermatologist level dermoscopy skin cancer classification using different deep learning convolutional neural networks algorithms. arXiv, arXiv:1810.10348
7. Foote M, Harvey J, Porceddu S, Dickie G, Hewitt S, Colquist S, Zarate D, Poulsen M (2010) Effect of radiotherapy dose and volume on relapse in merkel cell cancer of the skin. Int J Radiat Oncol 77:677–684
8. Qadir MI (2016) Skin cancer: etiology and management. Pak J Pharm Sci 29:999–1003
9. D'Orazio J, Jarrett S, Amaro-Ortiz A, Scott T (2013) UV radiation and the skin. Int J Mol Sci 14:12222–12248
10. Seebode C, Lehmann J, Emmert S (2016) Photocarcinogenesis and skin cancer prevention strategies. Anticancer Res 36:1371–1378
11. Houssein EH, Sayed A (2023) Dynamic candidate solution boosted beluga whale optimization algorithm for biomedical classification. Mathematics 11(3):707
12. Mohammed SS, Al-Tuwaijari JM (2021) Skin disease classification system based on machine learning technique: a survey. IOP Conf Ser Mater Sci Eng 1076:012045
13. Sharma V, Garg A, Thenmalar S (2020) A survey on classification of malignant melanoma and Benign Skin lesion by using machine learning techniques. Easy Chair Prepr 2611:2314–2516
14. Saherish F, Megha J (2020) A survey on melanoma skin cancer detection using CNN. Int J Sci Res Eng Manag (IJSREM) 4:1–4
15. Goswami T, Dabhi VK, Prajapati HB (2020) Skin disease classification from image-a survey. In: Proceedings of the 2020 6th international conference on advanced computing and communication systems (ICACCS). Coimbatore, India, pp 599–605
16. Sreelatha T, Subramanyam MV, Prasad MNG (2019) A survey work on early detection methods of melanoma skin cancer. Res J Pharm Technol 12:2589
17. DurgaRao N, Sudhavani G (2017) A survey on skin cancer detection system. J Eng Res Appl 7:59–64
18. Irum I, Sharif M, Raza M, Yasmin M (2014) Salt and pepper noise removal filter for 8-bit images based on local and global occurrences of grey levels as selection indicator. Nepal J Sci Technol 15:123–132
19. Sharif M, Irum I, Yasmin M, Raza M (2017) Salt and pepper noise removal from digital color images based on mathematical morphology and fuzzy decision. Nepal J Sci Technol 18:1–7
20. Reis HC, Turk V, Khoshelham K, Kaya S (2022) InSiNet: a deep convolutional approach to skin cancer detection and segmentation. Med Biol Eng Comput 60:643–662
21. Sikkandar MY, Alrasheadi BA, Prakash NB, Hemalakshmi GR, Mohanarathinam A, Shankar K (2021) Deep learning based an automated skin lesion segmentation and intelligent classification model. J Ambient Intell Humaniz Comput 12:3245–3255
22. Zghal NS, Kallel IK (2020) An effective approach for the diagnosis of melanoma using the sparse auto-encoder for features detection and the SVM for classification. In: Proceedings of the 2020 5th international conference on advanced technologies for signal and image processing (ATSIP). Sousse, Tunisia, pp 1–6

23. Guarracino MR, Maddalena L (2018) SDI+: a novel algorithm for segmenting dermoscopic images. IEEE J Biomed Health Inform 23:481–488
24. Nida N, Irtaza A, Javed A, Yousaf MH, Mahmood MT (2019) Melanoma lesion detection and segmentation using deep region based convolutional neural network and fuzzy C-means clustering. Int J Med Inf 124:37–48
25. Houssein EH, Abdelkareem DA, Emam MM, Hameed MA, Younan M (2022) An efficient image segmentation method for skin cancer imaging using improved golden jackal optimization algorithm. Comput Biol Med. 49 C. https://doi.org/10.1016/j.compbiomed.2022.106075
26. Mahum R, Aladhadh S (2022) Skin lesion detection using hand-crafted and DL-based features fusion and LSTM. Diagnostics 12:2974. https://doi.org/10.3390/diagnostics12122974
27. Dulhare UN, Gouse S Hands on MAHOUT—machine learning tool. Wiley. ISBN 9781119654834. Mach Learn Big Data 361–421. https://doi.org/10.1002/9781119654834.ch14
28. Dulhare UN, Mubeen A Ahmad K Hands-On H2O Machine Learning Tool. Wiley. ISBN 9781119654834. Booktitle = Machine Learning and Big Data, chapter 15, pp 423–453. https://doi.org/10.1002/9781119654834.ch15
29. Geetha S, Dulhare UN, Sivatha Sindhu SS (2018) Intrusion detection using NBHoeffding rule based decision tree for wireless sensor networks. In: 2018 Second international conference on advances in electronics, computers and communications (ICAECC). Bangalore, India, pp 1–5. https://doi.org/10.1109/ICAECC.2018.8479483
30. Arif F, Dulhare UN (2017) A machine learning based approach for opinion mining on social network data. In: Satapathy S, Bhateja V, Raju K, Janakiramaiah B (eds) Computer communication, networking and internet security. lecture notes in networks and systems, vol 5. Springer, Singapore. https://doi.org/10.1007/978-981-10-3226-4_13
31. Dulhare UN, Ghori I (2018) An efficient hybrid clustering to predict the risk of customer churn. In: 2018 2nd International conference on inventive systems and control (ICISC). Coimbatore, India, pp 673–677. https://doi.org/10.1109/ICISC.2018.8398883
32. Chandrahaas BV, Mohanty SN, Panda SK, Michael G (2023) An empirical study on classification of monkeypox skin lesion detection. EAI Endorsed Trans Perv Health Tech [Internet] 9:e4. https://publications.eai.eu/index.php/phat/article/view/3352; Ansari UB, Sarode T (2017) Skin cancer detection using image processing. Int Res J Eng Technol 4:2875–2881
33. Afza F, Sharif M, Khan MA, Tariq U, Yong H-S, Cha J (2022) Multiclass skin lesion classification using hybrid deep features selection and extreme learning machine. Sensors 22:799. https://doi.org/10.3390/s22030799
34. Irum I, Sharif M, Raza M, Mohsin S (2015) A nonlinear hybrid filter for salt and pepper noise removal from color images. J Appl Res Technol 13:79–85
35. Irum I, Sharif M, Yasmin M, Raza M, Azam F (2014) A noise adaptive approach to impulse noise detection and reduction. Nepal J Sci Technol 15:67–76
36. Khan MA, Sharif M, Akram T, Damaševičius R, Maskeliūnas R (2021) Skin lesion segmentation and multiclass classification using deep learning features and improved moth flame optimization. Diagnostics 11:811. https://doi.org/10.3390/diagnostics11050811
37. Malik S, Akram T, Awais M, Khan MA, Hadjouni M, Elmannai H, Alasiry A, Marzougui M, Tariq U (2023) An improved skin lesion boundary estimation for enhanced-intensity images using hybrid metaheuristics. Diagnostics (Basel) 13(7):1285. https://doi.org/10.3390/diagnostics13071285. PMID:37046503;PMCID:PMC10093272
38. Garg S, Jindal B (2021) Skin lesion segmentation using k-mean and optimized fire fly algorithm. Multimed Tools Appl 80:1–14. https://doi.org/10.1007/s11042-020-10064-8
39. Rasool S, Dulhare UN, Khan MN, Gangodkar D, Rana A, Kalra R (2022) Automated multiclass classification using deep convolution neural network on dermoscopy images. In: 2022 2nd International conference on technological advancements in computational sciences (ICTACS). Tashkent, Uzbekistan, pp 711–716. https://doi.org/10.1109/ICTACS56270.2022.9988394

40. Sengupta S, Mittal N, Modi M (2019) Improved skin lesion edge detection method using ant colony optimization. Skin Res Technol 25(6):846–856. https://doi.org/10.1111/srt.12744. Epub 2019 Jun 22. PMID: 31228313
41. Mirjalili S, Lewis A (2016) The whale optimization algorithm. Adv Eng Softw 95:51–67. https://doi.org/10.1016/j.advengsoft.2016.01.008

Chapter 6
Automatic Prediction of Non-alcoholic Liver Disease Using Deep Learning Models

Uma N. Dulhare and Essam H. Houssein

Abstract Non-alcoholic fatty liver disease (NAFLD) is a prevalent worldwide health issue without an effective medical solution. It can result in various types of liver diseases. Its frequency is increasing at a rapid rate due to the added situations of rotundity, diabetic cases and hypertension. These diseases occur where the liver damage is unconnected to alcohol consumption. There are various methods to detect this for quick treatment and recovery as in some cases this could be fatal. Hence, using deep learning convolutional neural network algorithms, which use colourful natural features of the illustration, useful in determining and prognosticating directly the case of non-alcoholic fatty liver disease. This approach proves to be efficient in addressing these concerns, within the framework of this design, we've used different features to determine the subjects of this complaint. The research work presented here proposes a novel method for liver cancer detection using neural network concepts. Feature extraction is done using wavelet and grey-level co-occurrence matrix from an input image. One approach is to categorise the dataset as normal or abnormal. Meanwhile, additional methods that rely on the size and shape of lesions have been created for identifying liver tumours. The proposed model obtained accuracy of 0.881, precision of 0.922, recall of 0.894, F1-score of 0.887 and AUC score of 0.951.

Keywords NAFLD · CNN · Preprocessing · Wavelet scattering · Grey-level co-occurrence

---

U. N. Dulhare (✉)
Department of Computer Science and Artificial Intelligence, Muffakham Jah College of Engineering & Technology, Hyderabad, Telangana, India
e-mail: prof.umadulhare@gmail.com

E. H. Houssein
Faculty of Computers and Information, Minia University, Minia, Egypt
e-mail: essam.halim@mu.edu.eg

U. N. Dulhare and E. H. Houssein (eds.), *Machine Learning and Metaheuristics: Methods and Analysis*, Algorithms for Intelligent Systems,
https://doi.org/10.1007/978-981-99-6645-5_6

1 Introduction

Non-alcoholic fatty liver disease (NAFLD) has developed into a significant public well-being health concern on a global scale [1, 2]. NAFLD prevalence is thought to be between 20 and 30%, and it is increasing. During the past 10 years, the predominance rate of NAFLD has increased [3]. The predominance of NAFLD, which affects 60–90%, 27–92% and 28–70% of individuals with obesity, hyperlipidemia and type 2 diabetes mellitus, respectively, is closely related to that of type 2 diabetes mellitus (T2DM) hyperlipidemia and other chronic metabolic disorders. NAFLD can lead to liver steatohepatitis, non-alcoholic steatohepatitis (NASH), liver fibrosis, cirrhosis and possibly liver tumour. NAFLD is the primary reason for the speediest rising kind of liver cancer, and NASH is currently the most widespread cause of liver failure in the nation. Adults in mainland China had a predominance of NAFLD of 20.09%, according to a large sample meta-analysis from 2014 [4]. In China, the prevalence of NAFLD has been steadily rising over the past few years and has intensified. As a result, it is crucial to do epidemiological research or large-scale cohort studies on NAFLD. The execution of the nationwide physical checkup inspires extensive study, but a rapid and simple technique to categorise NAFLD patients in general residents is still needed.

An invasive and sophisticated procedure called a histologic culture is the benchmark for identifying NAFLD. Ultrasound, CT and MRI are common diagnostic methods; however, when population screening is carried out on a large scale, imaging tests are very expensive. To help with NAFLD diagnosis, a number of tests have been created. In order to predict hepatic steatosis in the typical populace, the fatty liver index uses body mass index (BMI), waistline perimeter, serum triglyceride and gamma-glutamyl transferase (GGT) values [5, 6]. The NAFLD liver fat score [7] employs a formula that accounts for the metabolic disorder, type 2 diabetes, fasting blood insulin, aspartate aminotransferase (AST) and alanine aminotransferase (ALT) stages to determine the proportion of liver fat substance. The SteatoTest logistic regression model [8], which uses 12 predictive attributes, includes A2-macroglobulin (A2M), apolipoprotein A1 (ApoA1), haptoglobin, total bilirubin, GGT levels cholesterol, triglycerides, glucose, age, gender and BMI. A forecast prototype built on research data comprises the following six indicators: alanine aminotransferase, high-density lipoprotein cholesterol, triglyceride, haemoglobin A1c (HbA1c), leukocytes count and the existence of high blood pressure.

A challenge arises, though, from the difficulty in obtaining these prediction parameters. The use of the current NAFLD estimate prototypes has been extensive, but their use in extensive epidemiologic studies and many provinces of developing countries like China is still quite restrained.

In medicine, knowledge engineering is employed to develop precise prediction models. Machine learning outperforms traditional statistical methods in terms of better prediction accuracy, better modelling of complicated associations, better capacity to discover through various segments of data and sturdiness to data interference. The techniques are used to analyse fatty liver, meningitis, glaucoma, coronary

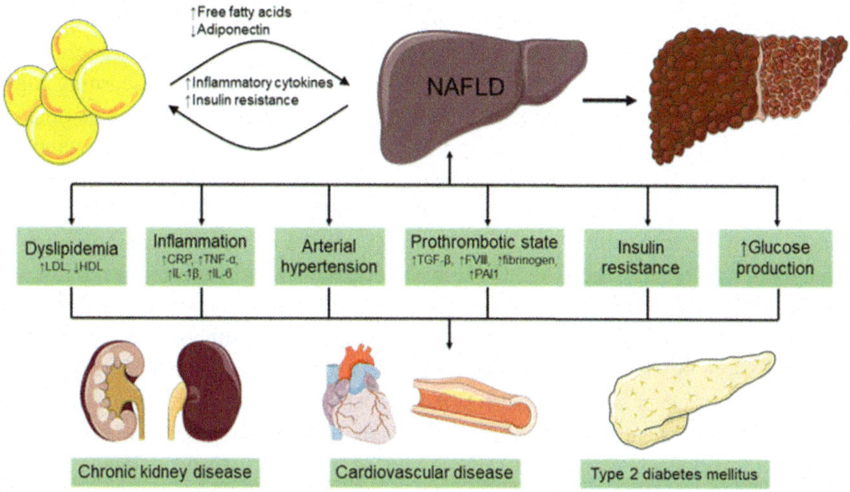

Fig. 1 Association of NAFLD with various diseases [17]

heart disease, cancer and other illnesses [9–16]. Our objective is to use knowledge engineering to analyse the data from 304,145 medical checkup of subjects.

These techniques have been used to make diagnoses for cancer, glaucoma, meningitis, fatty liver and other illnesses [9–16]. Figure 1 depicts the relationship among NAFLD and type 2 diabetes mellitus, chronic kidney disease and cardiovascular disease. Using deep learning to analyse the data, the authors aspire to build a simple NAFLD screening prototype without the need for laboratory-tested markers.

2 Related Work

The influence of AI [18–24], machine learning and deep neural networks (DNN) in health care on a range of medical specialties, including radiology, pathology and clinical specialty, has been the subject of numerous research. The current study intends to offer a thorough and synthesised review of DNN algorithms' applications in liver pathology across the literature, in the fields of metabolic and inflammatory as well as tumoral, and to discuss any bias that may exist as well as the opportunities that AI may present for liver pathologists.

The categorisation of fatty and normal liver sonography imageries has been the subject of the following studies in current times. The approaches for feature selection and feature extraction are used in this categorisation procedure.

Weidong et al. [25] used LASSO regression, obtained 12 non-zero coefficient characteristics and reduced from 17 to 12 indexes. A total of 58,654 subjects out of 304,135 subjects had NAFLD. The authors observed 17 types of attributes which were statistically ($p < 0.0001$) associated with NAFLD.

Garcia-Carretero et al. [26] assessed the occurrence of NASH in 2239 high blood pressure patients as well as the pertinent characteristics associated to high blood pressure and metabolic syndrome (MS) using supervised machine learning algorithms like least absolute shrinkage and selection operator (LASSO) and random forest classifier. The random forest approach was used to assess the importance of the characteristics in the LASSO regression model. In a univariate analysis, it was connected to metabolic syndrome, type 2 diabetes, Humulin defiance and Hyperlipidemia. With a true positive rate of 70%, true negative rate of 79% and area under the curve of 0.79 [26], the LASSO method correctly recognised serum ferritin and insulin.

Logistic regression (LR), k-nearest neighbour (kNN), support vector machine (SVM), Naive Bayes, Bayesian Network (BN), decision tree and K2 algorithm, which incorporates adaptive boosting (AdaBoost), bootstrap aggregating (bagging) and random forest, are some of the techniques employed by researchers in their work [10]. Hidden Naive Bayes (HNB) and aggregating one-dependence estimators (AODE) were added to the method. Ma et al. [10] examined the 11 machine learning techniques on 10,508 people to find the most precise investigative prototype for NAFLD. They asserted that the SVM prototype had the highest specificity and accurateness, with estimates of 0.946 and 0.725, respectively, and that the logistic regression (LR) model had an accuracy of 83.41%. The most sensitive model was the AODE one, with a value of 0.680 [10].

Yip et al. enrolled 922 patients to examine logistic regression, AdaBoost and ridge regression. Lastly, with an accuracy of 87–88% [9], the logistic regression model was able to collect six pertinent characteristics, like Humulin defiance, triglycerols or glutamate pyruvate transaminase.

Sorino et al. [27] paralleled eight distinct knowledge engineering algorithms, including the regularised multinomial classifier (using LR), boosting tree classifier (using the AdaBoost classifier), decision tree classifier, Naive Bayes classifier, K-nearest neighbours classifier, neural network classifier and SVM classifier. Employing the meta-learning technique, three prototypes were created: the FLI plus GLUCOSE plus SEX plus AGE, the stomach size index plus GLUCOSE plus GGT plus SEX plus AGE and the body roundness index plus GLUCOSE plus GGT plus SEX plus AGE. The SVM in the Python approach, according to the authors, was the most appropriate and performed best in the models under examination [27]. Model 3 had the maximum accuracy, at 77%, contrasted with Prototypes 2 and 1, each of which had an accuracy of 68%. Since Model 2 had less estimate miscalculations, it was thought to be the best model. [27].

Docherty et al. [28] used data from the National Institute of Diabetes and Digestive and Kidney Diseases (NIDDK) dataset that comprised both established NASH and non-NASH established liver anatomy findings [28] to develop a machine learning (ML) prototype to foresee NASH. The area under curve (0.82), sensitivity (81%) and accuracy (81%) of an extreme gradient boosting model (XGBoost) with 14 characteristics showed great performance in terms of predicting NASH [28]. Performance was marginally worse (0.79, 80 and 80%, respectively) [28] using an inadequate feature set of five variables. The entire model accurately predicted NASH in the Optum data

(AUC 0.76) [28]. The authors' NASH map model, which was validated on a huge, real-world patient dataset [28], was developed using liver biopsy data from patients of verified NASH and non-NASH.

For describing the texture of the liver and determining the risk of contracting fatty liver hepatitis under the symptosis class, Biswas et al. proposed a network based on CNN construct with various layers of convolution, pooling and inception. This algorithm gives this region in accordance with a gold truth after extracting its ROI attributes from background data. Only organs where the gold truth is present can be used in this approach. This clearly demonstrates the need for a professional using traditional techniques [29].

Essam et al. [30] proposed a moth flame optimisation algorithm (MFO) in their work to perform segmentation on liver images taken through MRI. MFO was used to cluster the abdominal image. The required clusters which represented the liver were picked up to get the initial segmented images. A group of 70 MRI imageries were taken to segment the liver and test the methodology. The results showed 95.6% accuracy [30].

Dulhare et al. [31] suggested a dual three-dimensional convolutional networks design for detection of the disease and ResNet for classification feature selection was performed using particle swarm optimisation. Using the expertly learnt features from particle swarm optimisation, classification was performed with the help of ResNet.

3 Methodology

The proposed model's architecture, which depicts the many processes employed in the model, is displayed in Fig. 2.

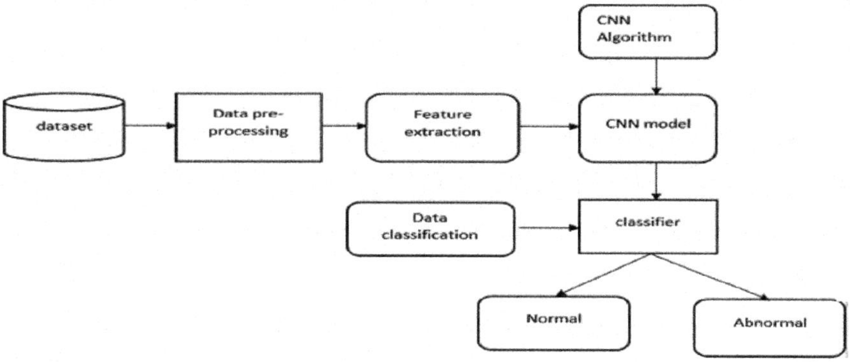

Fig. 2 Architecture showing the workflow the proposed model

3.1 Data Cleaning

NPE contains a huge volume of data, and the variables are disordered with many omitted values and outliers. Thus, data preparation is a crucial stage [32]. First, we eliminated 200 or so variables that didn't matter for our study. Then, after preprocessing the nulls and outliers, we removed the variables with more than 20% nulls and imputed the variables in their absence. In addition, categorical variables were occupied with the mode and continuous variables with the mean.

$$\text{Min}\, w1/2n \,||\, Xw - y||22 + \alpha \,||\, w||1 \tag{1}$$

where w is a regression model's coefficient vector, n is the number of samples, X is a matrix of subject features and $||w||1$ is the LASSO penalisation with the constant and the l1-norm of the coefficient vector $||w||1$ [33].

3.2 Data Imbalance Processing

In comparison to NAFLD (an unbalanced-class problem) individuals, there were more normal participants. Forecasting is often more difficult for people with fewer courses than for those with more disciplines. The embedded method's Penalised SVM algorithm was utilised to alleviate the harmful consequences of class imbalance. The main goal of the strategy is to raise the cost of incorrect classification in the minority class.

With the class_weight = 'balanced' parameter, errors made by the minority might be penalised to the extent that they are underrepresented during training.

3.3 Feature Extraction

Feature extraction is a method that converts unprocessed data into controllable numerical features while preserving the original dataset's information. It returns excellent outcomes when paralleled with utilising machine learning on the processed data straightforwardly. In this work, two techniques are applied for feature extraction, viz., wavelet scattering and grey co-occurrence matrix.

3.3.1 Wavelet Scattering Used for Feature Extraction

Wavelet scattering networks make it easier to extract low-variance information from signals and images for use in deep learning and machine learning applications [34].

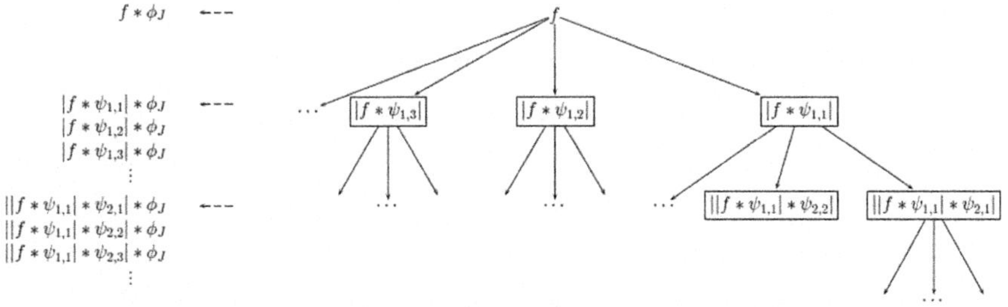

Fig. 3 Wavelet scattering coefficient modulus [35]

It can automatically find features that reduce differences within a class with the aid of scattering networks.

With the use of a wavelet scattering transform, data is processed in stages. The initial input for the next stage is the output of the previous stage. There are three operations in each stage. The input is simply averaged to determine the zeroth-order scattering coefficients. In this study, a wavelet scattering transform is employed to extract trustworthy features that are rotation- and deformation-invariant and stable to mild deformation. Number of years, sexual category, actions, smouldering, BMI, waistline perimeter, nutritional condition and liking of salty and fried foods were the characteristics that were extracted (Fig. 3).

Here, *j* is the scaling function, *j*, *k* are wavelets and f is the input data. For each *j*, *k* in the case of image data, there exists a variety of user-specified wavelet rotations. A path is a set of edges leading from the root to a node. The scalogram coefficients are the tree nodes. The scaling function J is convolved with the scalogram coefficients to produce the scattering coefficients. The prevalence characteristics produced from the data are the set of scattering coefficients. Information is lost during convoluting with the scaling function, which is substandard filtering. The data is nonetheless recovered when calculating the coefficients in the following step.

Before beginning to obtain characteristics from the facts, create and constitute the network using wavelet scattering (for time series data) or wavelet scattering2 (for picture data). You can alter the size of the invariance scale, the quantity of filter depositories and the total of wavelets per octave in each filter depository. The number of wavelet rotations can also be specified in wavelet Scattering2. To extract features from time series, use the wavelet scattering object's scattering Transform or feature Matrix operations. To extract characteristics from image data, use the scattering Transform or characteristics Matrix functions of the wavelet Scattering2 object.

The scattering transform is used to generate features iteratively. You must first convolute the facts with the scaling routine, *f J*, in order to derive S[0], the zeroth-order dispersion constants. Next, carry out the following:

1. Using every wavelet filter in the first filter bank, compute the wavelet transform of the input data.
2. Consider the modulus of every filtered output. Scalogram nodes are represented by U [1].

3. Utilising the scaling filter, average each modulus. The outcomes are the S [1] values for the first-order scattering coefficients.

Iterate the method at each node. The scattering Transform routine returns the scattering and scalogram coefficients. The feature Matrix function yields the scattering attributes.

3.3.2 Feature Extraction Using Grey Co-occurrence Matrix

It is a popular texture feature extracting methodology which lies on the foundation of statistical methods. This matrix is stated as the possibility of beginning from a pixel position with grey intensity i and dropping a static point with grey level j. The complete assessed estimates can be conveyed as a matrix. Figure 4 displays the grey co-occurrence matrix used in fatty liver feature extraction.

A point pair (x, y) is made up of every point in the imagery and any point that deviates from it $(x + a, y + b)$, and their combined grey value is (i, j). The values' frequency occurrences (i, j) across the entire image are calculated and then grouped in a square matrix. The grey co-occurrence matrix is created by normalising the total occurrence times to the probability of occurrence.

3.4 Classification Using CNN

CNNs, a subclass of deep neural networks, are commonly applied to the processing of visual images. CNNs are able to recognise and classify particular aspects in photos. Among their uses are natural language processing, image classification, video and image analysis for medical applications and image and video recognition [36–39]. Because of its high degree of accuracy, CNN is helpful for image recognition. Image recognition has several uses in a variety of industries, including phones, security, recommendation systems and medical image analysis, among others (Fig. 5).

There are two main parts to a CNN architecture. The process of feature extraction uses a convolution tool to separate and pinpoint the distinctive qualities of an image

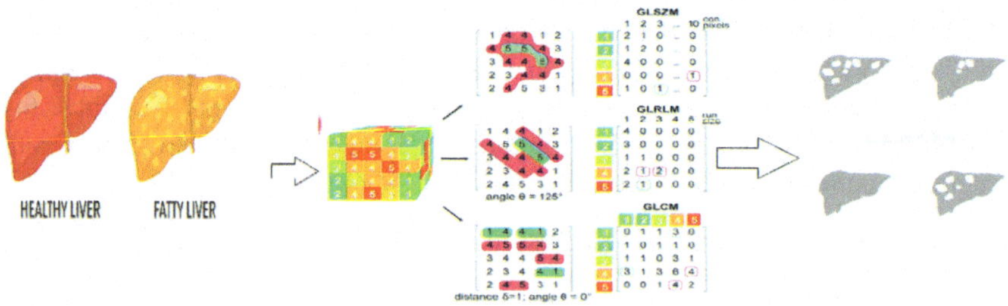

Fig. 4 Grey co-occurrence matrix applied for feature extraction

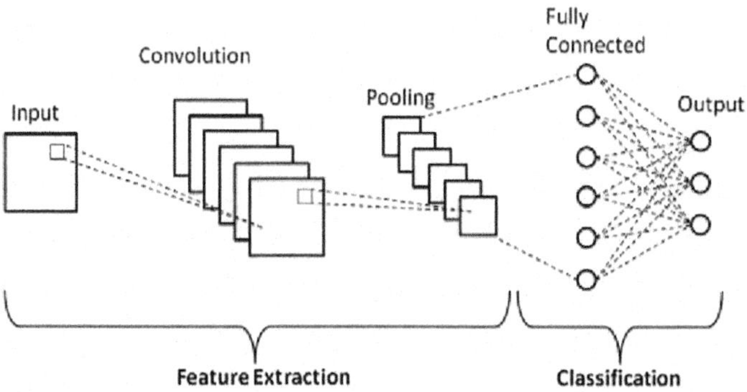

Fig. 5 Architecture of CNN

for analysis. The feature extraction network consists of many pairs of convolutional or pooling layers. A fully connected layer that utilises the results of the convolutional process and establishes the image class utilising the previously retrieved features. The goal of this CNN feature extraction model is to extract as few features from a dataset as possible. By integrating the existing features of an original collection of features into a single new feature, it develops new features. There are several CNN levels, as seen in the CNN architectural diagram (Figs. 6, 7, 8 and 9).

Evaluation standards for the accuracy of the results of applicable networks are different. The main criteria are accuracy, sensitivity and specificity.

Sensitivity: Sensitivity is termed as the ratio of elements accurately categorised in class 1 (True Positive (TP)) to all elements labelled in class 1 (True Positive (TP) + False Negative (FN)) and it is calculated as shown below:

$$\text{Sensitivity} = \frac{\text{TP}}{\text{TP} + \text{FN}} \tag{2}$$

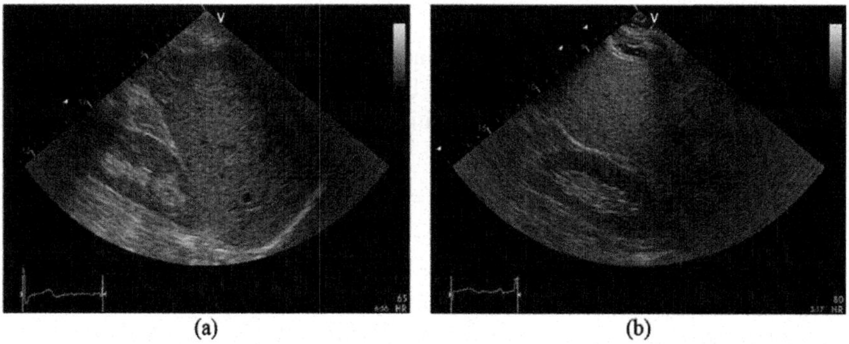

Fig. 6 Ultrasound image samples for patients: **a** normal liver (Index 3%), **b** susceptible to fatty liver (index 20%) [40]

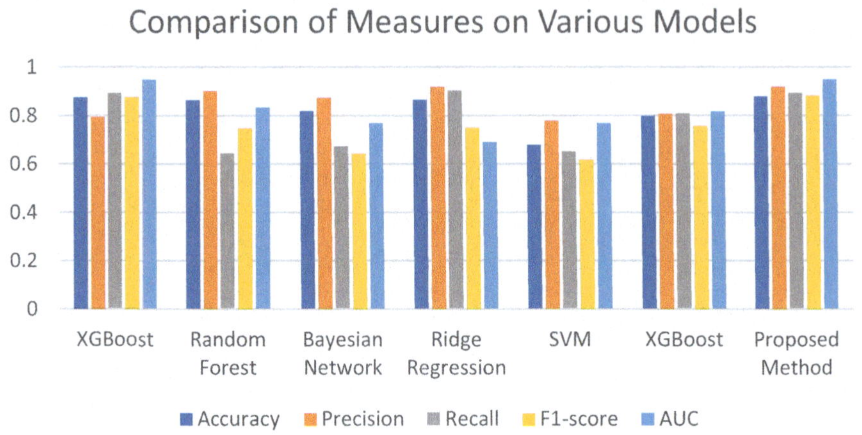

Fig. 7 Comparison of measures on various models

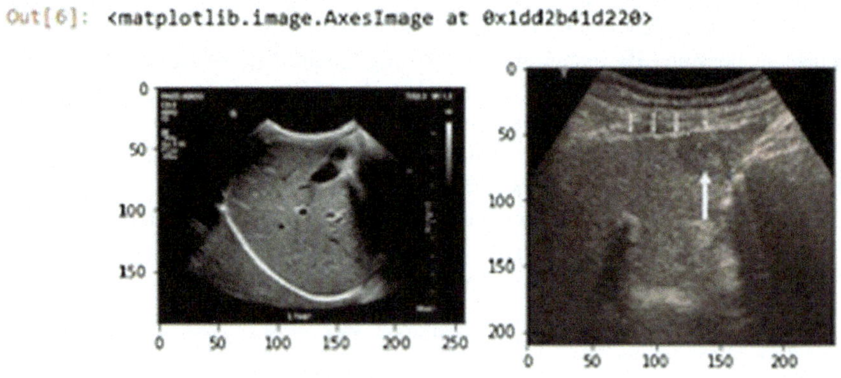

Fig. 8 Output image showing the classified NAFLD

Specificity: Specificity can also be defined by the fraction of elements categorised appropriately in class 2 (True Negative (TN)), in contrast to all existing factors in class 2 (True Negative (TN) + False Positive (FP)), and it can be shown as follows:

$$\text{Specificity} = \frac{\text{TN}}{\text{TN} + \text{FP}} \tag{3}$$

Accuracy: The accuracy value can be calculated as follows and is defined by the total sum of elements correctly categorised to all currently used features for categorisation.

$$\text{Accuracy} = \text{True Positive} + \text{True Negative} \tag{4}$$

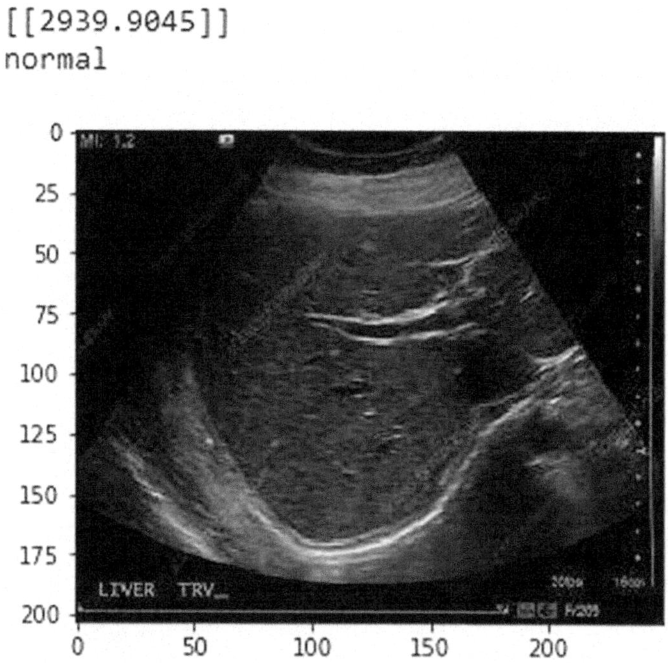

Fig. 9 Output image showing the normal liver

True Positive + True Negative + False Positive + False Negative

For the presented algorithm, the metrics of accuracy, sensitivity and specificity are calculated. The error matrix, in general, can be a suitable aggregate to convey the state of these factors for test data to contrast their functioning.

4 Results and Discussions

The CNN classification method was used to determine how well the work that was provided was performed. To assess the results of the pre-trained network approaches discussed above in the associated work, these outcomes are given. When CNN was used, the outcomes of the suggested system when compared to other models revealed better results (Tables 1 and 2).

Table 1 Comparison of original data and ratio and Penalised SVM

Dataset	Samples	Ratio	Description
Original data	245,490/58,654	4:1	Original data with all instances
Penalised SVM	243,300/243,300	1:1	Dataset balanced using Penalised SVM

Table 2 Summarised performance results for different networks

Model/Network	Accuracy	Precision	Recall	F1-score	AUC
XGBoost	0.879	0.800	0.892	0.881	0.950
Random forest	0.866	0.904	0.643	0.750	0.836
Bayesian network	0.821	0.876	0.672	0.645	0.769
Ridge regression	0.867	0.921	0.902	0.752	0.691
SVM	0.680	0.78	0.65	0.62	0.77
XGBoost	0.80	0.81	0.809	0.76	0.819
Proposed method	0.881	0.922	0.894	0.887	0.951

5 Conclusion

Over time, efforts have increased to more accurately discover and categorise fatty liver disease and the associated medical phases than doing it manually. To complete the work, majority of the work is focussed on extracting characteristics from processed imageries. A big step in the right direction is the use of various preprocessing and feature extraction methods and application of CNN for classifying.

To advance prospective future research, more effort must be invested into establishing problem-solving models and conducting randomised scientific investigations on a greater number of affected persons. The discoveries will support the advancement of comprehensible models in the future. To establish the most accurate phases of the images, more work must be put into processing images and extracting features, taking into consideration the structural differences.

References

1. Younossi ZM, Koenig AB, Abdelatif D, Fazel Y, Henry L, Wymer M (2016) Global epidemiology of nonalcoholic fatty liver disease-Meta-analytic assessment of prevalence, incidence, and outcomes. Hepatology 64:73–84. https://doi.org/10.1002/hep.28431
2. Rinella ME (2015) Nonalcoholic fatty liver disease: a systematic review. JAMA 313:2263–2273. https://doi.org/10.1001/jama.2015.5370
3. Wesolowski SR, Kasmi KC, Jonscher KR, Friedman JE (2017) Developmental origins of NAFLD: awomb with a clue. Nat Rev Gastroenterol Hepatol 14:81–96. https://doi.org/10.1038/nrgastro.2016.160

4. Li Z, Xue J, Chen P, Chen L, Yan S, Liu L (2014) Prevalence of nonalcoholic fatty liver disease in mainland of China: a meta analysis of published studies. J Gastroenterol Hepatol 29:42–51. https://doi.org/10.1111/jgh.12428
5. Kwok R, Tse YK, Wong GL, Ha Y, Lee AU, Ngu MC et al (2014) Systematic review with meta-analysis: non-invasive assessment of non-alcoholic fatty liver disease–the role of transient elastography and plasma cytokeratin-18 fragments. Aliment Pharmacol Ther 39:254–269. https://doi.org/10.1111/apt.12569
6. Wong VW, Wong GL (2014) When and how to use steatosis biomarkers? Aliment Pharmacol Ther 40:1359–1360. https://doi.org/10.1111/apt.12983
7. Bedogni G, Bellentani S, Miglioli L, Masutti F, Passalacqua M, Castiglione A et al (2006) The fatty liver index: a simple and accurate predictor of hepatic steatosis in the general population. BMC Gastroenterol 6:33. https://doi.org/10.1186/1471-230X-6-33
8. Kotronen A, Peltonen M, Hakkarainen A, Sevastianova K, Bergholm R, Johansson LM et al (2009) Prediction of non-alcoholic fatty liver disease and liver fat using metabolic and genetic factors. Gastroenterology 137:865–872. https://doi.org/10.1053/j.gastro.2009.06.005
9. Yip TC-F, Ma AJ, Wong VW-S, Tse Y-K, Chan HL-Y, Yuen P-C, Wong GL-H (2017) Laboratory parameter-based machine learning model for excluding non-alcoholic fatty liver disease (NAFLD) in the general population. Aliment Pharmacol Ther 46(4):447–456. https://doi.org/10.1111/apt.14172
10. Ma H, Xu C, Shen Z, Yu C, Li Y (2018) Application of machine learning techniques for clinical predictive modeling: a cross-sectional study on nonalcoholic fatty liver disease in China. Biomed Res Int 2018:e4304376. https://doi.org/10.1155/2018/4304376
11. Perveen S, Shahbaz M, Keshavjee K, Guergachi A (2018) A systematic machine learning based approach for the diagnosis of non-alcoholic fatty liver disease risk and progression. Sci Rep 8:2112. https://doi.org/10.1038/s41598-018-20166
12. Lélis VM, Guzmán E, Belmonte MV (2017) A statistical classifier to support diagnose meningitis in less developed areas of Brazil. J Med Syst 41:145. https://doi.org/10.1007/s10916-017-0785-5
13. Huang ML, Chen HY (2010) Glaucoma classification model based on GDx VCC measured parameters by decision tree. J Med Syst 34:1141–1147. https://doi.org/10.1007/s10916-009-9333-2
14. Gregori D, Bigi R, Cortigiani L, Bovenzi F, Fiorentini C, Picano E (2009) Non-invasive risk stratification of coronary artery disease: an evaluation of some commonly used statistical classifiers in terms of predictive accuracy and clinical usefulness. J Eval Clin Pract 15:777–781. https://doi.org/10.1111/j.1365-2753.2008.01034.x
15. Chao CM, Yu YW, Cheng BW, Kuo YL (2014) Construction the model on the breast cancer survival analysis use support vector machine, logistic regression and decision tree. J Med Syst 38:106. https://doi.org/10.1007/s10916-014-0106-1
16. Kate RJ, Nadig R (2017) Stage-specific predictive models for breast cancer survivability. Int J Med Inform 97:304–311. https://doi.org/10.1016/j.ijmedinf.2016.11.001
17. Kechagias S, Ekstedt M, Simonsson C, Nasr P (2022) Non-invasive diagnosis and staging of non-alcoholic fatty liver disease. Hormones (Athens). 21(3):349–368. https://doi.org/10.1007/s42000-022-00377-8. Epub 2022 Jun 4. PMID: 35661987; PMCID: PMC9464753
18. Hamad A, Houssein EH, Hassanien AE, Fahmy AA (2016) Feature extraction of epilepsy EEG using discrete wavelet transform. In: 12th international computer engineering conference (ICENCO). IEEE
19. Shaban H, Houssein EH, Pérez-Cisneros M, Oliva D, Hassan AY, Ismaeel AA, AbdElminaam DS, Deb S, Said M (2021) Identification of parameters in photovoltaic models through a runge kutta optimizer. Mathematics 9(18):2313
20. Abdelminaam DS, Said M, Houssein EH (2021) Turbulent flow of water-based optimization using new objective function for parameter extraction of six photovoltaic models. IEEE Access 9:35382–35398
21. Houssein EH, Hassaballah M, Ibrahim IE, AbdElminaam DS, Wazery YM (2022) An automatic arrhythmia classification model based on improved marine predators algorithm and convolutions neural networks. Expert Syst Appl 187: 115936

22. Houssein EH, Neggaz N, Hosney ME, Mohamed WM, Hassaballah M (2021) Enhanced Harris hawks optimization with genetic operators for selection chemical descriptors and compounds activities. Neural Comput Appl 33:13601–13618
23. Ahmed MM, Houssein EH, Hassanien AE, Taha A, Hassanien E (2018) Maximizing lifetime of wireless sensor networks based on whale optimization algorithm. In: Proceedings of the international conference on advanced intelligent systems and informatics 2017. Springer International Publishing, pp 724–733
24. Houssein EH, Sayed A (2023) Dynamic candidate solution boosted beluga whale optimization algorithm for biomedical classification. Mathematics 11(3):707
25. Ji W, Xue M, Zhang Y, Yao H, Wang Y (2022) Original research, a machine learning based framework to identify and classify non-alcoholic fatty liver disease in a large-scale population. Front Public Health 10:2296–2565. https://www.frontiersin.org/articles/https://doi.org/10.3389/fpubh.2022.846118
26. Garcia-Carretero R, Vigil-Medina L, Barquero-Perez O, Ramos-Lopez J (2019) Relevant features in nonalcoholic steatohepatitis determined using machine learning for feature selection. Metab Syndr Relat Disord 17(9):444–451. https://doi.org/10.1089/met.2019.0052
27. Sorino P, Caruso MG, Misciagna G, Bonfiglio C, Campanella A, Mirizzi A, Franco I, Bianco A, Buongiorno C, Liuzzi R, Cisternino AM, Notarnicola M, Chiloiro M, Pascoschi G, Osella AR, Group, M (2020) Selecting the best machine learning algorithm to support the diagnosis of non-alcoholic fatty liver disease: a meta learner study. PLoS One 15(10):e0240867. https://doi.org/10.1371/journal.pone.0240867
28. Docherty M, Regnier SA, Capkun G, Balp M-M, Ye Q, Janssens N, Tietz A, Löffler J, Cai J, Pedrosa MC, Schattenberg JM (2021) Development of a novel machine learning model to predict presence of nonalcoholic steatohepatitis. J Am Med Inform Assoc:JAMIA 28(6):1235–1241. https://doi.org/10.1093/jamia/ocab003
29. Biswas M, Kuppili V, Edla DR, Suri HS, Saba L et al (2018) Symtosis: a liver ultrasound tissue characterization and risk stratification in optimized deep learning paradigm. Comput Methods Progr Biomed 155:165–177. https://doi.org/10.1016/j.cmpb.2017.12.016
30. Said S, Mostafa A, Houssein EH, Hassanien AE, Hefny H (2018) Moth-flame optimization based segmentation for MRI liver images. In: Hassanien A, Shaalan K, Gaber T, Tolba M (eds) Proceedings of the international conference on advanced intelligent systems and informatics 2017. AISI 2017. Advances in intelligent systems and computing, vol 639. Springer, Cham. https://doi.org/10.1007/978-3-319-64861-3_30
31. Dulhare U, Mubeen A (2023) Detection and classification of rheumatoid nodule using deep learning models. Procedia Comput Sci 218:2401–2410. https://doi.org/10.1016/j.procs.2023.01.215
32. Ngiam KY, Khor IW (2019) Big data and machine learning algorithms for health-care delivery. Lancet Oncol 20(e26):3. https://doi.org/10.1016/S1470-2045(19)30149-4
33. Liu J, Sun D, Chen L, Fang Z, Song W, Guo D et al (2019) Radiomics analysis of dynamic contrast-enhanced magnetic resonance imaging for the prediction of sentinel lymph node metastasis in breast cancer. Front Oncol 9:980. https://doi.org/10.3389/fonc.2019.00980
34. Dulhare UN, Khaleed AM (2020) Taj-Shanvi framework for image fusion using guided filters. In: Sharma N, Chakrabarti A, Balas V (eds) Data Management, analytics and innovation. advances in intelligent systems and computing, vol 1016. Springer
35. https://in.mathworks.com/help/wavelet/ug/wavelet-scattering.html
36. Arathi B, Dulhare UN (2023) Classification of cotton leaf diseases using transfer learning-DenseNet-121. In: Reddy AB, Nagini S, Balas VE, Raju KS (eds) Proceedings of third international conference on advances in computer engineering and communication systems. Lecture notes in networks and systems, vol 612. Springer
37. Gouse S, Dulhare UN (2022) Automation of rice leaf diseases prediction using deep learning hybrid model VVIR." In: Rajagopal S, Faruki P, Popat K (eds) Advancements in smart computing and information security. ASCIS 2022. Communications in computer and information science, vol 1759. Springer, Cham

38. Dulhare UN, Gouse S (2022) Automation of rice cultivation from ploughing–harvesting with diseases, pests and weeds to increase the yield using AI. In: Kumar A, Mozar S (eds) ICCCE 2021. Lecture notes in electrical engineering, vol 828. Springer
39. Dulhare UN, Taj STA (2023) Water quality risk analysis for sustainable smart water supply using adaptive frequency and BiLSTM. In: Kannan RJ, Geetha S, Sashikumar S, Diver C (eds) International virtual conference on industry 4.0. IVCI 2021. Lecture notes in electrical engineering, vol 1003. Springer
40. Zamanian H, Mostaar A, Azadeh P, Ahmadi M (2021) Implementation of combinational deep learning algorithm for non- alcoholic fatty liver classification in ultrasound images. J Biomed Phys Eng 11(1):73–84. https://doi.org/10.31661/jbpe.v0i0.2009-1180. PMID:33564642; PMCID:PMC7859380

Chapter 7
Fuzzy Logic Controller-Based Off-Grid Solar Water Pumping System

Santosh S. Raghuwanshi, Animesh Kumar Masih, Hasina Adil, and M. Murali

Abstract A photovoltaic cell, also called a solar cell, directly transforms sunlight energy into DC electrical energy. PV systems have various applications such as lighting, off-grid systems, and pumping the water in remote places. Water irrigation systems have been developed and tested throughout India. This chapter discusses the design, modeling, and evaluation of an off-grid solar pumping system powered by a 3-Φ induction motor (IM). The size of the solar photovoltaic water pumping system (SPVWPS) can be calculated for any study area. The performance of all system components is analyzed under different solar irradiation conditions using various methods such as P&O, ICA, fuzzy logic, and neuronal network. The efficiency of the conventional controller (PI) and FLC are compared, with FLCs being better at handling nonlinearities in the system due to changes in solar radiation, temperature, and other environmental factors. Overall, Fuzzy controllers provide better results than PI controllers in terms of efficiency and stability.

Keywords FLC · Solar water pumping · P&O · ICA · Controller · Inverter

1 Introduction

SPVWPS is a sort of new technology in which solar panels are used to power a pump that pulls water from a well or other water source. These methods are frequently utilized in rural regions when power is scarce or non-existent. A solar panel array, a pump, a storage tank, and a control system are the fundamental components of an SPVWPS [1]. The solar panel array transforms solar energy into electrical energy, which powers the pump. The pump pulls water from the well or another source and pumps it into the storage tank. The control system regulates the pump's operation

S. S. Raghuwanshi (✉) · A. K. Masih · M. Murali
Department of Electrical Engineering, Medi-Caps University, Indore, Madhya Pradesh, India
e-mail: santoshsraghuwanshi@gmail.com

H. Adil
Department of Basic Science, Oriental University, Indore, Madhya Pradesh, India

U. N. Dulhare and E. H. Houssein (eds.), *Machine Learning and Metaheuristics: Methods and Analysis*, Algorithms for Intelligent Systems,
https://doi.org/10.1007/978-981-99-6645-5_7

and guarantees that the system runs within safe parameters [2]. PV water pumping systems have various benefits over traditional water pumping systems that rely on fossil fuels or grid power. They are ecologically beneficial since they emit no greenhouse gases or other pollutants. They are also less expensive because they do not use gasoline or power from the grid. They are also simple to install and maintain, making them suitable for isolated sites [3–5].

PV water pumping systems, however, have several limits. Because they rely on sunshine, weather variables such as clouds or rain can have an impact on their function. They also require a significant quantity of sunshine to function properly, which can be difficult in locations with little solar exposure [3, 4, 6–8]. Despite these restrictions, PV water pumping systems have grown in popularity in recent years due to their numerous advantages. These systems are becoming more efficient and dependable as new control systems like the FLC are developed, making them an appealing solution for a wide range of applications [9].

A number of research articles have been available on off-grid SPVWP systems. Modeling and controlling of SPVWPS for agriculture and lighting load at different solar irradiations and temperatures [10]. PV with wind turbine system is a more cost-effective arrangement to be capable to cover the farmhouse, irrigation system as well as city lighting system load [11, 12]. The numbers of intelligent algorithms (fuzzy logic and genetic algorithm) are available for optimal management of RE applied to water pumping systems [13]. The optimum SPVWP system is designed using soft computing techniques with cost-effective analysis [14]. In India, several water irrigation systems have been established in all states [15]. The water demand for agriculture and livestock is rising daily in rural areas [16]. In remote zones, diesel fuel is used for pumping the water. The key problem in the SPVWP system is to synchronize with load [17]. For more efficient use of PV energy system, MPPT methods are used [18]. Solar PV cells show nonlinear characteristics and provide very low efficiency. Thus, it becomes necessary to harvest maximum solar PV power using MPPT. There are a number of MPP techniques available, i.e., P&O, ICA, FLC, neural network, etc. [2, 19].

Motahhir et al. [20] designed a PV model based on an experimental and simulation program with integrated circuit emphasis (SPICE) coding in the Proteus tool. It proposed a low-cost MPPT controller after modification of the I&C algorithm. The modified controller has given better performance and is more efficient. Mamaghani et al. [21] proposed technically and economically based parameters of hybrid (PV/WT/DG) system configuration using the HOMER tool for different village areas of Colombia. Results validate that hybrid system are more efficient and cost-effective configurations at different locations. An economic evaluation of hybrid diesel-RES and grid-RES has been done for remote electrification in Malawi's rural villages using HOMER software [22]. In the first part economic evaluation is made on standalone (diesel and RE generation), while, in the second part, RE system is interfaced with utility grid power supply [23] A techno-economic evaluation was carried out [24, 25] to find out optimum solution among using a software tool. The PV/battery with DG model is the optimum solution for remote areas. Rural areas in Malaysia do not have utility grid connectivity.

These three cascaded fuzzy-predictive controllers for a PV/battery pumping system. The controllers include MPPT, power flow management, and duty ratio control. The results show that the proposed controllers provide better tracking, reduced ripples, regulated voltage, and high accuracy in speed tracking and torque reduction, and can protect the motor from overcurrent [26]. This reference [27] presents a study on a multifunction system consisting of PV power source connected to a battery, with a focus on improving the quality of energy produced and injected into the grid. The system utilizes a three-level inverter and two fuzzy logic controllers, one for MPPT and the other for DC bus stabilization. The results show improved energy quality and effectiveness of the proposed inverter.

The algorithm uses an adaptive an FLC and a supervisory power management strategy to maintain a stable power flow and prevent system component limits from being exceeded. The proposed energy management system (EMS) is modular and includes faster state of charge balancing and smaller DC bus voltage fluctuations. The algorithm also considers optimal energy flow between the PV system with battery system and load while reducing component constraints. Real-time simulation and experimentation using Matlab/Simulink and DSpace are used to validate the effectiveness of the proposed algorithm and EMS [28]. Altimania et al. [6, 7] propose a new application of PV system for water pumping using a 3-Φ IM while maximizing efficiency and harvested power. The proposed technique involves using MPPT of the PV system and the optimal firing pattern of the inverter for the motor through optimal flux control. The results show significant improvements in system performance, which are further enhanced using an adaptive neuro-fuzzy inference system (ANFIS) trained offline. Shetty et al. [29] discuss a study on using IM in solar water pumping systems for agricultural applications in India. The paper proposes a hybrid PSO-INC algorithm to track peak power efficiently in partial shading conditions and compares it to other MPPT techniques. The results show improvements in the motor's performance, including torque ripple and speed settling time.

The focus of this chapter is on the design, modeling, and assessment of an SPVWP system powered by a submersible motor pump, which has been tested under various weather conditions to determine its performance.

2 Solar Water Pumping System in India

SPVWPS is acquisition acceptance in India owing to their low-cost and environmental benefits. Solar panels are utilized to create electricity, which is subsequently used to power a water pump in these systems. The pump may take water from wells, boreholes, and other sources and distribute it for irrigation, drinking, and other uses. Solar water pumping systems have the benefit of not requiring either fuel or power from the grid, making them perfect for isolated places with limited access to electricity. They also need minimal maintenance and have a lengthy lifespan, making them a long-term cost-effective alternative [30].

Table 1 India SPVWPS cost and government subsidies [32]

Rating of pump	Pump cost (Rs.)	Subsidy (Indian Govt.) (Rs.)	Subsidy (State Govt.) (Rs.)	Employer part @ 10% (Rs.)
2.0 HP DC pump	184,955	86,500	80,056	18,496
5.0 HP (AC submersible pump)	439,000	163,000	232,300	43,900

In India, solar water pumping devices are widely employed for irrigation in rural regions. Farmers may utilize these devices to irrigate their crops without relying on costly diesel-powered pumps or grid power. This has aided in increasing agricultural output and decreasing reliance on fossil fuels. The Indian government has initiated a numeral of programs to inspire the usage of solar water pumping systems throughout the country [31]. Subsidies and incentives are available for farmers and other users who install these systems. Table 1 tabulates India SPVWPS cost and government subsidies.

There are three types of PV water pumping systems available in India.

2.1 Direct-Coupled Systems

In this system, the PV panels are linked directly to the pump, with no battery or controller in between. The pump only runs in the presence of sunshine, and the water is pushed straight to the storage tank or irrigation system. Direct-coupled SPVWPSs are simple and dependable, as no batteries or inverters are required. They are useful for outlying areas where grid electricity is unavailable or unstable. They can be used for irrigation, animal watering, home water supply, and a variety of other purposes [31].

The size of the PV module and pump is calculated by the volume of water required and the solar resources available at the location. A bigger array can generate more power and pump more water, although a smaller array may be enough for low-volume applications. Proper size and installation are crucial to ensuring the system's best performance and lifespan. Overall, direct-coupled PV water pumping systems provide an ecologically responsible and cost-effective alternative for off-grid access to clean water [33].

2.2 Battery-Coupled Systems

The PV panels in this system are linked to a battery bank through a charge controller. The battery stores surplus energy created during bright days and powers the pump during overcast or nighttime conditions. Battery-coupled SPVWPSs are a form of solar-powered water pumping system that generates energy from sunshine using PV panels. The PV panels create electricity, which is stored in batteries and utilised to power a water pump. The PV panels, batteries, charge controller, inverter, and water pump are among the system's components [34].

The charge controller regulates the amount of power transferred to the batteries and prevents them from being overcharged or undercharged. When there is no sunshine, the batteries collect the excess power of generator throughout the day and operate the water pump. The inverter turns the DC power stored in the batteries into AC electricity that the water pump can use. Depending on the use, the water pump can be either submersible or surface. It transports water from a well or other source to a storage tank or straight to where it is required. SPVWPSs powered by batteries are perfect for rural places where grid electricity is unavailable or unreliable. They are also ecologically beneficial because they emit no emissions and use renewable energy.

2.3 *Hybrid SPVWP Systems*

These systems mix direct and battery-coupled technologies. It combines solar electricity with grid or generator power to offer uninterrupted water delivery even on overcast days or at night. The battery bank stores surplus energy generated by the solar panels on bright days, while the grid or generator provides backup power as needed. Direct-coupled PV water pumping systems are solar-powered systems that generate electricity to operate a water pump using PV panels. A PV array, a motor, and a pump comprise the system [35]. The PV array transforms sunlight into DC power, which is subsequently sent to the motor. The motor transforms DC power into mechanical energy, which drives the pump to transport water from a source, such as a well or reservoir, to a storage tank or straight to the place of usage [36].

Overall, SPVWPS pumping systems offer a lot of potential in India as a sustainable and cost-effective way to address the country's expanding water demand.

3 Solar Water Pumping System

An SPVWPS pumping system is a renewable energy system that uses sunlight to power a water pump. Solar panels capture energy from the sun and turn it into electricity, which is then utilised to operate a water pump. The water pump is capable of pumping water from a well, borehole, or other water source to a storage tank or straight to a place of use. The regulated energy from the system is sent to the pump [37]. The pump then uses that energy to move water from the basis to a storing reservoir or point of use. The benefits of an SPVWPS are numerous. First, it is

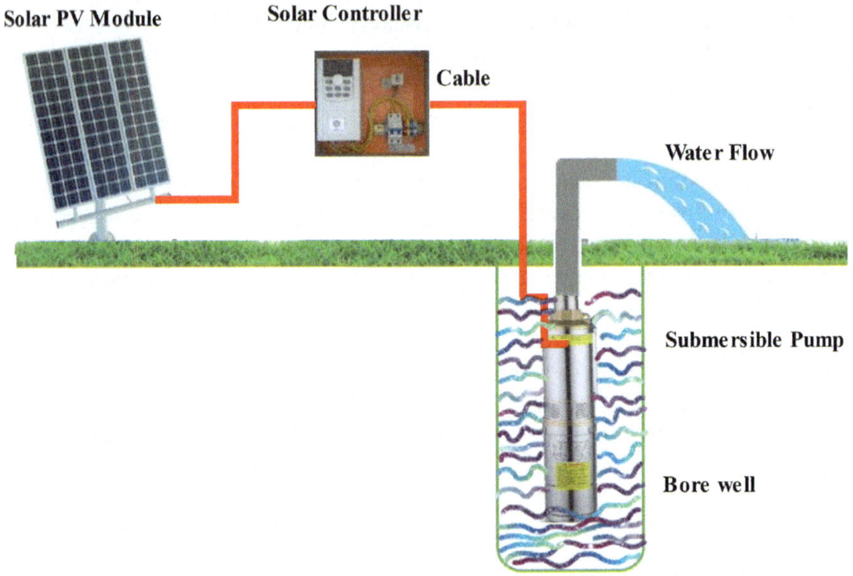

Fig. 1 Solar water pumping system

a renewable energy source, meaning it does not rely on fossil fuels or other non-renewable resources. Second, it is economical in the long run as it requires little maintenance and has low running costs.

Finally, it is environmentally friendly as it produces no emissions or pollutants. Overall, an SPVWPS is an efficient and sustainable way to provide water for agricultural, domestic, or commercial use in areas without grid access or with high electricity bills. An SPVWPS is shown in Fig. 1. The system typically includes a solar module, an inverter, a converter, and a monitoring system.

3.1 Photovoltaic Module

A PV module is expedient that changes sunshine into electricity. It consists of multiple interconnected solar cells enclosed in a protective housing. When sunlight hits a cell, it creates an electric field, creating a stream of electrons that can be converted into electrical energy. Photovoltaic modules are widely used in residential and commercial applications to provide clean, renewable energy for homes and businesses [38].

3.2 Solar Power Controller

A charge controller is an automatic expedient that regulates the amount of power transferred from a solar panel to a battery or other load. To avoid damaging the battery and reducing its lifespan, the controller's primary role is to prevent both overcharging and over discharging. Solar PV controllers monitor the voltage and current of solar panels and batteries. When the battery voltage level reaches a certain threshold, the controller will stop charging the battery to prevent overcharging. When the battery voltage drops below a certain threshold, the controller also stops discharging to prevent deep discharge. Various types of photovoltaic controllers are available on the market, such as PWM (Pulse Width Modulation) controllers and MPPT controllers. PWM regulators are cheaper, but less efficient than MPPT regulators [17].

MPPT controllers are expensive, but they can harvest more power from the solar panel by tracking the maximum power point. Overall, photovoltaic controllers are essential components in any photovoltaic system as they ensure that batteries are charged efficiently and safely.

3.3 Solar MPPT

A solar MPPT is used to optimize the performance of a solar module system. The performance of the solar modules is constantly monitored and the voltage and current are adjusted to draw maximum power from the modules. MPPT controllers regulate voltage and current using a technique called "pulse width modulation". Using this technology, the voltage can be switched on and off quickly, allowing precise control of the amount of current delivered to the battery or load. MPPT controllers are especially useful in situations where there is a mismatch between the solar panel voltage and the battery or load voltage. Matching voltages and currents ensure maximum power transfer between these components, improving efficiency and extending battery life. Overall, a solar MPPT controller is an integral part of a solar panel system as it helps maximize energy production and ensures reliable operation. P&O and ICA are the conventional MPPT techniques [39].

3.3.1 Perturb and Observe Algorithm

The P&O MPPT procedure is a generally used method for optimizing the power output of PV systems. It works by continually perturbating the working of the PV array and observing the resulting change in generated power. The method first measures the current and voltage of the PV system and calculates its output power. Then it disturbs the operating point by changing the voltage or current depending on what type of converter is used in the installation. A change in output power is observed, and as these increases, the disturbance continues in that direction until a

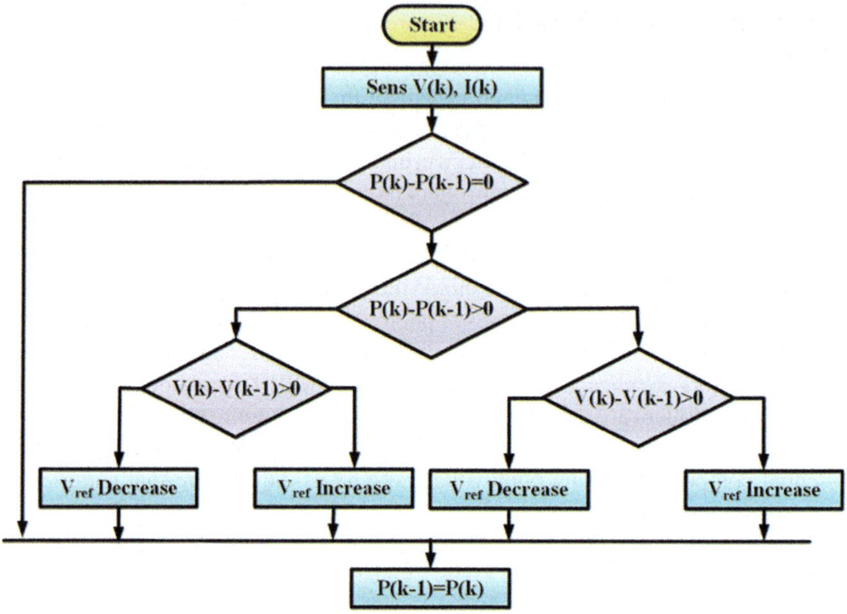

Fig. 2 P&O method [2]

further increase in power is observed. However, as the output power decreases, the noise reverses and continues in the opposite direction until the MPP is reached. Once an MPP is found, the algorithm continues to track, disturb, and monitor it regularly to ensure that any changes in environmental conditions or load requirements are taken into account. Although MPPT and P&O algorithms are easy to implement and require minimal computational resources, they can suffer from MPP jitter due to their dependence on small perturbations. Also, they may not perform as well as other MPPT algorithms when environmental conditions are changing rapidly or there is partial shading [2]. Figure 2 illustrates the diagram of P&O.

3.3.2 Incremental Conductance Algorithm

The ICA is a technique used in PV panel MPPT systems. The ICA uses a mathematical model of the panel's behavior to determine the optimal load resistance. The algorithm works by measuring the panel's voltage and current, and then calculating its incremental conductance, which is the change in current divided by the change in voltage. If the incremental conductance is positive, it means that increasing the load resistance will increase power output, so the algorithm increases the load resistance. If it is negative, decreasing the load resistance will increase power output, so the algorithm decreases it.

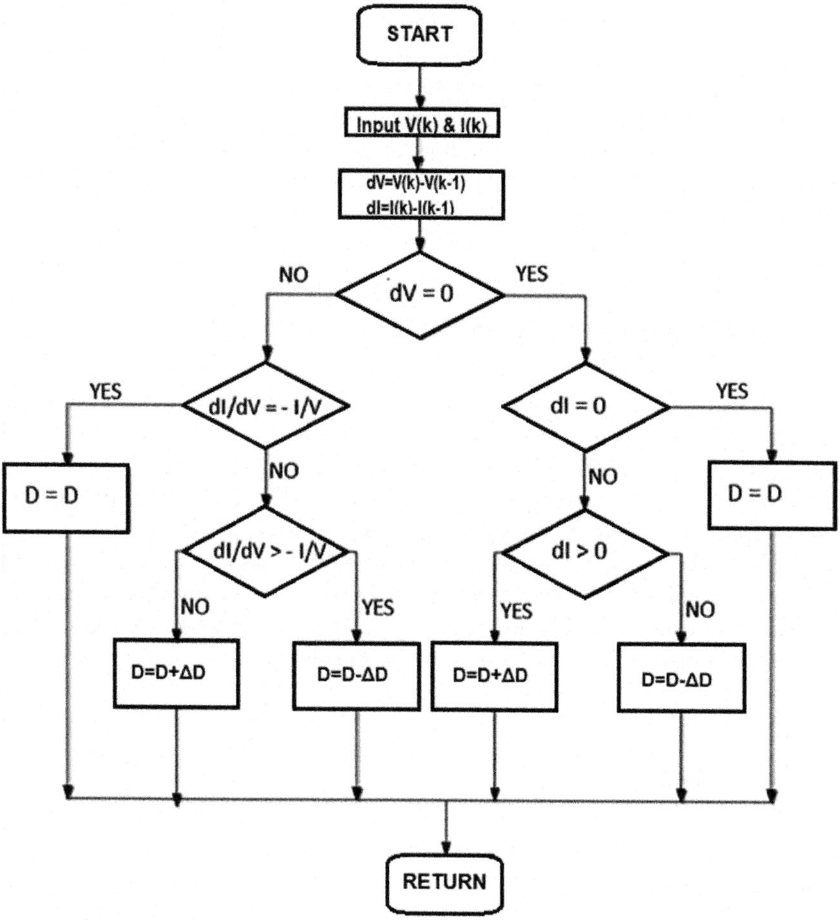

Fig. 3 ICA method [40]

The ICA considers variations in temperature and irradiance that may impact panel performance. The second derived of power w.r.t voltage is used to determine whether to increase or decrease load resistance in response to changes in these variables. Overall, the incremental conductivity algorithm is an effective method for MPPT as it can track changes in irradiance and temperature quickly and accurately, and recover maximum power from PV modules [40]. The ICA is illustrated in Fig. 3.

3.4 *Solar Inverter*

The function of a solar pump inverter is to convert the DC voltage into usable AC voltage. It is used in solar-powered water pump systems for irrigation, livestock watering, and other applications [41]. The inverter adjusts the V and F of the AC output according to the needs of the pump motor. It also has protection features such as overvoltage, overcurrent, and short circuit protection to ensure safe and reliable operation. Solar pump inverters are designed to work with different types of pumps such as centrifugal, submersible, and floating pumps [42]. They are an efficient and cost-effective solution for remote locations where mains power is unavailable or unreliable.

3.5 Solar Centrifugal Water Pump

A solar centrifugal water pump is a type of water pump that uses solar energy to operate. The centrifugal design of this pump allows it to generate high flow rates and high pressures, making it ideal for pumping water from wells, rivers, lakes, and other sources. Solar centrifugal water pumps are often used in remote areas where electricity is unavailable or unstable. It is also very popular in agriculture and irrigation systems as it can be used to transport large amounts of water over long distances. One of the main advantages of using a solar centrifugal water pump is that it is environmentally friendly and produces no harmful emissions. Additionally, it requires little maintenance and can last for years without the need for major repairs.

4 Controller

The existing SPVWPS uses conventional controllers such as P, PI, PD, and PID as per requirement. FLC is an intelligent controller that enhances the performance of SPVWPS. The motor pump flow rate is controlled using vector control method. This section is divided into three subsections, e.g., vector control drive, conventional controller, and FLC. The SPVWPS currently in use employs traditional control methods such as Proportional (P), Proportional Integral (PI), Proportional Derivative (PD), and Proportional Integral Derivative (PID) according to the specific needs.

4.1 Vector Control Drive

A Vector Control Drive (VCD) is a control method for AC underwater motors that uses advanced algorithms to precisely control the motion and force of the pump. This method is especially useful for AC submersible motors, which are commonly used in applications where precise control is essential, such as water treatment plants, oil and gas wells, and other industrial environments. A vector control drive measures motor current and voltage and uses this information to calculate motor speed and torque.

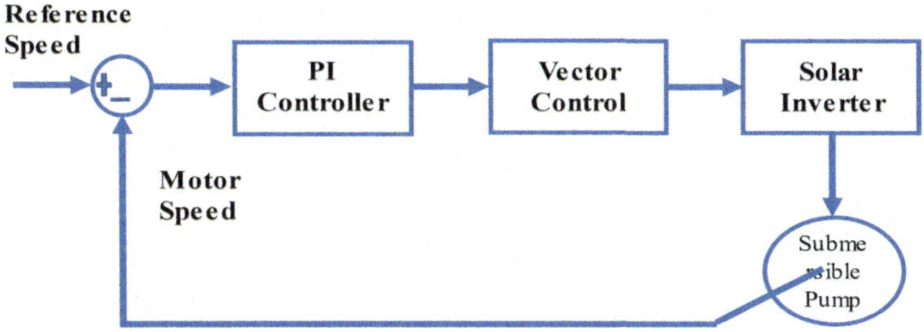

Fig. 4 Vector control method

The drive then regulate F and V of the power delivered to the motor to maintain the desired speed and torque [43].

A significant benefit of vector-controlled drives is that they offer accurate regulation across a broad range of speeds. This makes them ideal for applications requiring variable speed operation, such as pumps and fans that need to adjust to changing conditions. Another advantage of vector-controlled drives is their ability to provide high starting torque. This is essential for AC underwater motors that need to start under heavy loads. This reduces engine wear and extends life. Overall, vector control drives are an effective way to control AC underwater motors, providing precise speed and torque. The VCD is illustrated in Fig. 4.

4.2 Conventional Controller

4.2.1 P-controller

A P-controller in the solar energy sense is a system that adjusts the performance of a solar panel based on the difference between desired and actual performance.

4.2.2 PI Controllers

PI controllers are not only similar to P controllers for solar energy but also take into account the period they take for the system to reach the desired performance. For example, if the system eventually reaches 100 Watts, but it takes too long to reach it, the PI controller will adjust the power to reach 100 Watts sooner.

4.2.3 PD Controller

A solar energy-related PD controller is a system that adjusts the performance of a solar panel based on the difference between desired and actual performance and the rate of change of that difference. For example, if a sudden change in cloud cover causes the solar panel's output to drop, the PD controller will adjust quickly to maintain constant power generation [44].

4.2.4 PID Controllers

PID controllers related to solar energy combine all three elements (P, I, D) to create a more accurate control system. It considers the difference between desired and actual results (P), the time it takes to achieve those results (I), and how quickly those differences change (D) increase. This type of control system helps optimize power generation from solar panels by adapting to changing weather conditions and other factors that affect power generation.

4.3 FLC

FLC refers to a type of control system that utilizes fuzzy logic principles to make choices dependent on the inputs it receives. Fuzzy logic is a mathematical concept that allows for some degree of truth rather than just a true/false binary value. In FLC, input variables are mapped into fuzzy sets defined by membership functions that assign degrees of membership to each value of the input variables. After receiving inputs, FLC applies rules to determine the level of activation for each rule, based on how well the input variables match the fuzzy sets. The output of FLC is then determined by aggregating the activation level of each rule [30]. FLC is particularly useful in situations where the complexity of the system or the uncertainty of the input variables makes implementation of traditional control systems difficult. They are used in various applications such as robotics, process control, and automotive systems. An example application of FLC is an SPVWPS.

The system consists of solar panels that generate electricity to power the water pump. Inputs to the FLC include solar panel voltage, water flow rate, and storage tank water level. FLC maps these input variables into a fuzzy set and uses rules to determine the activation level for each rule. For example, if the solar panel voltage is high, the FLC may activate a rule to increase water flow [45]. If storage tank water indicates low, the FLC may activate rules to reduce water flow. The FLC's output is used to control the water pump, adjusting the flow rate to maintain the desired water indicated in the storage tank. By using fuzzy logic, FLC can handle uncertainties in input variables and make decisions based on truthiness rather than binary true/false values. Overall, FLC is a powerful tool for controlling complex systems with uncertain input variables. They are successfully used in a variety of applications and are predictable to endure to be important tools for control engineers in the forthcoming [31]. An SPVWPS can be modeled using FLC as follows:

Input Variables

1. Solar panel voltage
2. Water flow rate
3. Water level in the storage tank.

Output Variable

1. Pump speed.

Fuzzifier

1. Solar panel voltage: Low, Medium, High
2. Water flow rate: Low, Medium, High
3. Water level in the storage tank: Low, Medium, High
4. Pump speed: Slow, Medium, Fast.

Fuzzy Rules

1. If solar panel voltage is low OR water level in the storage tank is low, then impel speed is slow.
2. If solar panel voltage is medium AND water level in the storage tank is middle, then impel motion is medium.
3. If solar panel voltage is high AND water level in the storage tank is high, then impel motion is fast.
4. If water flow rate is low, then pump speed is slow.
5. If water movement amount is average, then impel speed is medium.
6. If water flow rate is high, then pump speed is fast.

Defuzzifier

The output of the FLC is defuzzified using the centroid method to obtain a crisp value for the pump speed. Figure 5 shows the FLC structure.

Control

The pump speed crisp value is used to control the water pump, adjusting the flow rate to maintain the desired water level in the storage tank. The FLC continuously monitors input variables and adjusts pump speed accordingly to ensure the system is operating efficiently. Overall, FLC provides an effective means of controlling an SPVWPS by considering the uncertainty of input variables and making truth-based decisions. This improves system performance and reliability. Additionally, the FLC can be programmed to include safety measures such as shutting off the pump if the water level in the storage tank reaches a critical low or high level [46]. This ensures

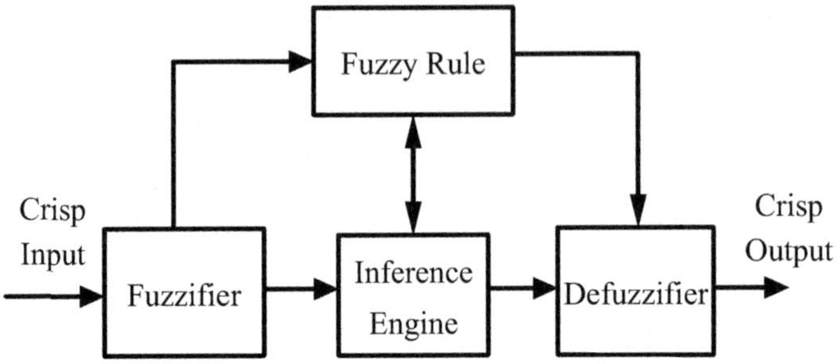

Fig. 5 FLC structure

that the system operates within safe limits and prevents damage to the pump or other components.

Furthermore, the FLC can be adapted to optimize the system's energy efficiency by adjusting the pump speed based on the solar panel voltage. For example, if the solar panel voltage is low, the FLC can reduce the pump speed to conserve energy and extend the system's operating time. Overall, the FLC provides a flexible and adaptive control solution for SPVWPS pumping systems that can improve performance, reliability, and energy efficiency while ensuring safe operation [47].

5 Design of Solar Water Pump System

An SPVWPS usually consists of a solar module, a pump, a controller, and a storage tank. Solar panels collect solar energy, convert it to DC, and use it to drive pumps. A pump is responsible for drawing water from a well or other source and pumping it into a storage tank {Formatting Citation}. The controller regulates the power flow to the pump to ensure efficient operation. Storage tanks are used to store water for later use. The design of an SPVWPS depends on aspects like the amount of water essential, water source, and the location and climate of the installation site [48]. Proper sizing of system components is critical so that they can meet water demand and operate efficiently. Additionally, the system may include a backup generator or battery bank given at non-sunshine hours. Monitoring systems can also be integrated to track system performance and identify problems. Overall, a well-designed SPVWPS can provide a reliable and sustainable water source for agricultural, residential, or commercial use.

SPVWPSs have several advantages over conventional pumps. It requires little maintenance, costs no fuel, and can be deployed in remote locations where mains power is not available They also have a long lifespan, typically 25–30 years [49]. It can be used for agricultural irrigation, livestock watering, and rural domestic water supply. Overall, the SPVWPS is a promising solution to meet community water demand in a sustainable and cost-effective manner. As technology continues to

Table 2 Solar panel (EGE-100W-36) parameters [50]

S. No	Name of parameters	Rating
1	Open circuit voltage	23.4 V
2	Short circuit current	5.55 A
3	PV module power	100 W
4	No. of module	20
5	Irradiation	1000 W/m^2
6	Dimensions	840 × 680 × 30 mm

improve and costs drop, it is expected that these systems will become even more widely used in future.

The overall design can be divided into five steps as given below:

Step one involves calculating the daily water requirement.

Step two involves determining the vertical height needed for water pumping.

In step three, the total hydraulic energy needed per day (Wh/day) to pump the required amount of water is determined.

Step four involves finding out the solar radiation available at the chosen location.

Step five includes figuring out the size and quantity of PV modules needed, selecting an appropriate motor, as well as determining the motor's rating, taking into account losses and efficiency.

5.1 Case Study

Design an SPVWPS which is required to draw 30,000 L of water every day from a depth of 15 m. Table 2 shows PV module's technical parameters.

Data required for calculation is as follows:

Water required per day = 30,000 l = 30 m^3

The vertical lift = 15 m + (Frictional losses- 5% of total perpendicular height) = 15.75 m.

Water density = 1000 kg/m^3

Acceleration gravity = 9.8 m/s^2

Operating factor = 0.75.

Pump efficiency = 30%

Mismatch factor = 0.85.

Calculation for SPVWPS

Step 1: To find the daily water demand,

daily water demand = 30 m^3

Step 2 To find the vertical height,

Total vertical head (TVH) = 15.75 m.

Step 3 To calculate hydraulic energy [51]

$$\begin{aligned} E_{\text{hydraulic}} &= \rho \times v \times g \times \text{TVH} \times 10^{-3}\,\text{MJ} \\ &= 4630\,\text{MJ} \end{aligned} \tag{1}$$

$$\text{We know that, } 3.6\,\text{MJ} = 1\text{kWh} = 1.286\,\text{kWh}$$

$$= 1.286\,\text{kWh}$$

Step 4 determine solar radiation data.
= 6 h/day (peak of 1000 W/m^2).
Step 5 determines the number of PV panels and pump sizes [52]:

$$\begin{aligned} \text{Total kW of PV panel} &= E_{\text{hydraulic}}/6\,\text{h} \\ &= 0.214\,\text{kW} \end{aligned}$$

$$\begin{aligned} \text{Consider system losses} &= \frac{\text{PV panel wattage}}{\text{pump efficiency} \times \text{mismatch factor}} \\ &= \frac{0.214\,\text{kW}}{0.3 \times 0.85} = 0.840\,\text{kW} = 840\,\text{W} \end{aligned} \tag{2}$$

Consider operational factor is 0.75.
Now,

$$\text{power} = \frac{840}{0.75} = 1120\,\text{W}$$

Now,

$$\text{Number of PV module required} = \frac{1120}{100} = 11.2 \cong 12\,\text{module}$$

PV system (PVS) efficiency (η_{pvs}) is [11]

$$\eta_{\text{pv}} = \frac{\text{Total PV required power}}{A_{\text{pvs}} \times G_{\text{irr}}} \tag{3}$$

$$\eta_{\text{pv}} = 16.37\%$$

where A_{pvs} *and* G_{irr} are PVS surface area and solar irradiation.

Inverter and converter efficiency are 95% and 93%, respectively.

Hydraulic pump efficiency is

$$\eta_{\text{hydraulic}} = \frac{2.725 \times Q \times \text{TVH}}{P_{\text{shaft}}} \tag{4}$$

$$Q = \frac{\text{water flow}}{\text{time}} = \frac{30}{5} = 6\text{m}^3/\text{h} \tag{5}$$

$$\eta_{\text{hydraulic}} = \frac{2.725 \times 6 \times 15.75}{746\,\text{W}} = 34.5\%$$

In this case study, 1hp motor has been considered.

where P_{shaft} *and* Q are motor shaft power and water flow rate, respectively.

The total overall system efficiency is [53]

$$\eta_{\text{sys}} = \eta_{\text{pvs}} \times \eta_{\text{converter}} \times \eta_{\text{inverter}} \times \eta_{\text{motor}} \times \eta_{\text{hydraulic}} \tag{6}$$

$$\eta_{\text{sys}} = (0.16 \times 0.95 \times 0.93 \times 0.34) \times 100 = 4.8\%$$

The performance of the motor pump drive with PI and fuzzy controllers is illustrated in Table 3. The hydraulic efficiency and flow rate averages of the PI controller are 34.05% and 6m3/hrs, respectively, with an overall efficiency of the solar water pump system of merely 4.8%. Comparatively, the FLC controller-based system exhibits an average hydraulic efficiency and flow rate of 35.95% and 6.25 m^3/h, respectively, thereby producing an efficiency of 5.07%.

The Q and $\eta_{\text{hydraulic}}$ of a PI controller system is depicted for different solar radiation levels in Fig. 6. The Q and $\eta_{\text{hydraulic}}$ of a system based on an FLC controller are depicted for different solar radiation levels in Fig. 7. It is evident from Fig. 7 that an upsurge in solar radiation results in a rise in both flow rate and pump efficiency.

These results indicate that the FLC controller system is more efficient than the conventional controller-based system.

Table 3 The performance of the drive controller is compared

Drive controller	Average flow rate, Q (m^3/h)	Average ($\eta_{\text{hydraulic}}$%)	Average (η_{system}%)
PI	6	34.05	4.8
Fuzzy	6.25	35.95	5.07

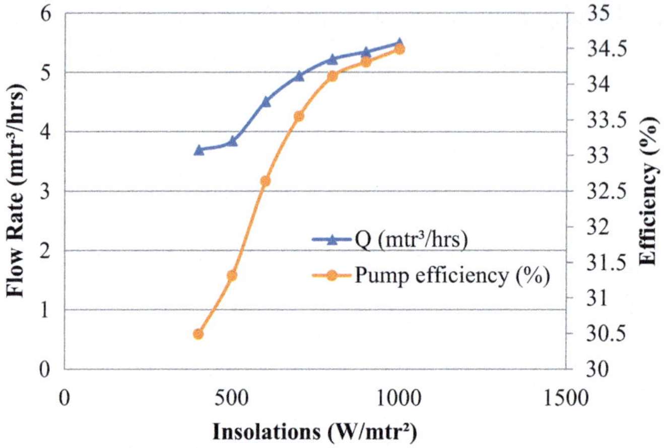

Fig. 6 PI-based flow rate and pump efficiency at different solar irradiations

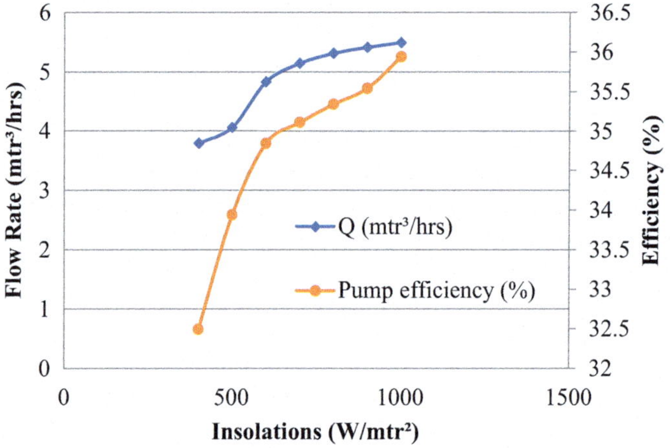

Fig. 7 FLC-based flow rate and pump efficiency at different solar irradiations

6 Comparison of Conventional Controller to FLC

Traditional PI controller is a widely used control strategy in various industrial applications such as SPVWPSs.

A PI controller computes the proportional and integral parts of the error signal and adds them together to produce the control signal. FLC, on the other hand, is a more advanced control strategy that uses fuzzy sets and rules to generate control signals. FLC can handle nonlinearities, uncertainties, and imprecise information in the system. Use linguistic variables instead of numbers to represent system inputs and outputs. In an SPVWPS, traditional controller and the FLC can be used to control the

Q of the pump. However, FLC has some advantages in performance and robustness over traditional PI controllers.

First, FLCs can handle nonlinearities in the system better than PI controllers. SPVWPSs are inherently nonlinear due to changes in solar radiation, temperature, and other environmental factors. FLC can adapt to these changes by adapting its rules based on fuzzy logic.

Second, FLCs can handle system uncertainty better than PI controllers. SPVWPSs can introduce uncertainties in pump characteristics and sensor readings. FLC uses fuzzy sets to represent these uncertainties and can adjust its rules accordingly.

Third, FLC can perform better than the PI controller in terms of accuracy and stability. FLC can generate smoother control signals that reduce overshoot and fluctuations in system response.

In summary, both traditional PI controllers and FLC can be used to control flow in SPVWPSs, but FLC handles nonlinearities and uncertainties better than traditional systems, which has several advantages over traditional PI controllers. The can controller offers excellent performance in terms of accuracy and stability.

7 Conclusion

The chapter delves into the creation, simulation, and assessment of an off-grid PV generator pumping system that uses a submersible motor pump system. The system components' efficiency is scrutinized in varying solar irradiation conditions. The research shows that IM-based pumping systems are more effective in different climates, and fuzzy controllers outperform PI controllers in terms of efficiency and stability. The PV water pumping system's size can be computed for any study area. Proper PV panel sizing and selection are also discussed to guarantee optimal performance and longevity of the system. Ultimately, the study concludes that off-grid PV generator pumping systems can provide a sustainable and clean water supply in remote areas, and further research is necessary to enhance their efficiency and reliability.

The chapter explores the development and evaluation of an off-grid PV generator pumping system that utilizes a submersible motor pump system, with a focus on analyzing the efficiency of system components under varying solar irradiation conditions. The study finds that IM-based pumping systems are more effective in diverse climates, and fuzzy controllers outperform PI controllers in terms of efficiency and stability. The sizing and selection of PV panels are also crucial for optimal performance and longevity. Overall, off-grid PV generator pumping systems have the potential to provide a sustainable and clean water supply in remote areas, but further research is necessary to enhance their efficiency and reliability. Additionally, government policies and incentives such as subsidies and tax credits can facilitate the adoption of these systems in rural communities by reducing the initial investment cost. In conclusion, successful implementation of off-grid SPVWPS requires proper system sizing and maintenance, government support, and further research and development to improve their effectiveness.

References

1. Farrar LW, Bahaj AS, James P, Anwar A, Amdar N (2022) Floating solar PV to reduce water evaporation in water stressed regions and powering water pumping: case study Jordan. Energy Convers Manage 260(April):115598. https://doi.org/10.1016/j.enconman.2022.115598
2. Raghuwanshi SS, Khare V (2018) Sizing and Implementation of photovoltaic water pumping system for irrigation. IAES Int J Artif Intell 7(1):54–62. https://doi.org/10.11591/ijai.v7.i1.pp54-62
3. Houssein EH, Helmy BED, Rezk H, Nassef AM (2021) An enhanced archimedes optimization algorithm based on local escaping operator and orthogonal learning for PEM fuel cell parameter identification. Eng Appl Artif Intell 103:104309
4. Houssein EH, et al (2021) A modified marine predator algorithm based on opposition based learning for tracking the global MPP of shaded PV system. Expert Syst Appl 183:115253
5. Saady I, Karim M, Bossoufi B, El Ghzizal N, Motahhir S (2023) Results in engineering optimization and control of photovoltaic water pumping system using kalman filter based MPPT and multilevel inverter fed DTC-IM. Results Eng 17: 100829. https://doi.org/10.1016/j.rineng.2022.100829
6. Altimania MR, Elsonbaty NA, Enany MA, Gamil MM, Alzahrani S, Alraddadi MH, Alsulami R, Alhartomi M (2023) Optimal performance of photovoltaic-powered water pumping system
7. Altimania MR et al (2023) Optimal performance of photovoltaic-powered water pumping system. Mathematics 11:731. https://doi.org/10.3390/math11030731
8. Shaban H, et al (2021) Identification of parameters in photovoltaic models through a runge kutta optimizer. Mathematics 9(18):2313
9. Mercedes AV, Francisco JS (2022) A new optimization approach for the use of hybrid renewable systems in the search of the zero net energy consumption in water irrigation systems. 195. https://doi.org/10.1016/j.renene.2022.06.060
10. Kashif M, Bhim S (2022) Engineering science and technology , an international journal design optimization with improved torque performance of a new flux- intensifying pmsm using multi-layer barriers for solar water pumps. Int J Eng Sci Technol 36:101134. https://doi.org/10.1016/j.jestch.2022.101134
11. Hilali A, Mardoude Y, Essahlaoui A, Rahali A (2022) Migration to solar water pump system: environmental and economic benefits and their optimization using genetic algorithm based MPPT. Energy Rep 8:10144–10153. https://doi.org/10.1016/j.egyr.2022.08.017
12. Hilali A, El Ouanjli N, Mahfoud S, Al-sumaiti AS (2022) Optimization of a solar water pumping system in varying logic and incremental conductance
13. Okakwu IK, Alayande AS, Akinyele DO, Olabode OE, Akinyemi JO (2022) Effects of total system head and solar radiation on the techno-economics of PV groundwater pumping irrigation system for sustainable agricultural production. Sci Afr 16:e01118. https://doi.org/10.1016/j.sciaf.2022.e01118
14. Biswas S, Tariq Iqbal M (2018) Dynamic modelling of a solar water pumping
15. Harish G, Varma K, Barry VR, Jain RK (2022) A total-cross-tied-based dynamic photovoltaic array reconfiguration for water pumping system. IEEE Access 10:4832–4843. https://doi.org/10.1109/ACCESS.2022.3141421
16. Dulhare UN, Taj STA (2023) Water quality risk analysis for sustainable smart water supply using adaptive frequency and BiLSTM. In: Kannan RJ, Geetha S, Sashikumar S, Diver C (eds) International virtual conference on industry 4.0. IVCI 2021. Lecture notes in electrical engineering, vol 1003. Springer, Singapore

17. Ibrahim MN, Rezk H, Al-Dhaifallah M, Sergeant P (2020) Modelling and design methodology of an improved performance photovoltaic pumping system employing ferrite magnet synchronous reluctance motors. Mathematic 8(9). https://doi.org/10.3390/MATH8091429
18. Ashraf U, Tariq Iqbal M (2020) Optimised design and analysis of solar water pumping systems for pakistani conditions. Energy Power Eng 12(10):521–542. https://doi.org/10.4236/epe.2020.1210032
19. Houssein EH, Sayed A (2023) Dynamic candidate solution boosted beluga whale optimization algorithm for biomedical classification. Mathematics 11(3):707
20. Motahhir S, Chalh A, El Ghzizal A, Derouich A (2018) Development of a low-cost PV system using an improved INC algorithm and a PV panel proteus model. J Clean Prod. https://doi.org/10.1016/j.jclepro.2018.08.246
21. Mamaghani AH, Escandon SAA, Najafi B, Shirazi A, Rinaldi F (2016) Techno-economic feasibility of photovoltaic, wind, diesel and hybrid electrification systems for off-grid rural electrification in Colombia. Renew Energy 97:293–305
22. Poudyal R, Loskot P, Parajuli R (2021) Techno-economic feasibility analysis of a 3-kW PV system installation in Nepal. Renewables Wind Water Sol 8(1). https://doi.org/10.1186/s40807-021-00068-9
23. Odoi-Yorke F, Woenagnon A (2021) Techno-economic assessment of solar PV/fuel cell hybrid power system for telecom base stations in Ghana. Cogent Eng 8(1). https://doi.org/10.1080/23311916.2021.1911285
24. Malanda C, Makokha AB, Nzila C, Zalengera C (2021) Techno-economic optimization of hybrid renewable electrification systems for Malawi's rural villages. Cogent Eng 8(1):2021. https://doi.org/10.1080/23311916.2021.1910112
25. Mayyas A, Chadly AA, Khaleel I, Maalouf M (2021) Techno-economic analysis of the Li-ion batteries and reversible fuel cells as energy-storage systems used in green and energy-efficient buildings. Clean Energy 5(2):273–287. https://doi.org/10.1093/ce/zkab009
26. Massaq Z (2023) Design of high—performance fuzzy—predictive controllers for a photovoltaic. Batter Pump Syst 13(1)
27. Mohan BM et al (2023) Multifunctional control for PV integrated battery energy storage system with enhanced power quality. Int J Res Appl Sci Eng Technol 11(3):659–667
28. Assem H, et al (2023) Adaptive fuzzy logic-based control and management of photovoltaic systems with battery storage. Int Trans Electr Energy Syst
29. Shetty D, Sabhahit JN, Mudlapur A, Hebbar P (2023) Hybrid PSO-incremental conductance mppt for induction motor based solar water pumping system under partial shading conditions. Adv Electr Comput Eng 23(1):87–94
30. Hannan MA, Member S, Ghani ZABD (2019) Fuzzy logic inverter controller in photovoltaic applications: issues and recommendations. IEEE Access 7:24934–24955. https://doi.org/10.1109/ACCESS.2019.2899610
31. Nejad HD, Nazari M, Nazari M (2022) Fuzzy state-dependent Riccati equation (FSDRE) control of the reverse osmosis desalination system with photovoltaic power supply. IEEE Access 10(September):95585–95603. https://doi.org/10.1109/ACCESS.2022.3204270
32. MNRE, Implementation of PM-KUSUM (2023) https://pib.gov.in/PressReleasePage.aspx?PRID=1881947#:~:text=As%20on%2031.10.,lakh%20pumps%20under%20the%20Scheme
33. Varma GKH, Barry VR, Jain RK (2022) Improved performance of PV water pumping system using dynamic reconfiguration algorithm under partial shading conditions. 7(2):206–15
34. Sharma R, Sarma S, Tiwari S (2020) Design optimization of solar PV water pumping system. Mater Today: Proc 21(xxxx):1673–79. https://doi.org/10.1016/j.matpr.2019.11.322
35. Raghuwanshi SS, Masih A, Raghuwanshi P (2022) Design and optimisation of stand-alone hybrid energy system for rural areas. Int J Ambient Energy 43(1):8707–8721. https://doi.org/10.1080/01430750.2022.2103185
36. Gualteros S, Rousse DR (2021) Solar water pumping systems : a tool to assist in sizing and optimization. Solar Energy 225(2020):382–398. https://doi.org/10.1016/j.solener.2021.06.053
37. Zelalem G (2017) Techno-economic analysis of photovoltaic pumping system for rural water supply in Ethiopia. Int J Sustain Energy 36(3):277–295

38. Singh B, Sharma U, Kumar S (2018) Standalone photovoltaic water pumping system using induction motor drive with reduced sensors. IEEE Trans Ind Appl 54(4):3645–3655. https://doi.org/10.1109/TIA.2018.2825285
39. Davey H, Ingram W, Memon FA (2018) An Integrated optimal approach for solar powered rural water distribution systems in the Gambia. 408–421. https://doi.org/10.4236/jwarp.2018.104022
40. Masih A, Verma HK (2022) Optimum sizing and simulation of hybrid renewable energy system for remote area. Energy Environ 33(5). https://doi.org/10.1177/0958305X211030112
41. Raghuwanshi SS, Khare V (2018) Sizing and modelling of stand-alone photovoltaic water pumping system for irrigation. https://doi.org/10.1177/0958305X17752739
42. Adle R, Renge M, Muley S (2020) Improvement in performance of series Z-source inverter with an application as solar PV fed water pump. Iran J Sci Technol Trans Electr Eng 44(3):1263–1279. https://doi.org/10.1007/s40998-020-00310-y
43. Menghala PM, Laxmi AJ (2016) Fuzzy based real time control of induction motor drive. Procedia Comput Sci 85:228–235
44. Sarr A, Diop L, Diatta I, Wane YD, Bodian A (2021) Baseline of the use of solar irrigation pump in the Niayes area in Senegal. 125–46. https://doi.org/10.4236/nr.2021.125010
45. Zand SJ, Mobayen S, Member S, Gul HZAD (2022) Optimized Fuzzy controller based on cuckoo optimization algorithm for maximum power-point tracking of photovoltaic systems. IEEE Access 10(May):71699–71716. https://doi.org/10.1109/ACCESS.2022.3184815
46. Parimala V, Ganeshkumar D, Divya M (2020) Implementation of Solar Water Pumping System Using Fuzzy Logic Controller. In: 2020 6th International conference on advanced computing and communication systems. ICACCS, pp 967–970. https://doi.org/10.1109/ICACCS48705.2020.9074427
47. Rai N, Rai B (2018) Control of fuzzy logic-based PV-battery hybrid system for stand-alone DC applications. J Electr Syst Inf Technol 5(2):135–143. https://doi.org/10.1016/j.jesit.2018.02.007
48. Errouha M, Derouich A, El Ouanjli N, Motahhir S (2020) High-performance standalone photovoltaic water pumping system using induction motor. Int J Photoenergy. https://doi.org/10.1155/2020/3872529
49. Antón JM, Carricondo MA, Jiménez Bello J, Juárez M, Royuela A, Tomas. (2023) Optimization of an isolated photovoltaic water pumping system with technical—economic criteria in a water users association. Irrig Sci. https://doi.org/10.1007/s00271-023-00859-6
50. Eco green energy (2023) https://www.eco-greenenergy.com/product/solar-module-ege-100m-36-monocrystalline-module/
51. Kumar A, Kumar P (2021) Power quality improvement for grid-connected PV system based on distribution static compensator with fuzzy logic controller and UVT/ADALINE-based least mean square controller. 9(6):1289–99. https://doi.org/10.35833/MPCE.2021.000285
52. Said SM, Ali A, Hartmann B (2020) Tie-line power flow control method for grid- connected microgrids with smes based on optimization and fuzzy logic 8(5):941–50. https://doi.org/10.35833/MPCE.2019.000282
53. Raghuwanshi SS, Raghuwanshi P, Masih A, Singh P (2023) Modeling and optimization of hybrid renewable energy with storage system using flamingo swarm intelligence algorithms. Energy Storage March:1–21. https://doi.org/10.1002/est2.470

Chapter 8
Renewable Energy Optimization System Using Fuzzy Logic

Pawan Whig, Bhupesh Bhatia, Ashima Bhatnagar Bhatia, and Pavika Sharma

Abstract Contrary to the current decline in the use of fossil fuels, the globe is now paying close attention to the need for energy, global warming, and the ongoing rise in oil prices. Since life cannot exist without energy, the recently developed renewable energy technologies hold out some promise for at least partially alleviating the issue caused by an energy shortage or an imbalance in the distribution of energy between and within nations. In recent decades, hybrid renewable energy systems have gained popularity and are increasingly being used to electrify isolated rural regions throughout the world where grid extension is difficult and uneconomical. These systems combine one or more renewable energy sources, such as solar photovoltaic, wind, microhydro, biomass, and geothermal energy, and they may also include conventional backup generators. This book chapter assembles renewable energy systems along with their benefits and drawbacks, hybrid wind and solar energy systems with various hybrid energy system components to identify the best possible combination of energy components for a typical rural community in order to minimize the total net present cost of the system over its lifetime. In this chapter, a case study using fuzzy logic including a few simulation technique tools is also described, along with some of the component highlights.

Keywords Fuzzy logic · Simulation · Economical · Geothermal · Global warming hybrid

---

P. Whig (✉) · A. B. Bhatia
Vivekananda Institute of Professional Studies—TC, Delhi, India
e-mail: pawan.whig@vips.edu

A. B. Bhatia
e-mail: Ashima.bhatia@vips.edu

B. Bhatia
NSUT, Delhi, India
e-mail: bhupeshbhatia@wilp.bits-pilani.ac.in

P. Sharma
Department of ECE, BPIT, Delhi, India
e-mail: pavikasharma@bpitindia.com

U. N. Dulhare and E. H. Houssein (eds.), *Machine Learning and Metaheuristics: Methods and Analysis*, Algorithms for Intelligent Systems,
https://doi.org/10.1007/978-981-99-6645-5_8

Acronyms

PV	Photovoltaic
EV	Electric Vehicle
REOS	Renewable Energy Optimization Systems
I/O	Input
O/P	Output
AI	Artificial Intelligence
IoT	Internet of Things

1 Introduction

The world is facing an ever-increasing demand for energy, driven by population growth, urbanization, and industrialization [1]. However, traditional energy sources such as fossil fuels are finite, and their continued use is unsustainable due to environmental concerns. This has led to the adoption of renewable energy sources as a sustainable alternative, but their intermittent nature and variability pose significant challenges for energy system operators [2].

Renewable energy sources such as solar, wind, hydro, and biomass offer a significant opportunity to meet the growing energy demand while reducing carbon emissions. However, the integration of these sources into the existing grid is challenging due to their intermittent and variable nature [3–6]. This leads to problems such as system instability, grid congestion, and high operational costs, which limit their effectiveness in meeting the growing energy demand [7].

To overcome these challenges, an optimization system is required that can manage the variability of renewable vigor bases and ensure their efficient addition to the existing grid. Fuzzy logic is a promising optimization technique that can be used to develop such a system. This chapter provides an overview of renewable energy optimization systems using fuzzy logic [8–10].

2 Background and Motivation

Renewable energy sources have gained significant attention due to the environmental concerns associated with traditional energy sources such as coal, oil, and gas. The shift toward renewable energy sources is driven by the need to reduce carbon emissions and mitigate the impacts of climate change. The adoption of renewable energy sources has increased in recent years, and it is expected to continue in the future [11–15].

However, the variability of renewable energy sources poses significant challenges for their integration into the existing grid. The intermittent nature of solar and wind power makes it difficult to predict their output, which can lead to system instability, grid congestion, and high operational costs [16–18]. These challenges limit the effectiveness of renewable energy sources in meeting the growing energy demand.

To address these challenges, an optimization system is required that can manage the variability of renewable energy foundations and ensure their efficient addition to the existing grid. Fuzzy logic is a promising optimization technique that can be used to develop such a system [19, 20].

3 Overview of Renewable Energy

Renewable energy sources are energy sources as shown in Fig. 1. that are replenished naturally and can be used without depleting the source. Solar, wind, hydro, and biomass are examples of renewable energy sources [21–25]. These sources offer a significant opportunity to meet the growing energy demand while reducing carbon emissions.

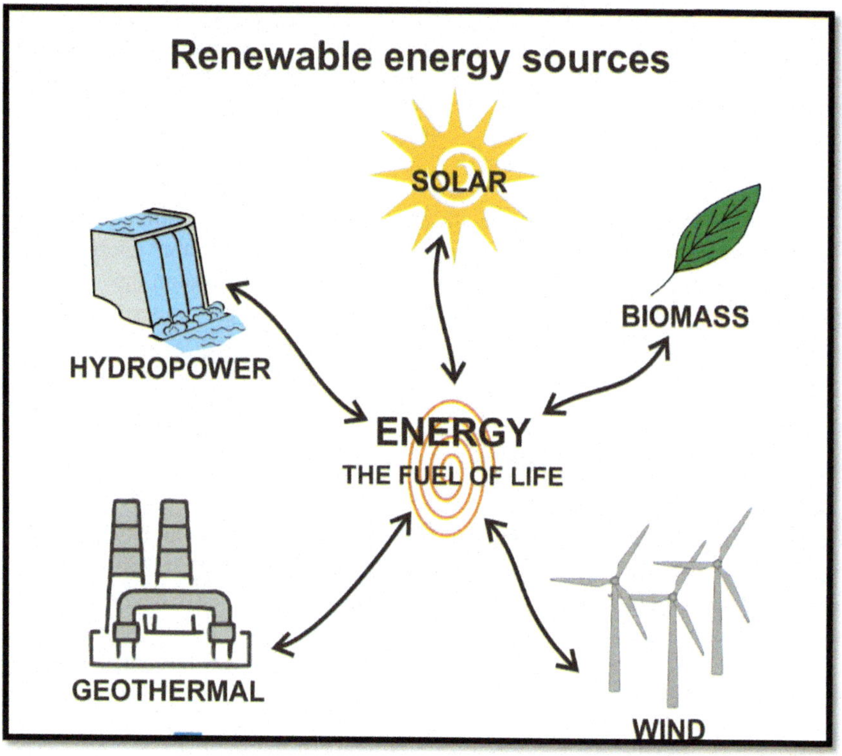

Fig. 1 Renewable energy sources

Table 1 Comparison of previous study

Study title	Authors	Year	Methodology	Key findings
"Optimal sizing of solar photo voltaic (PV) systems for residential buildings"	Smith and Johnson [26]	2018	Mathematical optimization models	The optimal sizing of solar PV systems can significantly reduce electricity costs and improve system efficiency
"Wind farm layout optimization using genetic algorithms"	Zhang and Wang [27]	2019	Genetic algorithms	Genetic algorithms can effectively optimize the layout of wind farms, leading to increased energy production
"Optimal dispatch of hybrid energy systems in remote areas"	Brown and Davis [28]	2020	Linear programming and economic analysis	Optimal dispatch strategies for hybrid energy systems can improve cost efficiency and reduce reliance on fossil fuels
"Multi-objective optimization of tidal energy converter arrays"	Chen and Li [29]	2021	Evolutionary algorithms and numerical simulations	Multi-objective optimization can identify optimal tidal energy converter array designs considering energy production and environmental impact
"Optimal placement of electric vehicle charging stations"	Kim and Lee [30]	2022	Geographic information systems and optimization models	Optimized placement of electric vehicle (EV) charging stations can reduce infrastructure costs and enhance accessibility

However, the variability of renewable energy sources poses significant challenges for their integration into the existing grid. The output of renewable energy sources varies with time and is difficult to predict, which can lead to system instability, grid congestion, and high operational costs. The comparison of previous study is shown in Table 1.

4 Overview of Fuzzy Logic

Fuzzy logic is an exact method that can grip doubt and fuzziness in data as shown in Fig. 2. It is a powerful tool for modeling complex systems that involve uncertainty, and it has been used in various applications such as control systems, decision-making, and optimization.

Fuzzy logic is based on fuzzy set theory, which allows for the representation of uncertain and imprecise data. Fuzzy logic operations such as fuzzy inference, fuzzy reasoning, and fuzzy control can be used to develop optimization systems that can manage the variability of renewable energy bases and ensure their efficient addition to the existing network [31].

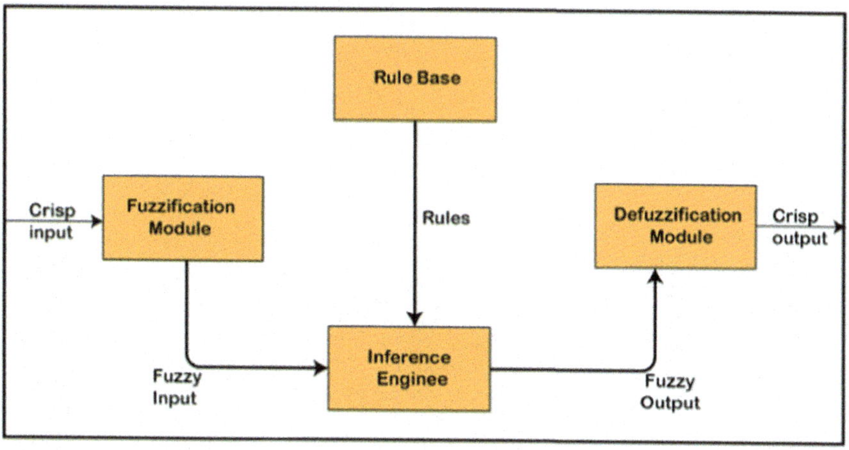

Fig. 2 Overview of fuzzy logic

Renewable energy sources offer a significant opportunity to meet the growing energy demand while reducing carbon emissions. However, the variability of renewable energy sources poses significant challenges for their integration into the existing grid [32].

Fuzzy logic is a promising optimization technique that can be used to develop such a system. This chapter provides an overview of renewable energy optimization systems using fuzzy logic. The following chapters will provide a detailed explanation of renewable energy systems, fuzzy logic, optimization techniques, and renewable energy optimization systems. The final chapter will discuss future directions and challenges in renewable energy optimization using fuzzy logic and provide concluding remarks [33].

5 Renewable Energy Systems

Renewable energy systems are systems that generate energy from renewable sources. These systems offer a significant opportunity to meet the growing energy demand while reducing carbon emissions [34–36]. However, their integration into the existing grid is challenging due to their intermittent and variable nature. An example of renewable energy system is shown in Fig. 3.

Renewable energy systems consist of several components such as generators, inverters, batteries, and control systems. These components work together to generate, store, and distribute energy efficiently.

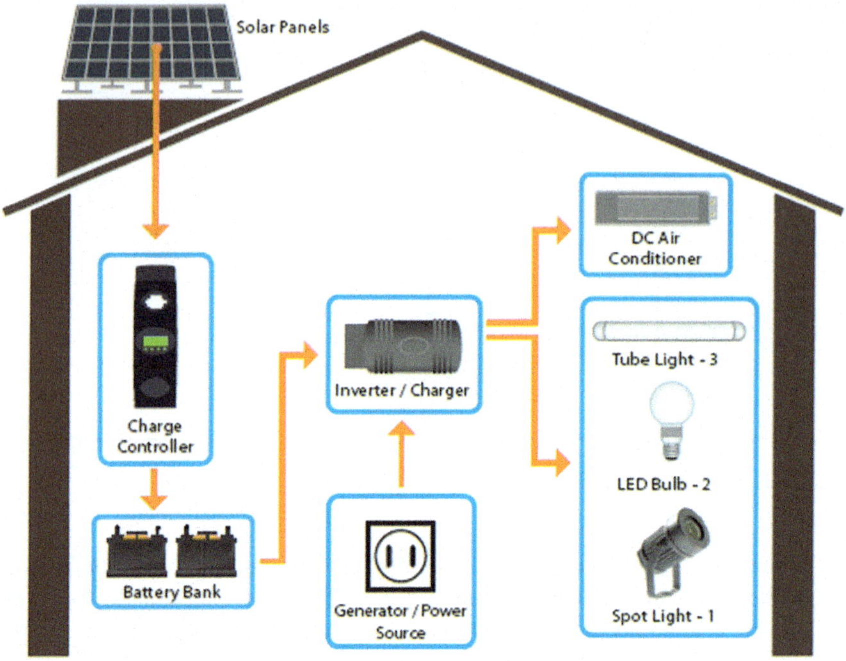

Fig. 3 Example of renewable energy system

5.1 *Advantages and Disadvantages of Renewable Energy Systems*

Renewable energy systems offer several advantages over traditional energy sources such as coal, oil, and gas. They are cleaner, more sustainable, and have a lower carbon footprint.

However, renewable energy systems also have several disadvantages. They are intermittent and variable, which can lead to system instability, grid congestion, and high operational costs. They also require significant investment in infrastructure and technology, which can be a barrier to their adoption.

Fuzzy Logic

It is a powerful tool for modeling complex systems that involve uncertainty, and it has been used in various applications such as control systems, decision-making, and optimization.

Fuzzy logic is based on fuzzy set theory, which allows for the representation of uncertain and imprecise data. Fuzzy logic operations such as fuzzy inference, fuzzy reasoning, and fuzzy control can be used to develop optimization systems that can manage variability.

Optimization Techniques

Optimization techniques [38–50] are mathematical approaches employed to discover the optimal solution to a problem while considering specific limitations. Various optimization techniques exist, such as genetic algorithms, particle swarm optimization, and ant colony optimization. Genetic algorithms mimic the principles of natural selection and can be applied to optimize intricate systems. Particle swarm optimization draws inspiration from the collective behavior of social insects like birds and can be utilized for optimizing systems with numerous variables. Ant colony optimization is inspired by the behavior of ants and is employed to optimize systems that involve numerous constraints.

6 Renewable Energy Optimization System

A renewable energy optimization system using fuzzy logic can be developed to manage variability. This system can consist of several components such as sensors, controllers, and actuators.

The fuzzy logic control system can be used to manage the variability of renewable energy sources by adjusting the energy output of generators, batteries, and inverters. The system can also optimize energy distribution to reduce grid congestion and ensure efficient use of energy.

Case studies and simulations can be used to evaluate the performance of the renewable energy optimization system. The system can be compared to existing optimization systems to determine its effectiveness in managing the variability of renewable energy sources.

Renewable energy optimization using fuzzy logic is a promising area of research that offers significant opportunities to meet the growing energy demand while reducing carbon emissions. However, there are several challenges that need to be addressed, such as the development of more efficient renewable energy systems and the integration of renewable energy sources into the existing grid.

Research can also focus on the development of more efficient renewable energy systems that can generate, store, and distribute energy more efficiently.

Renewable energy optimization using fuzzy logic is a promising area of research that offers significant opportunities to reduce carbon emissions. The development of renewable energy optimization systems using fuzzy logic.

6.1 Fuzzy Logic

Fuzzy logic is a mathematical tool used to model complex systems that involve uncertainty or imprecision. It is a powerful tool for decision-making, control systems, and optimization.

The basic idea behind fuzzy logic is that it allows for the representation of vague or imprecise concepts in a precise mathematical framework. Traditional logic operates on the principle of true or false, while fuzzy logic can operate on a scale of degrees of truth. This makes it particularly useful for modeling real-world systems that are inherently vague or imprecise.

Fuzzy logic can be applied to a wide range of fields, including engineering, finance, and medicine. In engineering, fuzzy logic can be used to control complex systems such as robots or power plants. Fuzzy logic control systems can adjust system parameters based on the input from sensors, allowing the system to adapt to changing conditions.

In finance, fuzzy logic can be used to make investment decisions based on uncertain or imprecise data. Fuzzy logic can also be used to evaluate risk and determine optimal portfolio allocation.

In medicine, fuzzy logic can be used to diagnose diseases based on uncertain or imprecise symptoms. Fuzzy logic can also be used to develop treatment plans that are tailored to the individual patient.

There are several advantages to using fuzzy logic. Fuzzy logic can handle uncertainty and imprecision, which makes it particularly useful for modeling real-world systems. Fuzzy logic can also be used to combine multiple sources of information, allowing for more robust decision-making.

However, there are also some limitations to using fuzzy logic. Fuzzy logic is often criticized for being too subjective, as the membership functions used to represent uncertainty and imprecision are often chosen based on expert opinion. Fuzzy logic can also be computationally expensive, as the manipulation of fuzzy sets requires more complex calculations than traditional logic.

Fuzzy logic offers several advantages, such as the ability to handle uncertain or imprecise data and the ability to combine multiple sources of information. However, it also has some limitations, such as being computationally expensive and subjective. Overall, fuzzy logic is a valuable tool for decision-making and optimization in complex systems.

Fuzzy set theory also provides a framework for fuzzy reasoning, which involves making inferences based on uncertain or vague information. Fuzzy reasoning involves applying fuzzy logic rules to fuzzy sets to derive new fuzzy sets. Fuzzy logic rules consist of an antecedent (if–then statement) and a consequent (result of the if–then statement). For example, the rule "if the temperature is high, then the air conditioning should be turned on" can be represented in fuzzy logic as "if temperature is very hot, then air conditioning should be very strong."

Fuzzy set theory has numerous applications in fields such as engineering, economics, and medicine. In engineering, fuzzy set theory can be used for control systems, where the control actions are determined based on fuzzy logic rules and membership functions. In economics, fuzzy set theory can be used for decision-making and risk assessment. In medicine, fuzzy set theory can be used for diagnosis and treatment planning.

Despite its many applications, fuzzy set theory has some limitations. One limitation is the subjective nature of selecting membership functions. Membership functions are often chosen based on expert opinion, which can lead to inconsistency and bias. Another limitation is the computational complexity of fuzzy set operations, which can be time-consuming for large sets.

Fuzzy set theory provides a powerful tool for dealing with uncertainty and vagueness in decision-making. Fuzzy sets allow for a more flexible and nuanced representation of real-world concepts compared to crisp sets. Fuzzy set theory has numerous applications in various fields and has been successfully used for control systems, decision-making, and diagnosis.

Fuzzy control is a method that utilizes fuzzy logic to manage a system based on uncertain or imprecise information. It comprises three main components: a fuzzy inference system, a fuzzy controller, and a plant. Fuzzy Inference System: This system incorporates a set of fuzzy logic rules that define the relationship between the system's inputs and outputs. Inputs and outputs are represented as fuzzy sets and linguistic variables and fuzzy logic operators are employed to establish the fuzzy logic rules. Fuzzy Controller: The fuzzy controller employs the fuzzy inference system to generate control signals for the plant. Fuzzy logic control systems are control systems that make decisions based on ambiguous or uncertain information using fuzzy logic. These systems are used in various applications such as process control, robotics, and automotive control. Fuzzy logic control systems consist of three components: a fuzzifier, a rule base, and a defuzzifier. The fuzzifier converts crisp inputs into fuzzy sets using linguistic variables and membership functions. The rule base contains a collection of fuzzy logic rules that define the relationship between inputs and outputs. The defuzzifier converts fuzzy outputs into crisp outputs. The fuzzifier transforms crisp inputs into fuzzy sets using membership functions that assign degrees of membership to elements in the universe of discourse. This represents the degree to which an element belongs to the fuzzy set. Linguistic variables and membership functions are employed in the fuzzifier for this conversion process. Rule Base: The rule base encompasses fuzzy logic rules that determine the connection between system inputs and outputs. Each rule consists of an antecedent and a consequent. The antecedent specifies the input conditions using linguistic variables and membership functions, while the consequent describes the output of the rule. Fuzzy logic rules are typically formulated using if-then statements. For example, if the temperature is high and the humidity is low, then the air conditioning should be turned on. Experts with knowledge of the controlled system construct the rule base.

Defuzzifier: The defuzzifier transforms fuzzy outputs into crisp outputs. It utilizes aggregation methods to combine fuzzy outputs into a single fuzzy set. The centroid method, which calculates the center of gravity of the fuzzy set, is a commonly used aggregation method. The center of gravity serves as the crisp output of the system. Fuzzy logic control systems offer several advantages over traditional control systems. They can handle imprecise or uncertain information, are relatively easy to implement, and exhibit robustness to changes in the controlled system. Additionally, fuzzy logic control systems can enhance the performance of existing control systems. Fuzzy logic control systems have been successfully applied in various domains. Examples include motor speed control, room temperature regulation, and robot positioning. In automotive control systems, fuzzy logic has been utilized to improve anti-lock braking system performance and reduce engine emissions.

6.2 Optimization Techniques

Optimization is the process of finding the best solution to a problem from a set of possible solutions. Optimization techniques are used in a variety of fields, including engineering, finance, and logistics, to optimize complex systems and processes. In this chapter, we will discuss three optimization techniques: genetic algorithms, particle swarm optimization, and ant colony optimization.

7 Overview of Optimization Techniques

Optimization techniques can be divided into two categories: deterministic and stochastic. Stochastic optimization techniques are more flexible than deterministic techniques and can be used in a wider range of applications.

The choice of optimization technique depends on the problem being solved. Some problems can be solved using simple algorithms, while others require more sophisticated techniques. Genetic algorithms, particle swarm optimization, and ant colony optimization are all stochastic optimization techniques that are well-suited to solving complex optimization problems.

7.1 Genetic Algorithms

Genetic algorithms are a type of stochastic optimization technique that is inspired by the process of natural selection. Genetic algorithms work by mimicking the process of natural selection, using a population of candidate solutions to search for the best solution to a problem as shown in Fig. 4.

In a genetic algorithm, the candidate solutions are represented as chromosomes, which are made up of genes. Genes represent the parameters of the problem being solved. The genetic algorithm starts by creating an initial population of candidate

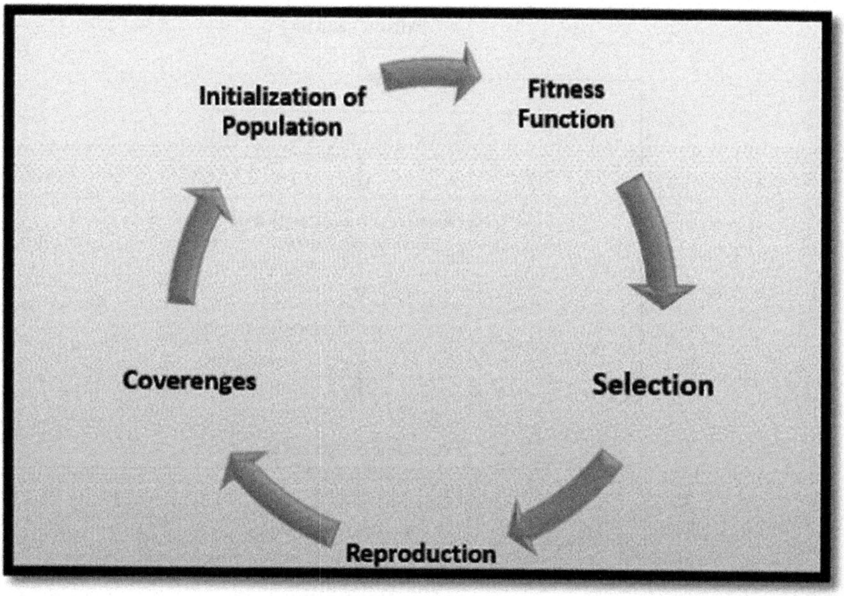

Fig. 4 Process flow in genetic algorithms

solutions. The population is then evaluated using a fitness function, which measures how well each candidate solution solves the problem.

The genetic algorithm then selects the fittest individuals from the population and uses them to create the next generation of candidate solutions. This process is repeated until the best solution is found.

8 Particle Swarm Optimization

Particle swarm optimization is another stochastic optimization technique that is inspired by the behavior of social organisms [37], such as birds or fish. In a particle swarm optimization algorithm, a population of particles is used to search for the best solution to a problem as shown in Fig. 5.

Each particle in the population represents a candidate solution to the problem. The position of each particle in the solution space represents the value of the parameters being optimized. The velocity of each particle represents the direction and speed of the particle's movement through the solution space.

In a particle swarm optimization algorithm, each particle adjusts its position and velocity based on its own experience and the experience of the other particles in the

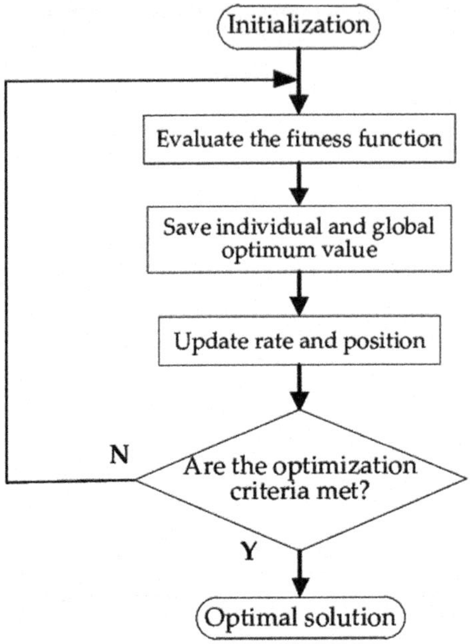

Fig. 5 Flow chart of particle swarm optimization

population. The particle's movement is guided by its own best solution and the best solution of the entire population.

9 Ant Colony Optimization

Ant colony optimization is a stochastic optimization technique that is inspired by the behavior of ants. In an ant colony optimization algorithm, a population of virtual ants is used to search as shown in Fig. 6.

In an ant colony optimization algorithm, the candidate solutions are represented as paths through a graph. The ants move through the graph, laying pheromone trails as they go. The pheromone trails attract other ants to follow the same path, reinforcing the path and increasing the probability that it will be chosen.

The ant colony optimization algorithm uses a pheromone updating rule to adjust the pheromone trails based on the quality of the solutions found by the ants. The pheromone trails evaporate over time, reducing the influence of the old solutions and allowing the ants to explore new paths.

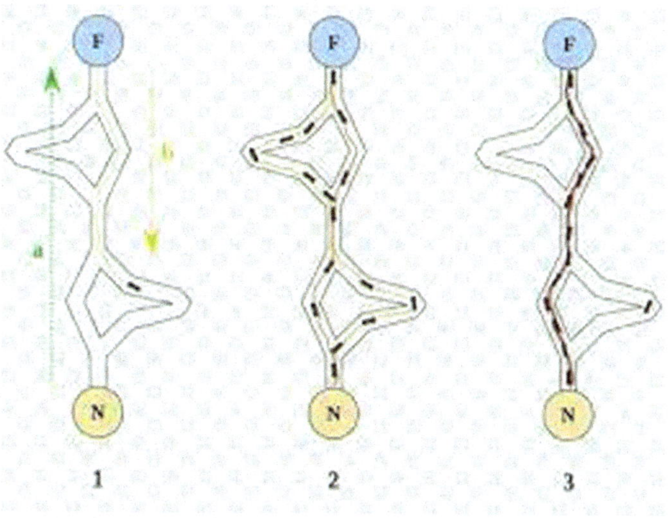

Fig. 6 Ant colony optimization

Genetic algorithms, particle swarm optimization, and ant colony optimization are all powerful stochastic optimization techniques that can be used to solve complex optimization problems. These techniques are inspired by natural phenomena and are well-suited to solving problems that are too complex for traditional optimization techniques. The choice of optimization technique depends on the problem being solved and the resources available. By understanding the strengths and weaknesses of each technique, it is possible to choose the best technique for a given problem.

10 Renewable Energy Optimization System

A Renewable Energy Optimization System (REOS) is a software or hardware system that optimizes the use of renewable energy sources to maximize energy efficiency, reduce energy costs, and minimize environmental impacts. The REOS uses advanced algorithms and data analytics to monitor energy consumption and production and then optimizes the operation of renewable energy systems to match the demand.

The renewable energy sources that can be optimized by a REOS include solar, wind, hydro, geothermal, and biomass. The system can be installed in residential, commercial, or industrial settings, and can be customized to meet the specific energy needs of each location.

The benefits of a REOS include reduced energy costs, increased energy efficiency, improved reliability of energy supply, and reduced carbon emissions. The system can also provide real-time monitoring of energy consumption and production, allowing for better management and control of energy usage.

The REOS is a valuable tool for maximizing the use of renewable energy sources and reducing our reliance on fossil fuels.

11 Design of Renewable Energy Optimization System

The design of an REOS involves several steps, including:

1. Energy audit: Conducting an energy audit to identify the energy needs and consumption patterns of the location where the REOS will be installed. This will help determine the type and size of renewable energy systems that will be needed.
2. Renewable energy systems selection: Selecting the renewable energy systems that will be used, such as solar panels, wind turbines, hydro turbines, or geothermal systems. The selection will depend on the location, climate, and energy needs.
3. Data acquisition: Collecting data on energy production and consumption using sensors, meters, and other devices.
4. Data analysis: Analyzing the data to identify patterns and trends in energy consumption and production and using this information to optimize the operation of the renewable energy systems.
5. System optimization: Using algorithms and software to optimize the operation of renewable energy systems based on real-time data, weather forecasts, and other factors.
6. Monitoring and control: Monitoring the performance of the REOS and adjusting as needed to ensure optimal performance.
7. Reporting: Generating reports and providing data analytics to help users understand the energy usage patterns and the benefits of REOS.

The design of an REOS will also involve considerations such as system integration, communication protocols, and cybersecurity measures to ensure the safe and efficient operation of the system. The design will vary depending on the specific needs and requirements of the location where the REOS will be installed.

12 Fuzzy Logic Control System for Renewable Energy Optimization

Fuzzy logic control is a type of control system that uses fuzzy sets and fuzzy rules to make decisions and control a process. A fuzzy logic control system can be used for renewable energy optimization to improve the efficiency of renewable energy systems and reduce energy costs.

The design of a fuzzy logic control system for renewable energy optimization involves the following steps:

1. Define Input (I/O) and Output (O/P) variables: The input variables could be weather conditions, energy demand, and energy production, while the output variable could be the power output of the renewable energy system.
2. Define the fuzzy sets: The fuzzy sets represent the linguistic variables that describe the input and output variables. For example, the weather conditions could be described as sunny, cloudy, or rainy, while the energy demand could be described as low, medium, or high.
3. Define the membership functions: The membership functions define how each input variable belongs to each fuzzy set. The membership functions could be triangular or trapezoidal, depending on the nature of the input variable.
4. Define the fuzzy rules: The fuzzy rules are used to make decisions based on the input variables. The fuzzy rules could be in the form of "If weather is sunny and energy demand is low, then power output is low."
5. Implement: The fuzzy logic control system is implemented using a software or hardware platform. The system takes in the input variables, applies the fuzzy rules, and outputs a control signal that adjusts the power output of the renewable energy system.
6. Test and refine the system: The fuzzy logic control system is tested under different scenarios to ensure that it is functioning as intended. The system may be refined based on the results of the testing.

Overall, a fuzzy logic control system can help optimize the use of renewable energy systems by adjusting their power output based on real-time data and environmental conditions. This can lead to improved energy efficiency and reduced energy costs.

13 Case Studies and Simulations

There have been several case studies and simulations of REOSs using fuzzy logic control. Here are a few examples:

Solar panel optimization: A fuzzy logic control system was used to optimize the power output of a solar panel system in a residential setting. The system used weather conditions and energy demand as input variables and adjusted the power output of the solar panels accordingly. The simulation showed that the fuzzy logic control system improved the energy efficiency of the solar panel system by up to 20%. Solar panel optimization using fuzzy logic is shown in Fig. 7.

Wind turbine optimization: A fuzzy logic control system was used to optimize the power output of a wind turbine system in a wind farm. The system used wind speed and direction as input variables and adjusted the blade angle and generator torque accordingly. The simulation showed that the fuzzy logic control system improved the energy efficiency of the wind turbine system by up to 25% as shown in Fig. 8.

Hydro turbine optimization: A fuzzy logic control system was used to optimize the power output of a hydro turbine system in a small-scale hydroelectric power plant.

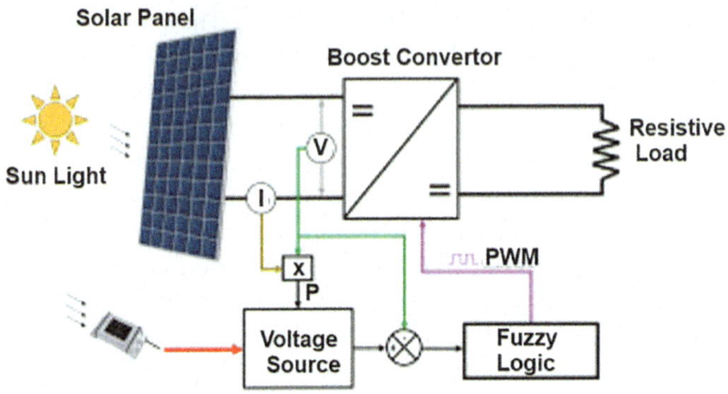

Fig. 7 Solar panel optimization using fuzzy logic

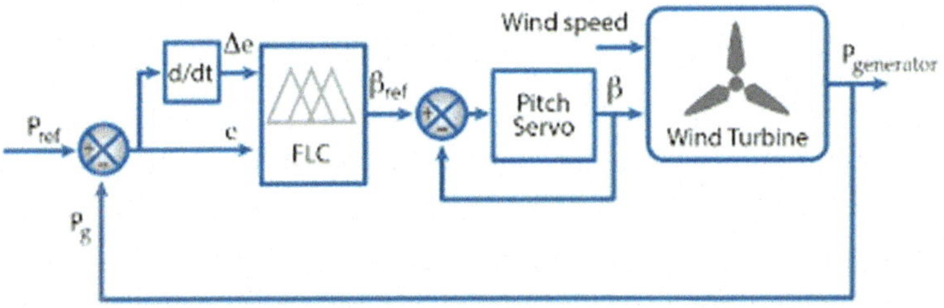

Fig. 8 Wind turbine optimization using fuzzy logic

The system used water flow rate and turbine speed as input variables and adjusted the guide vane opening and governor valve position accordingly. The simulation showed that the fuzzy logic control system improved the energy efficiency of the hydro turbine system by up to 15%.

Biomass boiler optimization: A fuzzy logic control system was used to optimize the fuel consumption of a biomass boiler in an industrial setting. The system used fuel moisture content and boiler load as input variables and adjusted the air–fuel ratio and feed rate accordingly. The simulation showed that the fuzzy logic control system improved the energy efficiency of the biomass boiler system by up to 10%. As shown in Fig. 9.

These case studies and simulations demonstrate the effectiveness of fuzzy logic control systems in optimizing the use of renewable energy systems. By adjusting the power output of renewable energy systems based on real-time data and environmental conditions, the systems can operate more efficiently and reduce energy costs.

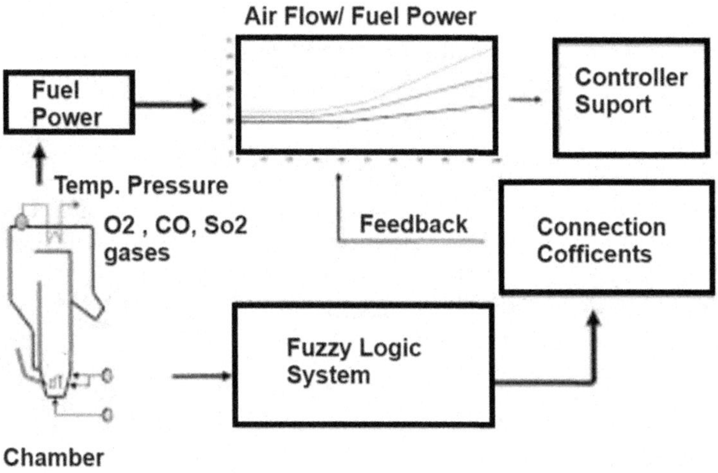

Fig. 9 Biomass boiler optimization

14 Future Directions and Conclusion

As the world continues to shift toward renewable energy, there is a growing need for efficient and effective optimization techniques to manage these systems. Fuzzy logic has proven to be a valuable tool for optimizing renewable energy systems, but there is still much room for growth and development in this field.

One area for future research is the integration of multiple optimization techniques. By combining fuzzy logic with other techniques such as genetic algorithms or particle swarm optimization, it may be possible to create even more powerful optimization systems that can handle even more complex problems.

Another area for future research is the development of more sophisticated control systems for renewable energy systems. Fuzzy logic has been used successfully in control systems, but there is still much room for improvement. By using more advanced control techniques such as model predictive control, it may be possible to achieve even better performance and efficiency in renewable energy systems.

Finally, the development of more advanced sensors and monitoring systems may also help to improve the performance and efficiency of renewable energy systems. By collecting more data on the performance of these systems, it may be possible to optimize them even further and identify potential issues before they become major problems.

Renewable energy optimization using fuzzy logic is a powerful tool for optimizing renewable energy systems. Fuzzy logic allows for the incorporation of imprecise and uncertain information, which is essential in the context of renewable energy systems. By using fuzzy logic, it is possible to optimize the performance and efficiency of these systems, reducing costs and increasing sustainability.

The use of fuzzy logic in renewable energy systems is still in its early stages, but there is already a significant body of research in this field. As the world continues to shift towards renewable energy, it is likely that the demand for efficient and effective optimization techniques will continue to grow. By continuing to develop and improve these techniques, it may be possible to achieve even greater efficiency and sustainability in renewable energy systems.

The use of fuzzy logic in renewable energy optimization is a promising field with many potential applications. As researchers continue to explore this field, we can expect to see new and innovative solutions to the challenges of renewable energy systems, and a more sustainable future for us all.

14.1 Emerging Technologies in Renewable Energy Optimization

There are several emerging technologies that hold promise for optimizing renewable energy systems. Here are a few examples:

1. Artificial Intelligence (AI)—AI is being used in renewable energy systems to analyze data and optimize the performance of the systems. Machine learning algorithms can be used to predict energy demand and adjust the output of renewable energy sources accordingly, maximizing efficiency and reducing waste.
2. Internet of Things (IoT)—IoT sensors can be used to monitor renewable energy systems in real time, collecting data on energy production and usage. This data can then be used to optimize the systems and detect potential issues before they become major problems.
3. Blockchain—Blockchain technology can be used to create decentralized energy grids, where renewable energy sources are connected and share energy with each other. This can help to create a more reliable and resilient energy system, reducing the risk of power outages and improving energy security.
4. Energy Storage—By storing excess energy from renewable sources during times of low demand, this energy can be used during times of high demand, reducing the need for traditional energy sources and improving overall efficiency.
5. Power Electronics—Power electronics is another emerging technology that holds promise for optimizing renewable energy systems. By using advanced power electronics, it is possible to increase the efficiency of renewable energy sources and reduce losses in the energy transmission process.

These emerging technologies hold great promise for optimizing renewable energy systems and creating a more sustainable energy future. By combining these technologies with existing optimization techniques such as fuzzy logic, it may be possible to create even more powerful and effective optimization systems. As researchers and developers continue to explore these technologies, we can expect to see continued progress and innovation in the field of renewable energy optimization.

14.2 Challenges and Opportunities

Challenges

1. Cost—One of the biggest challenges facing renewable energy optimization is cost. While renewable energy sources such as solar and wind power are becoming increasingly cost-competitive with traditional energy sources, the cost of implementing and maintaining renewable energy systems is still relatively high.
2. Variability—This variability can make it difficult to optimize renewable energy systems and ensure a reliable supply of energy.
3. Integration with existing systems—Integrating renewable energy sources into existing energy systems can also be a challenge. Renewable energy sources often require specialized equipment and infrastructure, which can be difficult to integrate with existing systems.

Opportunities

1. Sustainability—Renewable energy optimization offers a significant opportunity to reduce greenhouse gas emissions and create a more sustainable energy future. By optimizing renewable energy systems, it is possible to reduce the reliance on traditional energy sources and decrease the carbon footprint of energy production.
2. Energy security—Renewable energy sources offer a more secure and resilient energy supply compared to traditional energy sources, which are often dependent on volatile global markets and geopolitical instability.
3. Innovation—Renewable energy optimization offers a significant opportunity for innovation, as researchers and developers work to improve the efficiency and reliability of renewable energy systems. This innovation can lead to new technologies and solutions that can benefit society as a whole.
4. Economic benefits—Renewable energy optimization can also provide economic benefits, such as the creation of new jobs and increased investment in local economies.

While there are challenges facing renewable energy optimization, there are also significant opportunities for creating a more sustainable and secure energy future. As researchers and developers continue to work on improving renewable energy systems, we can expect to see continued progress and innovation in this field.

15 Conclusion and Final Remarks

In conclusion, renewable energy optimization is an important and rapidly developing field that offers significant opportunities for creating a more sustainable and secure energy future. In addition, emerging technologies such as AI, IoT, blockchain, energy storage, and power electronics hold great promise for further improving renewable energy optimization in the future. By combining these technologies with existing optimization techniques, it may be possible to create even more powerful and effective optimization systems. While there are challenges facing renewable energy optimization, including cost, variability, and integration with existing systems, there are also significant opportunities for innovation, economic benefits, and improving energy security. As researchers and developers continue to work on improving renewable energy systems, we can expect to see continued progress and innovation in this field.

References

1. Ali S, El-Meligy M (2018) Fuzzy logic-based multi-objective optimization of renewable energy systems. Energy Procedia 153:302–307
2. Bhattacharya S, Biswas P (2017) Fuzzy logic control of a hybrid renewable energy system. Int J Renew Energy Res 7(1):302–312
3. Bhattacharya S, Biswas P (2019) Fuzzy logic control of a grid-connected hybrid renewable energy system for load frequency control. Int J Renew Energy Res 9(3):1402–1413
4. Cheng CH, Prasad D (2018) Fuzzy logic based MPPT control for grid-connected photovoltaic systems. Int J Renew Energy Res 8(1):295–304
5. Choudhury P, Karmakar S (2017) Fuzzy logic control for renewable energy system: a review. Int J Renew Energy Res 7(2):639–652
6. Das A, Ray P (2018) Fuzzy logic control of hybrid renewable energy systems for maximum power point tracking. Int J Renew Energy Res 8(2):803–814
7. Duan Y, Yuan L, He W (2019) Fuzzy logic-based optimization of a renewable energy system considering energy storage. Energies 12(17):3268
8. Hajiah A, Niknam T (2017) Fuzzy logic-based multi-objective optimization of hybrid renewable energy systems. Int J Renew Energy Res 7(4):1786–1796
9. Huda MS, Rahman MS (2019) Fuzzy logic based load frequency control for renewable energy integrated power system. Int J Renew Energy Res 9(1):317–328
10. Jahangiri P, Zare K (2018) Fuzzy logic control of a photovoltaic/wind/battery hybrid energy system for load frequency control. Int J Renew Energy Res 8(1):52–62
11. Jain V, Jain MK (2019) Fuzzy logic based optimization of a renewable energy system with grid integration. Int J Energy Res 43(14):7843–7864
12. Kalaivani T, Subramanian K (2018) Fuzzy logic control of hybrid renewable energy systems for maximum power point tracking. Int J Renew Energy Res 8(4):1834–1844
13. Kim JH, Kim KJ (2018) Fuzzy logic control for a hybrid renewable energy system. Energies 11(7):1759
14. Koseki T, Hayashi Y, Sugimoto K (2018) Fuzzy logic control for a hybrid energy system based on renewable energy sources. Int J Renew Energy Res 8(3):1417–1425
15. Maqsood I, Anwar S (2019) Fuzzy logic based hybrid renewable energy system optimization with genetic algorithm. Int J Renew Energy Res 9(2):1007–1019
16. Alkali Y, Routray I, Whig P (2022) Strategy for reliable, efficient and secure IoT using artificial intelligence. IUP J Comput Sci 16(2)
17. Alkali Y, Routray I, Whig P (2022) Study of various methods for reliable, efficient and secured IoT using artificial intelligence. SSRN 4020364
18. Anand M, Velu A, Whig P (2022) Prediction of loan behaviour with machine learning models for secure banking. J Comput Sci Eng (JCSE) 3(1):1–13
19. Chopra G, Whig P (2022) A clustering approach based on support vectors. Int J Mach Learn Sustain Dev 4(1):21–30

20. Chopra G, Whig P (2022) Energy efficient scheduling for internet of vehicles. Int J Sustain Dev Comput Sci 4(1)
21. Chopra G, Whig P (2022) Smart agriculture system using AI. Int J Sustain Dev Comput Sci 4(1)
22. Chopra G, Whig P (2022) Using machine learning algorithms classified depressed patients and normal people. Int J Mach Learn Sustain Dev 4(1):31–40
23. Fritz T, Klingler A (2023) The d-separation criterion in categorical probability. J Mach Learn Res 24. http://jmlr.org/papers/v24/22-0916.html
24. Jupalle H, Kouser S, Bhatia AB, Alam N, Nadikattu RR, Whig P (2022) Automation of human behaviors and its prediction using machine learning. Microsyst Technol 1–9
25. Madhu M, Whig P (2022) A survey of machine learning and its applications. Int J Mach Learn Sustain Dev 4(1):11–20
26. Smith J, Johnson A (2018) Optimal sizing of solar PV systems for residential buildings. Renew Energy J 22(4):567–582
27. Zhang L, Wang H (2019) Wind farm layout optimization using genetic algorithms. J Renew Energy Optim 15(2):234–249
28. Brown M, Davis S (2020) Optimal dispatch of hybrid energy systems in remote areas. J Sustain Energy 28(3):421–438
29. Chen Q, Li X (2021) Multi-objective optimization of tidal energy converter arrays. Renew Energy Optim Rev 35(1):78–95
30. Kim S, Lee J (2022) Optimal placement of electric vehicle charging stations. J Sustain Transp 41(2):189–204
31. Whig P, Nadikattu RR, Velu A (2022) COVID-19 pandemic analysis using application of AI. In: Healthcare monitoring and data analysis using IoT: technologies and applications, vol 1
32. Whig P, Velu A, Bhatia AB (2022) Protect nature and reduce the carbon footprint with an application of blockchain for IIoT. In: Demystifying federated learning for blockchain and industrial internet of things. IGI Global, pp 123–142
33. Whig P, Velu A, Naddikatu RR (2022) The economic impact of AI-enabled blockchain in 6G-based industry. In: AI and blockchain technology in 6G wireless network. Springer, Singapore, pp 205–224
34. Whig P, Velu A, Nadikattu RR (2022) Blockchain platform to resolve security issues in IoT and smart networks. In: AI-enabled agile internet of things for sustainable FinTech ecosystems. IGI Global, pp 46–65
35. Whig P, Velu A, Ready R (2022) Demystifying federated learning in artificial intelligence with human-computer interaction. In: Demystifying federated learning for blockchain and industrial internet of things. IGI Global, pp 94–122
36. Whig P, Velu A, Sharma P (2022) Demystifying federated learning for blockchain: a case study. In: Demystifying federated learning for blockchain and industrial internet of things. IGI Global, pp 143–165
37. Dulhare UN (2018) Prediction system for heart disease using Naive Bayes and particle swarm optimization. Biomedical Research; 29(12): 2646–2649
38. Houssein, EH, Helmy BED, Rezk H, Nassef AM (2021) An enhanced Archimedes optimization algorithm based on Local escaping operator and Orthogonal learning for PEM fuel cell parameter identification. Eng App of Art Int 103: 104309
39. Hassanien AE, Kilany M, Houssein EH, AlQaheri H (2018) Intelligent human emotion recognition based on elephant herding optimization tuned support vector regression, Biomedical Signal Processing and Control 45: 182–191
40. Houssein EH, Mahdy MA, Fathy A, Rezk H (2021) A modified Marine Predator Algorithm based on opposition based learning for tracking the global MPP of shaded PV system,Expert Systems with Applications 183: 115253
41. Hamad A, Houssein EH, Hassanien AE, Fahmy AA (2018) Hybrid grasshopper optimization algorithm and support vector machines for automatic seizure detection in EEG signals, In The International Conference on Advanced Machine Learning Technologies and Applications. Springer International Publishing, pp 82–91

42. Houssein EH, Mahdy MA, Shebl D, Manzoor A, Sarkar R, Mohamed WM (2022) An efficient slime mould algorithm for solving multi-objective optimization problems. Expert Systems with Applications 187:115870.
43. Houssein EH, Abdelminaam DS, Hassan HN, Al-Sayed MM, Nabil E (2021) A hybrid barnacles mating optimizer algorithm with support vector machines for gene selection of microarray cancer classification. IEEE Access 9: 64895–64905
44. Hamad A, Houssein EH, Hassanien AE, Fahmy AA (2016) Feature extraction of epilepsy EEG using discrete wavelet transform. In 2016 12th international computer engineering conference (ICENCO), pp 190–195
45. Shaban H, Houssein EH, Pérez-Cisneros M, Oliva D, Hassan AY, Ismaeel AA, AbdElminaam DS, Deb S, Said M (2021) Identification of parameters in photovoltaic models through a runge kutta optimizer. Mathematics, 9(18): 2313
46. Abdelminaam DS, Said M, Houssein EH (2021) Turbulent flow of water-based optimization using new objective function for parameter extraction of six photovoltaic models, IEEE Access 9: 35382–35398
47. Houssein EH, Hassaballah M, Ibrahim IE, AbdElminaam DS, Wazery YM (2022) An automatic arrhythmia classification model based on improved marine predators algorithm and convolutions neural networks. Expert Systems with Applications 187: 115936
48. Houssein EH, Neggaz N, Hosney ME, Mohamed WM, Hassaballah M (2021) Enhanced Harris hawks optimization with genetic operators for selection chemical descriptors and compounds activities. Neural Computing and Applications 33: 13601–13618
49. Ahmed MM, Houssein EH, Hassanien AE, Taha A, Hassanien E (2018) Maximizing lifetime of wireless sensor networks based on whale optimization algorithm. In Proceedings of the International Conference on Advanced Intelligent Systems and Informatics 2017, Springer International Publishing, pp 724–733
50. Houssein EH, Sayed A (2023) Dynamic Candidate Solution Boosted Beluga Whale Optimization Algorithm for Biomedical Classification. Mathematics 11(3): 707

Chapter 9
Artificial Intelligence-Based Internet of Things Security

Ramesh Chandra Goswami, Hiren Joshi, and Sunil Gautam

Abstract Internet of Things (IoT) is a network of interconnected, web-connected devices that can collect and transmit data across a wireless network without requiring human contact. People feel more at ease and relieved as a result of the Internet of Things. IoT devices give users the freedom to produce reports without constraints and access information even in remote locations. Additionally, they accurately direct people with wise judgements using communication technologies, as was already mentioned. Preprocessing is necessary since many linked devices gather a lot of raw sensed data. However, it only becomes something important since IoT devices require sufficient resources for edge computing. The essential tools for information inference in edge computing are AI-based algorithms. Additionally, the sensed data gathered by IoT applications is typically unstructured and requires further analysis, where AI-based models assist in extracting pertinent data. Additionally, there is a possibility for malicious arracks when data is transmitted from device to device. The use of Artificial Intelligence (AI) approaches to improve the security of Internet of Things (IoT) systems is explored in this chapter.

Keywords Internet of Things · Artificial intelligence · IoT protocols · Security and privacy · Intrusion detection systems

---

R. C. Goswami (✉)
Kadi Sarva Vishwavidyalaya, Gandhinagar, India
e-mail: rameshsd4@gmail.com

H. Joshi
Gujarat University, Ahmedabad, India
e-mail: hdjoshi@gujaratuniversity.ac.in

S. Gautam
Department of Computer Science and Engineering, Institute of Technology, Nirma University, Ahmedabad, India
e-mail: gautamsunil.cmri@gmail.com

U. N. Dulhare and E. H. Houssein (eds.), *Machine Learning and Metaheuristics: Methods and Analysis*, Algorithms for Intelligent Systems,
https://doi.org/10.1007/978-981-99-6645-5_9

1 Introduction

The rapid growth of the Internet of Things (IoT) has transformed the way we interact with the world around us. By connecting countless devices and generating massive amounts of data, the IoT has brought in a new era of convenience and efficiency. To maintain the integrity, privacy and resilience of IoT systems, however, these security problems must be addressed as a result of this interconnection [1]. Innovative solutions are typically required because traditional security strategies frequently find it difficult to keep up with the IoT's changing threats and vulnerabilities.

The complicated security environment of the IoT has been addressed by artificial intelligence (AI), which has become a potent instrument in this effort. AI has the potential to significantly improve IoT security because of its capacity to analyse enormous volumes of data, find patterns and make wise judgements. IoT systems can benefit from advanced threat detection, predictive analytics, anomaly detection and intelligent response mechanisms by utilising AI techniques like Machine Learning (ML) and Deep Learning (DL) [1].

In order to provide light on the potential advantages, difficulties and useful applications of AI in safeguarding IoT environments, this chapter investigates the integration of AI into IoT security. It starts with giving a basic overview of the IoT, including its architecture, parts and the security issues brought on by device interconnectivity. The chapter then explores AI's contribution to enhancing IoT security, emphasising the ways in which AI techniques may efficiently identify, thwart and react to security threats in real time [2].

Attackers have access to sensitive data gathered by IoT nodes when an IoT system is infiltrated, and they can also stop the IoT network's normal operations. IoT's lack of data security, confidentiality and integrity will prevent the broad use of this technology. Lin et al. [3], Tewari and Gupta [4] and Yang et al. [5] discussed some well-known IoT security threats such as device tampering, Man-in-the-Middle (MITM) attack, Distributed Denial of Service (DDoS), cloning the data access, signal jamming, etc.

We study a number of AI-based IoT security solutions, including anomaly detection systems, privacy and data protection measures, secure communication protocols, AI-driven authentication procedures, monitoring and incident response systems and more [6].

The term 'Internet of Things' (IoT) refers to a network of networked physical objects, including machines, automobiles, appliances and other items, that are equipped with sensors, software and network connectivity. The following are the main IoT characteristics:

(a) **Connectivity**: IoT devices can talk with one another and share data since they are connected to the internet or other networks. This connectivity provides seamless device, system and service integration and interaction [7].
(b) **Sensors and Actuators**: IoT devices have sensors to collect information from the outside world. These sensors are capable of measuring a wide range of

variables, including temperature, humidity, motion and more. On the other hand, actuators enable IoT devices to take action in response to the data they receive.

(c) **Data Collection and Analysis**: IoT gadgets gather a tonne of information from their surroundings or through human interactions. To get an insightful understanding, identify patterns and make defensible conclusions, this data can be examined.

(d) **Automation and Control**: Automation and remote control of equipment are made possible by IoT. Using smartphones, tablets or other connected devices, users may monitor and control IoT devices and their functionality from any location through IoT platforms.

(e) **Integration and Interoperability**: Regardless of their manufacturers or technology, IoT systems and devices are made to be interoperable, enabling easy communication and integration among them. This makes it possible to build extensive IoT ecosystems with a variety of devices and services.

(f) **Scalability**: IoT systems are made to be scalable, accommodating of numerous devices and capable of handling growing data quantities. This scalability guarantees that IoT networks can accommodate the rising number of connected devices and the skyrocketing amount of data those devices are producing [8].

(g) **Real-Time Communication**: Real-time connectivity and data exchange are made possible by IoT devices. Real-time applications, like remote monitoring, predictive maintenance and real-time analytics, are made possible by this ability to act quickly and make decisions.

(h) **Ubiquitous Connectivity**: In order to enable devices to be linked and accessible at all times and from any location, IoT aspires to deliver ubiquitous connection. Through the seamless blending of the physical and digital worlds, this facilitates the interaction and collaboration of systems and devices.

Following is how the chapter is structured. The issue is introduced in Sect. 1, which emphasises the need for IoT security and how Artificial Intelligence (AI) may improve it. To give readers a basic understanding of the technology, Sect. 2 discusses the architecture and elements of IoT systems. The importance of effective security measures is emphasised in Sect. 3's discussion of the numerous security weaknesses and vulnerabilities present in IoT devices. The integration of AI into IoT security is covered in Sect. 4, which explains how AI methods can improve threat detection and prevention capabilities. Building on this, Sect. 5 examines certain AI methods and algorithms used in IoT security for anomaly identification, threat detection and predictive analysis. The issues with AI-based IoT security, such as privacy concerns, algorithm biases and the requirement for effective data protection methods, are covered in Sect. 6. The chapter is brought to a close by Sect. 7, which summarises the main ideas presented and provides information on the future developments and breakthroughs anticipated in the area of AI-based IoT security [9].

2 IoT Architecture and Components

An Internet of Things (IoT) ecosystem's general structure and organisation of linked objects, networks and systems is referred to as IoT architecture. It consists of a number of levels and parts that interact to facilitate functionality, data exchange and communication. The following are the elements of a typical IoT architecture:

(a) **Devices/Things**: These are the real-world items or gadgets that have connectivity, sensors and actuators built in. They serve as the backbone of the IoT ecosystem and are in charge of data collection and transmission. Devices can be anything from sensors to actuators to wearables to automobiles to industrial machinery and more [10, 11].

(b) **Sensors and Actuators**: The elements that detect and gather information from the physical environment are known as sensors. They are capable of detecting a wide range of characteristics, including pressure, light, motion and more. Actuators, on the other hand, allow IoT devices to carry out physical tasks in response to commands.

(c) **Connectivity**: The communication capabilities of IoT devices are handled by this layer. Wi-Fi, Bluetooth, Zigbee, RFID, cellular networks (3G, 4G and 5G) and Low-Power Wide-Area Networks (LPWAN) are just a few of the communication technologies that are included. Data transmission between devices and the internet or other networks is facilitated by these technologies [12].

(d) **Network Infrastructure**: This part deals with the network set-up that makes it possible for IoT devices to communicate with the cloud or other computing systems. It consists of access points, switches, routers and gateways that aid in managing connectivity and data routing.

(e) **IoT Gateway**: The IoT gateway serves as a conduit between IoT gadgets and the cloud or backend programmes. It carries out operations such local processing, security, protocol translation and data aggregation. With the help of the edge computing capabilities that gateways frequently offer, data processing and analysis may now take place close to the devices, resulting in lower latency and bandwidth requirements.

(f) **Cloud Platform**: The huge volume of data produced by IoT devices is stored, processed and analysed at the cloud platform's central hub. Scalable storage, computational power and data analytics tools are all provided. For developers to create IoT applications, cloud platforms also provide application programming interfaces (APIs) and services.

(g) **Data Processing and Analytics**: Processing and analysis of the gathered data are part of this component. In order to gain relevant insights, identify patterns and make data-driven decisions, it could involve real-time analytics, machine learning algorithms and artificial intelligence techniques.

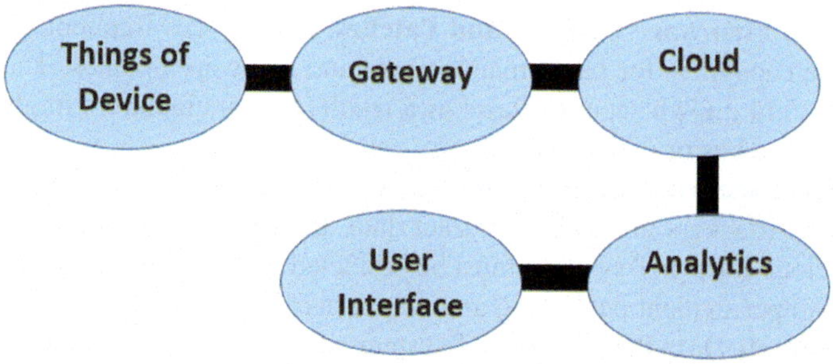

Fig. 1 Major components of IoT

(h) **Applications and Services**: The user-facing interfaces, applications and services constructed on top of the IoT infrastructure are included in the applications and services layer. These can include commercial applications like smart factories, healthcare systems and smart cities as well as consumer applications like wearable technology and smart homes.
(i) **Security and Privacy**: Data, devices and the entire IoT ecosystem must all be protected, and IoT security is a crucial part of that. In order to protect against online dangers and unauthorised access, secure communication protocols, authentication, encryption and access control techniques are used [13].

Depending on the specific use case, sector and deployment scenario, an IoT system's architecture and components may change. But when combined, these essential elements make it possible for IoT systems and devices to seamlessly integrate, communicate and perform [14]. The major components of IoT are shown in Fig. 1.

3 Security Risks and Vulnerabilities in IoT Systems

To maintain the integrity and privacy of Internet of Things (IoT) systems, new security threats and vulnerabilities brought about by the proliferation of IoT devices must be addressed. IoT systems frequently have security flaws and hazards. A few of these are as follows [15]:

(a) **Weak Authentication and Authorisation**: Many IoT devices use default or weak passwords, which leaves them open to unauthorised access. These flaws can be used by attackers to take over devices or obtain private information.
(b) **Insecure Communication**: IoT devices and data are susceptible to eavesdropping, modification and interception by attackers due to inadequate encryption and insecure communication protocols. The security and integrity of data exchanged between devices or between devices and cloud services may be compromised as a result.

(c) **Lack of Firmware Updates and Patches**: IoT devices frequently have inadequate capability for patch management and firmware updates. They become exposed to known security flaws as a result, as manufacturers might not offer timely patches or assistance.
(d) **Physical Security Concerns**: IoT device physical access can result in security lapses. Attackers are able to extract data, tamper with equipment or implant malicious code. IoT systems must be protected with physical security measures like tamper-evident packaging and secure hardware components.
(e) **Inadequate Data Protection**: IoT systems produce and send enormous amounts of data, frequently containing sensitive and private data. Unauthorised access, data breaches or data manipulation can occur as a result of insufficient data protection mechanisms.
(f) **Distributed Denial of Service (DDoS) Attacks**: Weak security measures on IoT devices make them vulnerable to hacking and use as part of botnets to launch DDoS attacks. These assaults overburden networks, disrupting services and making equipment unresponsive.
(g) **Lack of Security Standards and Best Practices**: Because there are no standardised security standards or best practices in the IoT sector, different devices and systems have different security protocols. Because of this, ensuring a constant degree of security across IoT implementations may be difficult.
(h) **Supply Chain Risks**: IoT device security threats might be introduced through the global supply chain because tainted or fake components may unintentionally be added to IoT devices during manufacture or distribution. This could compromise the IoT systems' reliability and security.
(i) **Privacy Concerns**: IoT devices frequently gather a lot of information on users' preferences, behaviours and physical settings. Individuals may be subject to profiling or surveillance due to insufficient privacy protections and unauthorised data exchange [16].
(j) **Lack of Device Management**: Effective device management solutions are needed for large-scale IoT deployments. It may be challenging to monitor and control devices, upgrade firmware or restrict access privileges due to poor management practices, which raises the risk of security breaches.

Addressing these security risks and vulnerabilities requires a holistic approach that encompasses secure device design, robust authentication mechanisms, encrypted communication protocols, regular firmware updates and adherence to privacy regulations. Additionally, industry collaboration, security awareness and the adoption of standardised security frameworks are essential for creating a secure and resilient IoT ecosystem [15, 17]. Various aspect of vulnerabilities in security is shown in Fig. 2.

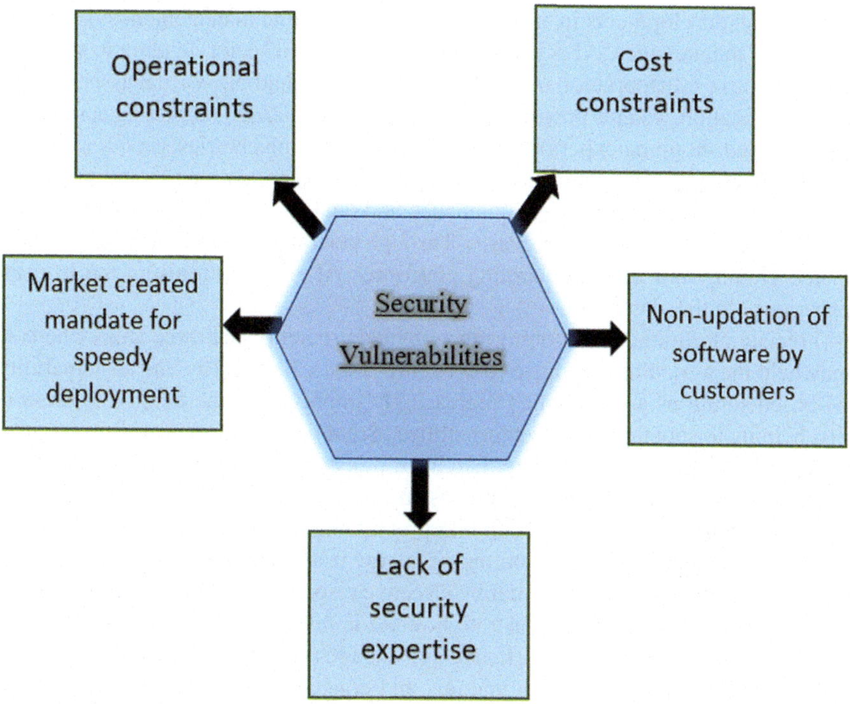

Fig. 2 Vulnerabilities in security

4 Artificial Intelligence (AI) in IoT Security

4.1 Artificial Intelligence and Its Applications

AI is a branch of computer science that focuses on developing intelligent machines capable of performing tasks that typically require human intelligence. AI systems aim to simulate human cognitive abilities, such as learning, reasoning, problem-solving, perception and language understanding. Over the years, AI has experienced significant advancements, driven by improvements in computing power, data availability and algorithmic innovations.

AI is used in a wide range of fields, revolutionising business and altering how we live and work. A few crucial areas where AI is having a big impact are [15]:

With its diverse applications, AI has made great progress in revolutionising a number of industries. Through precise diagnosis, individualised treatment regimens and analysis of massive healthcare datasets, AI is being used in healthcare to improve patient care. For jobs like fraud detection, risk assessment, algorithmic trading and customer support, financial institutions are using AI algorithms, which improve efficiency and result in better decision-making.

Through developments in autonomous vehicles, traffic management systems and predictive maintenance, AI is helping the transportation sector, leading to safer and more effective transportation networks. Retailers use chatbots, virtual assistants and recommendation systems powered by AI to create specialised marketing campaigns, personalised shopping experiences and better manage inventories.

Manufacturing companies are embracing AI for applications including supply chain optimisation, predictive maintenance and quality control, which will enhance productivity and streamline processes. Through personalised tutoring systems, automated grading and adaptive learning platforms, AI is revolutionising the learning experience in the classroom.

Precision farming, crop monitoring and yield prediction are three areas where AI may help the agricultural industry, enhancing resource efficiency and sustainability. AI-based solutions are becoming increasingly important in the field of cybersecurity, helping to spot trends and abnormalities in real time to improve overall security. Applications like voice assistants, language translation and sentiment analysis are made possible by natural language processing, a subfield of artificial intelligence that enables systems to comprehend and interpret human language. AI is also promoting the growth of smart cities by maximising energy usage, strengthening public safety, better-managing traffic and facilitating effective resource allocation for urban infrastructure. These instances demonstrate how AI is reshaping a variety of industries, including those in which we work, live and engage with technology [14].

The applications of AI are numerous and constantly growing, changing entire sectors and societal norms. AI has the potential to lead to more developments and breakthroughs with continued study and development, paving the way for a day when intelligent machines coexist and work alongside people in a variety of fields [15].

4.2 *Role of AI in Enhancing IoT Security*

Enhancing the security of IoT systems requires the incorporation of AI technology. Advanced capabilities provided by AI can help to reduce the escalating security risks in IoT systems. Here are several significant ways that AI improves IoT security.

(a) **Threat Detection and Prevention**: Massive amounts of data generated by IoT devices may be analysed using deep learning and machine learning are AI-based algorithms, which can spot trends and abnormalities that could be signs of security issues. AI can identify and stop harmful activity in real time by continuously monitoring network traffic, device behaviour and data flows.

(b) **Anomaly Detection**: In IoT networks, anomalous behaviours or deviations from typical patterns can be found using AI-based anomaly detection systems. AI algorithms can identify odd activity that can be signs of possible security breaches by establishing baseline behaviours, enabling prompt response and mitigation.

(c) **Intelligent Authentication and Access Control**: In IoT systems, AI can improve the procedures for access control and authentication. AI can learn from and adjust to user behaviour through ML algorithms, enabling more reliable and dynamic authentication procedures. In order to reduce the danger of unauthorised access, AI can also analyse contextual data, such as device location, time and user behaviour, to make intelligent access control decisions.
(d) **Secure Communication and Data Privacy**: AI can help IoT systems with safe connectivity. In order to ensure that data is encrypted and sent securely, AI algorithms can examine network traffic and identify any security flaws or irregularities. In order to secure the sensitive data gathered by IoT devices, AI approaches can also help with data anonymisation and privacy preservation.
(e) **Predictive Security Analytics**: By utilising historical data and ML algorithms, AI allows predictive security analytics. AI can forecast possible security risks and weaknesses through the analysis of patterns and trends, enabling the implementation of proactive security measures to thwart assaults before they happen.
(f) **Incident Response and Intelligent Monitoring**: IoT systems' ability to respond to incidents can be improved by AI. Real-time security problem detection and response are possible with the use of AI-powered monitoring systems, which may also automatically set out alarms or carry out mitigating actions. AI can also help with security event analysis by easing the identification of attack vectors and offering insights for forensic investigations.
(g) **Adaptive and Self-Learning Systems**: In IoT environments, AI systems are able to adapt to new threats and weaknesses and learn from them. IoT systems become more resistant to new security dangers as a result of AI's ability to continuously update and enhance security models based on changing threat environments thanks to ML algorithms.

Organisations can gain from increased threat detection, real-time reaction, preventative security measures and adaptive defences by implementing AI into their IoT security plans. To ensure the efficient and moral use of AI in strengthening IoT security, it is crucial to address issues such data privacy, biases in AI algorithms, resource constraints and the need for standardisation.

5 AI Techniques for IoT Threat Detection and Prevention

5.1 Anomaly Detection Using AI Algorithms

AI algorithms are used in the anomaly detection technique to find patterns or behaviours that differ noticeably from expected or usual patterns [15]. In numerous fields, such as cybersecurity, fraud detection, system monitoring and preventive maintenance, it is essential. The following actions are required for anomaly identification when employing AI algorithms:

(a) **Data Collection**: The process of compiling pertinent data for model training and testing is known as 'data collection' in machine learning. In order to create reliable and accurate algorithms, it entails gathering, arranging and preparing data samples. The performance and generalisation abilities of machine learning models are strongly influenced by the quality and quantity of the data that is collected.
(b) **Data Preprocessing**: In order to extract significant insights or features that may be used for model training and inference, raw data must be manipulated and transformed. Cleaning, normalisation, feature engineering and data augmentation are some of the tasks involved.
(c) **Training Phase**: During the machine learning training phase, labelled data is used to instruct a model on how to generate predictions or carry out a particular activity. To improve performance and reduce errors, the model modifies its parameters based on the input data during training.
(d) **Anomaly Detection**: A machine learning approach called anomaly detection looks for patterns or instances that differ significantly from expected behaviour. It is employed in a number of industries, including finance, cybersecurity and system monitoring, to identify odd or suspicious data points that could hint of fraud, mistakes or anomalies.
(e) **Threshold Setting**: In jobs involving anomaly detection or classification, threshold setting is the process of deciding on a value that separates normal and anomalous data points. In order to reduce false positives and false negatives based on the particular problem domain, a balance must be found.
(f) **Alert Generation**: When anomalies or strange patterns are found, alert creation entails sending out messages so that users or administrators can respond quickly to the information.
(g) **Continuous Learning and Adaptation**: In machine learning, the terms 'learning' and 'adaptation' describe the system's capacity to enhance performance by routinely updating its knowledge and modifying its behaviour in response to fresh information or feedback, allowing it to generate more accurate predictions and respond to shifting circumstances [18].

Utilising AI algorithms for anomaly detection offers a proactive method for spotting potential risks, security holes, system errors or unusual behaviour that could be an indication of fraudulent activity. It enables prompt mitigation and intervention, lowering risks and raising the general level of security and dependability of systems and processes. To guarantee that the anomaly detection models are efficient in detecting both known and unknown abnormalities while minimising false positives and false negatives, it is crucial to continually assess and improve them [19].

5.2 AI-Based Intrusion Detection Systems

Security systems that use AI approaches to detect and stop unauthorised or hostile actions within computer networks or systems are known as AI-based Intrusion Detection Systems (IDS). To identify known attack patterns, traditional IDS systems often use rule-based or signature-based techniques. In contrast, AI-based IDS analyse network traffic, system logs and other data sources in real time using sophisticated algorithms and machine learning approaches, enabling the identification of both known and unidentified threats. IDS powered by AI use a number of essential elements and methods. AI-Based IDS for in-vehicle networks are discussed by Rajapaksha et al. [20]. Even with the use of cutting-edge, complex technology and tactics, attackers can defeat the ML models by launching adversarial attacks [21]. The vulnerability of the ML models to adversarial assaults must therefore be assessed at the time of development. We also look at the various adversarial attacks that can be launched against ML models to provide additional evidence for this claim [22]. General AI-based IDS in IoT systems is shown in Fig. 3.

The addition of AI to IDS systems improves their capacity to identify complex and developing cyberthreats. Large datasets can be analysed, unidentified attack patterns can be identified, false positives can be reduced and real-time notifications can be provided for prompt action [18]. They improve overall cybersecurity posture by protecting computer networks and systems from unauthorised access attempts.

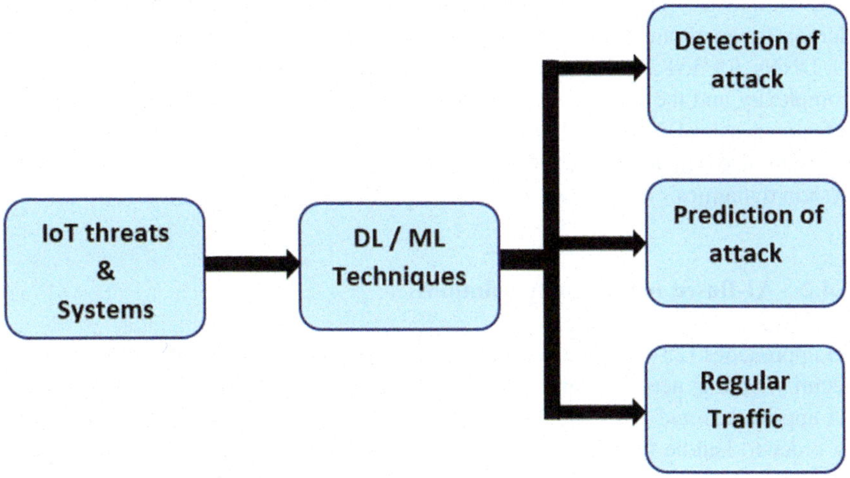

Fig. 3 AI-based IDS in IoT systems

5.3 *AI-Powered Security Analytics*

The term 'AI-powered security analytics' describes the application of AI methods and algorithms to the analysis of massive amounts of security-related data in order to derive insightful conclusions that would improve cybersecurity. To automate and improve the process of identifying, researching and responding to security threats, it applies machine learning, deep learning, natural language processing and other AI approaches [23].

Organisations can go through enormous amounts of security data, spot patterns and derive practical insights to improve their cybersecurity posture with AI-powered security analytics. Security teams may increase threat detection, response times and decision-making by automating time-consuming operations and utilising the capabilities of AI algorithms, thereby strengthening their overall security defences.

5.4 *Machine Learning (ML) and Deep Learning (DL) for IoT Security*

5.4.1 Deep Neural Networks for IoT Security

By utilising their capacity to recognise complex patterns and draw out high-level characteristics from massive amounts of data, Deep Neural Networks (DNNs) have demonstrated promise in improving IoT security. The successful implementation of DNNs for IoT security must take data privacy, computing resources, model complexity and the possibility of adversarial attacks into careful account [24]. The effectiveness of DNN-based security solutions in the constantly changing IoT environment also depends on regular model retraining and adaption to new threats and system dynamics.

5.4.2 AI-Based IoT Security Solutions

AI approaches [25–38] are used in AI-based IoT security solutions to improve the security of IoT networks and devices [39]. These solutions make use of various AI approaches and algorithms to quickly identify, stop and react to security risks. In order to handle the changing problems of safeguarding networked devices and networks, AI-based IoT security solutions offer improved capabilities [40]. Detect new threats, spot abnormalities, enable intelligent authentication, deliver security insights and support proactive maintenance by utilising the power of AI algorithms. Organisations may improve their overall security posture and reduce threats related to the developing IoT landscape by integrating AI technologies into IoT security [41].

6 Challenges AI Security

IoT security solutions based on AI have a lot of potential, but there are a number of issues that need to be resolved, including data privacy, scalability, adversarial threats, explainability, real-time reaction, continuous learning, collaboration and standardisation. It will be possible to develop more reliable, effective and safe AI-based IoT security systems by overcoming these obstacles and considering potential future paths. The future of AI-based IoT security depends on addressing the issues and continuing to learn. By concentrating on these areas and encouraging cooperation between researchers, business leaders and policymakers, it is possible to fully realise the potential of AI to improve the security and resilience of the globalised economy [42].

7 Conclusion

Security for the Internet of Things (IoT) enables a more reliable and secure connected world. AI algorithms and methodologies enhance threat identification, anomaly detection, and intelligent decision-making processes. Real-time monitoring, predictive analysis and adaptive reaction mechanisms make AI-based IoT security solutions superior to conventional ones. IoT security is more resilient and effective because of AI's ability to continuously learn from evolving threats and data. However, addressing privacy concerns, algorithmic biases and robust data protection protocols is crucial for responsible adoption. Due to scalability and interoperability difficulties, ongoing research and standardisation operations are also required to assure the long-term sustainability and effectiveness of AI-based IoT security solutions. AI-based IoT solutions provide concrete advantages in terms of efficacy and cost-savings, from improved security. Organisations may streamline operations, better manage resources and get a better return on investment by automating security processes and lowering their need on human interaction. The ethical and privacy issues related to AI-based IoT security must be addressed. To promote confidence and broad use of these technologies, it is crucial to strike the correct balance between data collecting for security objectives and individual privacy. Thus, the concerns around privacy and ethics in AI-based IoT security must be addressed. It's critical to strike the correct balance between data collecting for security objectives and individual privacy in order to promote trust and broad adoption of these technologies.

References

1. Chander B, Pal S, De D, Buyya R (2022) Artificial intelligence-based internet of things for industry 5.0. In: Artificial intelligence-based internet of things systems, pp 3–45
2. Khilar R et al (2022) Artificial intelligence-based security protocols to resist attacks in internet of things. Wirel Commun Mob Comput 2022
3. Lin J, Yu W, Zhang N, Yang X, Zhang H, Zhao W (2017) A survey on internet of things: architecture, enabling technologies, security and privacy, and applications. IEEE Internet Things J 4(5):1125–1142
4. Tewari A, Gupta BB (2020) Security, privacy and trust of different layers in internet-of-things (IoTs) framework. Futur Gener Comput Syst 108:909–920
5. Yang Y, Wu L, Yin G, Li L, Zhao H (2017) A survey on security and privacy issues in internet-of-things. IEEE Internet Things J 4(5):1250–1258
6. Awotunde JB, Misra S (2022) Feature extraction and artificial intelligence-based intrusion detection model for a secure internet of things networks. In: Illumination of artificial intelligence in cybersecurity and forensics. Springer, pp 21–44
7. Mishra S, Tyagi AK (2022) The role of machine learning techniques in internet of things-based cloud applications. In: Artificial intelligence-based internet of things systems, pp 105–135
8. Gao X, Li Q, Liu F (2021) Research on the new normal technology and application of artificial intelligence in the internet of things. J Phys: Conf Ser 42062
9. Prasad M, Tripathi S, Dahal K (2020) Unsupervised feature selection and cluster center initialization based arbitrary shaped clusters for intrusion detection. Comput Secur 99:102062
10. Pal S, De D, Buyya R (2022) Artificial intelligence-based internet of things systems. Springer
11. Dulhare UN, Rasool S Harnessing the IoT-based activity trackers and sensors for cognitive assistance in COVID-19. In: Principles and applications of socio-cognitive and affective computing. IGI Publisher, pp 64–92
12. Geetha S, Dulhare UN, Sivatha Sindhu SS (2018) Intrusion detection using NBHoeffding rule based decision tree for wireless sensor networks. In: 2018 second international conference on advances in electronics, computers and communications (ICAECC), Bangalore, India, pp 1–5. https://doi.org/10.1109/ICAECC.2018.8479483
13. Dulhare UN, Rasool S IOT evolution and security challenges in cyber space: IOT security. In: Countering cyber attacks and preserving the integrity and availability of critical systems. IGI Publisher, pp 99–127
14. Seng KP, Ang LM, Ngharamike E (2022) Artificial intelligence internet of things: a new paradigm of distributed sensor networks. Int J Distrib Sens Netw 18(3):15501477211062836
15. Sasikumar A, Ravi L, Kotecha K, Saini JR, Varadarajan V, Subramaniyaswamy V (2022) Sustainable smart industry: a secure and energy efficient consensus mechanism for artificial intelligence enabled industrial internet of things. Comput Intell Neurosci 2022
16. Rasool S, Dulhare UN (2018) Evolution of indian railways through IOT. In: Innovative applications of big data in the railway industry, chap 12. IGI Global, pp 269–290
17. Nirmala P, Ramesh S, Tamilselvi M, Ramkumar G, Anitha G (2022) An artificial intelligence enabled smart industrial automation system based on internet of things assistance. In: 2022 international conference on advances in computing, communication and applied informatics (ACCAI), pp 1–6
18. Krishnan R, Rajpurkar P, Topol EJ (2022) Self-supervised learning in medicine and healthcare. Nat Biomed Eng 1–7
19. Shah D, Singh A, Prasad SS (2022) Sentimental analysis using supervised learning algorithms. In: 2022 3rd international conference on computation, automation and knowledge management (ICCAKM), pp 1–6
20. Rajapaksha S, Kalutarage H, Al-Kadri MO, Petrovski A, Madzudzo G, Cheah M (2023) Ai-based intrusion detection systems for in-vehicle networks: a survey. ACM Comput Surv 55(11):1–40
21. Donalek C (2011) Supervised and unsupervised learning. In: Astronomy colloquia, USA, p 8

22. Sagar R, Jhaveri R, Borrego C (2020) Applications in security and evasions in machine learning: a survey. Electronics (Basel) 9(1):97
23. Dulhare UN, Ali H (2023) Underwater human detection using faster R-CNN with data augmentation. Mater Today: Proc 80:1940–1945. ISSN 2214-7853
24. Ullah I, Mahmoud QH (2021) Design and development of a deep learning-based model for anomaly detection in IoT networks. IEEE Access 9:103906–103926
25. Houssein EH, Helmy BE, Rezk H, Nassef AM (2021) An enhanced Archimedes optimization algorithm based on local escaping operator and orthogonal learning for PEM fuel cell parameter identification. Eng Appl Artif Intell 103:104309
26. Hassanien AE, Kilany M, Houssein EH, AlQaheri H (2018) Intelligent human emotion recognition based on elephant herding optimization tuned support vector regression. Biomed Signal Process Control 45:182–191
27. Ismaeel AAK, Elshaarawy IA, Houssein EH, Ismail FH, Hassanien AE (2019) Enhanced elephant herding optimization for global optimization. IEEE Access 7:34738–34752
28. Houssein EH, Mahdy MA, Fathy A, Rezk H (2021) A modified marine predator algorithm based on opposition based learning for tracking the global MPP of shaded PV system. Expert Syst Appl 183:115253
29. Hamad A, Houssein EH, Hassanien AE, Fahmy AA (2018) Hybrid grasshopper optimization algorithm and support vector machines for automatic seizure detection in EEG signals. In: The international conference on advanced machine learning technologies and applications (AMLTA2018). Springer International Publishing, pp 82–91
30. Houssein EH, Mahdy MA, Shebl D, Manzoor A, Sarkar R, Mohamed WM (2022) An efficient slime mould algorithm for solving multi-objective optimization problems. Expert Syst Appl 187:115870
31. Houssein EH, Abdelminaam DS, Hassan HN, Al-Sayed MM, Nabil E (2021) A hybrid barnacles mating optimizer algorithm with support vector machines for gene selection of microarray cancer classification. IEEE Access 9:64895–64905
32. Hamad A, Houssein EH, Hassanien AE, Fahmy AA (2016) Feature extraction of epilepsy EEG using discrete wavelet transform. In: 2016 12th international computer engineering conference (ICENCO). IEEE, pp 190–195
33. Shaban H, Houssein EH, Pérez-Cisneros M, Oliva D, Hassan AY, Ismaeel AA, AbdElminaam DS, Deb S, Said M (2021) Identification of parameters in photovoltaic models through a Runge Kutta optimizer. Mathematics 9(18):2313
34. Abdelminaam DS, Said M, Houssein EH (2021) Turbulent flow of water-based optimization using new objective function for parameter extraction of six photovoltaic models. IEEE Access 9:35382–35398
35. Houssein EH, Hassaballah M, Ibrahim IE, AbdElminaam DS, Wazery YM (2022) An automatic arrhythmia classification model based on improved marine predators algorithm and convolutions neural networks. Expert Syst Appl 187:115936
36. Houssein EH, Neggaz N, Hosney ME, Mohamed WM, Hassaballah M (2021) Enhanced Harris hawks optimization with genetic operators for selection chemical descriptors and compounds activities. Neural Comput Appl 33:13601–13618
37. Ahmed MM, Houssein EH, Hassanien AE, Taha A, Hassanien E (2018) Maximizing lifetime of wireless sensor networks based on whale optimization algorithm. In: Proceedings of the international conference on advanced intelligent systems and informatics 2017. Springer International Publishing, pp 724–733
38. Houssein EH, Sayed A (2023) Dynamic candidate solution boosted beluga whale optimization algorithm for biomedical classification. Mathematics 11(3):707
39. Alloghani M, Al-Jumeily D, Mustafina J, Hussain A, Aljaaf AJ (2020) A systematic review on supervised and unsupervised machine learning algorithms for data science. In: Supervised and unsupervised learning for data science, pp 3–21
40. Ahmad Z et al (2021) Anomaly detection using deep neural network for IoT architecture. Appl Sci 11(15):7050

41. de Assis MVO, Carvalho LF, Rodrigues JJPC, Lloret J, Proença ML Jr (2020) Near real-time security system applied to SDN environments in IoT networks using convolutional neural network. Comput Electr Eng 86:106738
42. Zhao Z, Liu H (2007) Spectral feature selection for supervised and unsupervised learning. In: Proceedings of the 24th international conference on machine learning, pp 1151–1157

Chapter 10
Memristors: A Missing Element is a Boon Toward the Development of Neuromorphic Computing and AI

Shubhro Chakrabartty, AlaaDdin Al-Shidaifat, Hanjung Song, and Ayush Kumar Singh

Abstract The advancement and research in memristor technology are proving to be highly beneficial in various fields, particularly in the realm of Artificial Intelligence and in assisting patients with autism disorders. Memristors possess unique capabilities such as simultaneous programming and data storage, emulation of brain cell behavior, and power-independent stability, making them a promising component. The chapter aims to delve into the manufacturing process of memristors and their crossbar structure. It is likely to discuss the utilization of heterojunction layers in the development of nanowire (NW) HfO_2 memristors and acknowledge the device's remarkable endurance for nearly two decades. However, along with the growing research in this field, there are challenges that must be addressed, such as enhancing conductivity and storage capacity. The chapter is also expected to cover the fabrication methods employed to evaluate multi-layered memory stores and the investigation of the device's surface morphology using the FEGSEM technique. Overall, the chapter will provide insights into the potential of memristors and their role in advancing various fields, particularly in the realm of neuromorphic computing. The chapter appears to discuss the properties and applications of two types of memristor devices: TiO_2 nanoparticles and Gd-doped HfO_2 nanoparticles. Through X-ray diffraction analysis,

S. Chakrabartty (✉)
School of Computer Science Engineering and Applications, DY Patil International University, Pune, India
e-mail: shubhro.chakrabartty87@gmail.com

A. Al-Shidaifat · H. Song
Department of Nanoscience and Engineering, Inje University, Gimhae-si, Korea
e-mail: alaaddin@inje.ac.kr

H. Song
e-mail: hjsong@inje.ac.kr

A. K. Singh
Department of Electrical and Electronics Engineering, National Institute of Technology Goa, Goa, India
e-mail: ayushsingh@nitgoa.ac.in

U. N. Dulhare and E. H. Houssein (eds.), *Machine Learning and Metaheuristics: Methods and Analysis*, Algorithms for Intelligent Systems,
https://doi.org/10.1007/978-981-99-6645-5_10

it has been observed that TiO_2 nanoparticles possess a crystalline nature, and the presence of oxygen vacancies increases with heat application. Various studies have examined the Gd-doped HfO_2 nanoparticles memristor device, including annealing, energy-dispersive X-ray, and photoluminescence examinations. The findings indicate device was subjected to an annealing process at a temperature of 600 °C degrees and demonstrates a notable enhancement in the terms of leakage current and interference state density factor. Additionally, the chapter delves into the potential applications of memristor devices in advanced neurocomputing, specifically as synaptic simulators connecting pre- and postsynaptic neurons. The chapter examines the utilization of memristor devices incorporating TiO_2 nanoparticles with two different Ag and Au electrodes by highlighting their properties and applications. The nanoparticle morphology verified using atomic force microscopy (AFM). Memristor behavior does indeed resemble that of a synaptic emulator, as it can modify its resistance based on the electrical signal history passing through it. This characteristic positions memristors as a promising candidate for constructing artificial neural networks and other neuromorphic computing systems. Regarding the comparison between memristors with Ag and Au contacts, measurements of capacitor–voltage curves have shown that Ag-based memristors exhibit superior charge-storage capabilities. This is attributed to Ag's lower work function compared to Au, enabling efficient electron exchange with other materials. As a result, Ag serves as an excellent electrode material for memristors, facilitating efficient charge storage and release. Memristor devices with Ag electrodes have been proposed as pre- and post-neuronal systems and have demonstrated comparable performance to Au electrode memristors. Additionally, these devices may offer significant advantages in the field of neuromorphic computing by potentially enhancing performance and energy efficiency. However, further research is necessary to fully comprehend the capabilities and limitations of Ag electrode memristors for neuromorphic computing applications.

Acronyms

NW	Nanowires
AFM	Atomic Force Microscopy
I–V	Current–Voltage
M	Memristance
Au	Gold
MIM	Metal–Insulator–Metal
TiO_2	Titanium Dioxide
Ag	Silver
RSM	Resistive Switch Non-volatile Memory
CMOS	Complementary Metal–Oxide–Semiconductor
Si	Silicon
Ge	Germanium
RRAM	Resistive Random-Access Memory

FEGSEM	Field Emission Gun Scanning Electron Microscopy
EDX	Energy-Dispersive X-ray Spectroscopy
GPU	Graphics Process Technology
NVM	Non-volatile Memory
HfO_2	Hafnium Oxide
ATT	Angle Toppling Technique
TF	Thin Film
PL	Photoluminescence
FEGSEM	Field Emission Gun Scanning Electron Microscopy

1 Introduction

Memristors, a passive electronic component, possess a remarkable ability to store the charge that has previously flowed through them. This distinctive characteristic positions memristors as highly promising components for various applications, including non-volatile memory and neuromorphic computing. Unlike other components, the resistance of a memristor can be manipulated by applying a voltage, and even after the voltage is removed, the resistance remains unchanged. This exceptional quality allows memristors to retain their resistance state even in the absence of power, making them exceptionally well suited for memory applications.

Additionally, the interconnectedness among the four fundamental two-terminal circuit elements (resistor, capacitor, inductor, and memristor) is widely recognized as Chua's circuit, honoring the memristor's discoverer, Leon Ong Chua. Chua's circuit represents a nonlinear system that exhibits intricate dynamics, symbolizing the intricate relationships between these fundamental elements in circuit theory (Fig. 1).

The behavior of a memristor in terms of electrical signal can be effectively depicted through its current–voltage (I–V) curve. This curve demonstrates the correlation between the current passing through the memristor and the voltage applied across it. Notably, the I–V curve of a memristor showcases nonlinearity, indicating that its resistance fluctuates in response to variations in the applied voltage or current. When a voltage is administered to a memristor, its resistance undergoes modification, and the current flowing through it becomes proportional to the rate at which the voltage changes. This characteristic is commonly referred to as "memristance," which quantifies the alteration in resistance caused by the historical pattern of voltage or current imposed on the device. Depending on the direction of the applied voltage or current, the memristance can exhibit either a positive or negative value (Fig. 2).

There is currently a significant interest in the search for the next generation of universal memory, capable of replacing traditional forms of memory such as SRAM, DRAM, and non-volatile storage [1]. In recent times, a novel device called the memristor has emerged as the first practical realization of the long-awaited fundamental circuit component. In the year 1971, Prof. Chua had initial theoretical concept

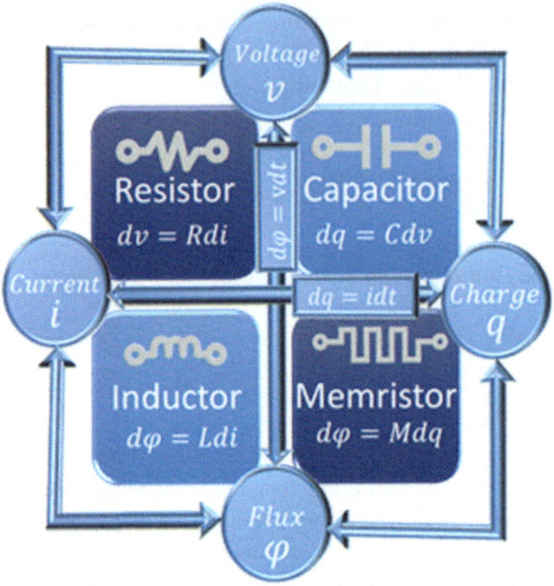

Fig. 1 The fundamental components of electronics

Fig. 2 Inherent non-volatile memristor

of the memristor. In late 2008, a team led by Williams et al. [2] introduced the electrical characteristics of nanoscale TiO_2 memristor, igniting significant interest in nanoscale devices [3–5].

The essential components of a circuit are resistance, capacitance, and inductance. Resistance relates voltage and current ($dv = R.di$), capacitance relates charge and voltage ($dq = C.dv$), and inductance relates flux and current ($d\varphi = L.di$), as established by [6–8]. However, there was a missing link between flux and charge. Figure 1 depicts this absent connection, known as memristance (M) [9]. Memristance $M(q)$ is defined by the Eq. (1) [10, 11].

$$M(q) = d\varphi/dq \tag{1}$$

Inductance, on the other hand, is represented as

$$W(\varphi) = dq/d\varphi \tag{2}$$

From Eqs. (1) and (2), it can be observed that the reciprocal of memductance corresponds to memristance. Additionally, it is evident that

$$v = M(q).i \quad ; \quad i = W(\varphi).v$$

The voltage–current characteristics displayed by memristance, which is similar to resistance, especially in linear scenarios, and is commonly denoted as the resistance.

Hence, memristance shares the same unit (Ohm) as resistance, and $M(q)$ can be regarded as a resistance that can be manipulated or regulated by the charge, and similarly the memductance is quantified in conductance units. It should be noted that memristor-based memory is unsuitable for conventional memories like CMOS memory due to its reliance on resistance values, which lack bistable properties [12, 13].

The memristor possesses inherent memory capabilities and has shown significant progress in the development of advanced neurocomputing paradigms over the years. These advanced systems utilize memristor devices to serve as junction emulators for facilitating information processing between presynaptic and postsynaptic neurons [14]. Similar to biological neurons, communication between artificial neurons occurs through synapses, where the strength of the transmitted signal is associated with the junction weight. In 1943 [15], McCulloch and Pitts introduced the first time-independent model of artificial neurons, demonstrating the intricate patterns of the brain using interconnected basic units known as neurons [16]. Later, in 1952, Hodgkin and Huxley implemented an electronic circuit to illustrate the electrical properties of neurons [17]. Since then, researchers have made remarkable advancements in emulating neurons and developing synthetic paradigms [18].

Illustrated in Fig. 1 is a comprehensive diagram of neuro-electronics computing, showcasing the transmission of signals from presynaptic neurons to the receptors of postsynaptic neurons via memristor devices [19]. These memristors are composed of conductor and gold (Au) electrodes, offering a promising neural interface approach for establishing connections between pre- and postsynaptic neurons and opening possibilities for innovative neural prostheses [20]. Functioning as two-terminal devices, memristors serve as essential circuit elements with metal–insulator–metal (MIM) structures, capable of retaining their resistance state even after the applied switching voltage or current is removed [21].

In recent years, extensive research has been directed toward metal memristor oxides, particularly ZnO, NiO, and CuO [2, 22–25], and particularly titanium dioxide (TiO_2), due to their rapid resistive switching speed and versatile applications in various fields, including solar cells, gas sensors, and memristors [26]. The metal–oxide sandwich structure shows promise for non-volatile memory devices which have been investigated extensively [27]. Resistive switching takes place at the interface between the metal electrode and the oxide region, creating a conductive path known as the interface path. The presence of a continuous flow of oxygen vacancies near the interface layer reduces the digital barrier in the low resistance state [28–32].

Memristors exhibit desirable characteristics such as minimum power consumption, simple composition, and compatibility with existing processing techniques. These attributes position them as excellent candidates for novel logical devices and artificial neuromorphic systems with superior performance [33, 34]. In this study, deposition technique in physical process was employed to design the top contact with two different electrode silver (Ag) and gold (Au) [35–37]. Comparative studies of these memristor layers were conducted to explore their charge-storage capabilities under various sweeping voltage conditions, with a particular focus on their suitability for neuromorphic computing applications [38–40].

In recent times, resistive switch non-volatile memory (RSM) devices have opened new avenues of analysis in the semiconductor industry, offering a wide range of potential mechanisms [41–45]. Researchers are particularly intrigued by the appealing qualities of these devices, as they possess a simple composition and the intriguing characteristics of these devices have captivated researchers, as they possess a straightforward composition and are seamlessly well suited for CMOS devices. They can be seamlessly integrated into the complementary metal–oxide–semiconductor (CMOS) fabrication process, as indicated by previous studies [46, 47]. While two-terminal devices like memristors have been extensively explored using oxides, metal oxides, organic polymers, and solid electrolytes, these hold promise as next-generation solutions for scaling down existing semiconductor memories such as flash [48–51]. However, the predominant use of silicon (Si) and germanium (Ge) semiconductors in memory device fabrication, due to their sensitivity to the thermal process, has limited the performance of storage devices [52–54].

Oxides, on the other hand, are versatile compounds with applications in various sectors such as photocatalysts, solar cells, supercapacitors, and energy storage devices. In recent years, titanium dioxide (TiO_2) has emerged as a highly promising candidate for diverse applications including resistive random-access memory (RRAM), memristors, reconfigurable analog integrated circuits, and neuromorphic computing [55–60]. Many research teams have extensively utilized metal nanoparticles like platinum, copper, and gold on TiO_2 films to enhance the oxygen-related defect states for memory storage applications [60–63]. However, in the long run, this approach fell short of meeting the requirements. To preserve the performance of the parent device, memory with a compound-thin film coated by metal oxide nanoparticles is now replacing the use of metal nanoparticles [64].

This study focuses on the preparation of two layers of TiO_2 thin film/TiO_2 nanoparticles on an n-type silicon substrate with a conductive layer of atomic number 47 [65, 66]. The impact of the conductive layer is investigated through I–V, switching, and C–V measurements. Additionally, this work demonstrates that the memory switch can be operated in various TiO_2 (amorphous)/TiO_2 nanoparticle (crystalline) geometries. The characterization of these layers is performed using field emission gun scanning electron microscopy (FEGSEM), energy-dispersive X-ray spectroscopy (EDX), photoluminescence, Raman spectroscopy, and X-ray diffraction, with detailed descriptions provided in Sections A, B, and C, respectively [67, 68].

In recent times, there has been significant interest in using Memory Process Systems (MPU) and Graphics Process Technology (GPU) for the identification of human tissue layers. The various edge detection techniques, including Sobel, Laplace, Canny, and others, are being utilized [69].

MPU involves non-volatile devices (NVM) that play a crucial role in artificial tissue layer and brain computing applications. Non-volatile memory (NVM) offers desirable features like high density, flexibility, low power consumption, and stability, making it suitable for resistance in various applications [70–72]. Two examples of NVM devices are Random-Access Memory (RRAM) devices and memristor devices. These devices possess inherent memory, enabling them to retain data and facilitate parallel computing paradigms. In 1962, Hickmott reported the resistive switch phenomena in materials like SiO_x, Al_2O_3, Ta_2O_5, ZrO_2, and TiO_2. Further exploration has been conducted on various transition metal oxides to enhance resistive switch performance [73, 74].

Hafnia, a recently discovered superior non-volatile device, in conjunction with the complementary metal–oxide-semiconductor (CMOS) device, showcases remarkable characteristics, including high stability, a wide bandgap, excellent electrical properties, high dielectric permittivity (ranging from $\varepsilon = 15$ to 27), high durability, and imposing overall performance [75].

The non-volatile device, in conjunction with the CMOS device, showcases remarkable characteristics, including high stability, a wide bandgap, excellent electrical properties, high dielectric permittivity (ranging from $\varepsilon = 15$ to 27), high endurance, and impressive overall performance [75]. Researchers have developed similar devices using different materials and electrode combinations. The necessary investigation was conducted by R.F. et al. on titanium nitride/titanium/Hafnia/titanium nitride device, while C. et al. [76] had shown the performance of HfO_2 with Ti, Pt, TiN electrodes for the resistive switch. The memory performance in Tantalum/Hafnia/Platinum combination was shown by Jiang et al. However, challenges such as excessive leakage current, viability concerns, and scalability issues continue to be prevalent in this research domain. Nevertheless, some teams have made efficient efforts to address these memory leakage problems in the mathematician filter. This draft proposes alternative systems to overcome this challenge [77–80].

Furthermore, various attempts have been made to improve retention characteristics by incorporating metal nanocrystals such as nickel, iridium, platinum, and gold with HfO_2 films [81, 82]. Currently, dopants like lutetium, aluminum, and zirconium are being added to HfO_2.

Nevertheless, there are still challenges to address, including the formation of oxygen-related defect states, the requirement for higher switching voltages, and the need to achieve optimal dopant concentrations to enhance non-volatile memory (NVM) devices [83]. To overcome these obstacles and enhance endurance by increasing oxygen defect-related traps, researchers have fabricated Gd-doped HfO_2 nanoparticles using deposition process in vacuum physical mode [84–86]. In addition to the previously mentioned studies, crossbar memory exhibits notable advantages in parallel computing, such as increased storage capacity and improved processing speed. It has good capacity and faster processing, which can be beneficial for image-processing applications. Different fabrication methods, such as atomic layer deposition, sol–gel method, and RF tube sputtering, have been incorporated to grow HfO_2 films [87].

In this context, a three × three non-volatile memory-based crossbars employing the Sobel filter are designed to evaluate its effectiveness and strength in determining the quality of neuro-artificial computing. The achievement of optimal charge storage in non-volatile memory (NVM) devices based on Gd-doped HfO_2 nanoparticles is a crucial objective [88–90].

To fabricate Gd-doped HfO_2 nanoparticles on an SiO_x insulating layer, a technique that does not involve chemical vapor deposition and ensures precise control of the body and stability of the angle toppling is utilized. The resulting samples are then exposed to various temperatures and analyzed using field emission gun scanning electron microscopy (FEGSEM, JEOL, JSM-7600F) for characterization purposes.

In 1962, Hickmott conducted a comprehensive investigation into the performance of resistance switches in various materials, such as ZrO_2, TiO_2, SiO_x, and different transition metal oxides. In the present day, there is a growing interest in exploring the potential of Hafnia (HfO_2) as a metallic compound due to its prime bandgap, exceptional electrical properties, extraordinary dielectric permittivity, robustness, low cost, and outstanding performance in creating CMOS devices [91–94].

The basic structure of tissue layer cells consists of five layers: [95] photoreceptor cells, which form the outer boundary, followed by horizontal cells, bipolar cells, amacrine cells, and neural cells, which form the inner boundary. The neural cells are connected to the photoreceptor cells, where the photoreceptor cells serve as processing units. They receive signals in the form of spikes and transfer them to the neural cells [96]. The spike frequency is inversely proportional to the illumination level, meaning it decreases as the illumination level increases, and vice versa. This effect occurs because illumination shifts the threshold of the photoreceptor cells to higher levels. Figure 1 illustrates the cells present in the tissue layer and the sensory system [97]. The field of artificial vision systems is an area of computer science that enables computers and systems to derive meaningful information from digital images, videos, and various visual inputs. In this research paper, we aim to develop the Ag/HfO_2 NPS/Al memristor for the advancement of the artificial vision system.

2 Experimental Details

Our research team has successfully developed an advanced electron beam evaporator system integrated with catalytic free-angle deposition equipment for the precise fabrication of optimized nanostructures, specifically targeting the development of a memristor. This groundbreaking achievement has been recognized through the acquisition of an international Korean patent [98].

The Angle Toppling Technique (ATT) is a method employed in physical vapor deposition, which exploits controlled substrate motion to create nanostructures [99].

Fig. 3 Manufactured angle deposition technique

Originally introduced by Smith, Knorr, and Hoffman in 1959 [100], ATT involves depositing atoms or molecules from a gas phase onto a vacuum substrate at an oblique angle. This results in the formation of columnar film morphology, where the tilt of the columns is determined by the angle of incidence, thereby influencing the film's shadowing and porosity. Unlike conventional deposition methods, the vapor flux trajectory in ATT is not parallel to the substrate surface but instead occurs at oblique angles. In 1999, Liu et al. demonstrated the successful fabrication of a porous film with evenly spaced columns using the ATT technique [101]. By adjusting the angular frequency and position of the substrate relative to the evaporation source, precise control over the film's topography can be achieved. Furthermore, by employing varying frequencies of substrate rotation, it is possible to create thin films with intricate structures. The ATT technique also allows for the growth of multi-component nanostructures by utilizing different source materials [102]. To provide a visual representation of the ATT apparatus, Fig. 3 presents a schematic diagram illustrating its configuration and components.

2.1 Theory Behind the Process

The ATT technique, when applied to flat substrates, involves a nucleation process that results in the formation of dispersed layers composed of columns. In the year 2006, Gall and Kesapragada successfully generated a periodic arrangement of columns by patterning the substrate with seed particles [103]. These patterned sites serve as nucleation points for the columns, effectively preventing growth on the surrounding substrate due to the shadowing effect. Ensuring that the diameter of the individual sites on the substrate closely matches the diameter of the growing structures is crucial to minimize the formation of multiple structures on each site. The extent of the

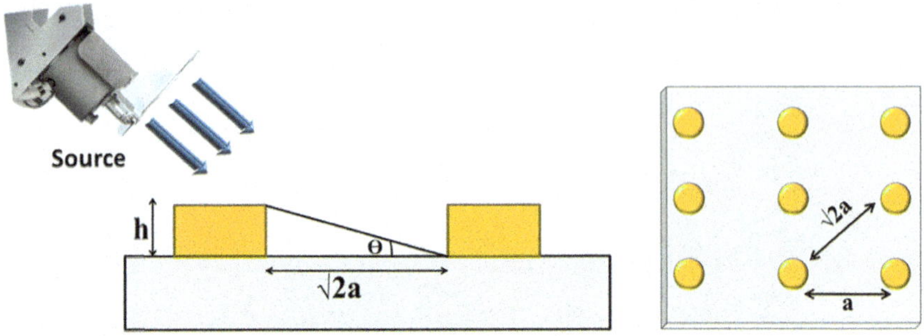

Fig. 4 Substrate patterning and its nucleation sites

shadowing effect is heavily influenced by the spacing between the individual sites on the substrate, as depicted in Fig. 4. The maximum spacing between sites, in relation to a specific deposition angle, significantly impacts the overall outcome which can be determined using the equation $\tan(\theta) = h/\sqrt{2}a$, where "$a$" represents the "lattice constant" of the array.

The Angle Toppling Technique (ATT) has shown great promise as a flexible and powerful method for producing thin films with customized nanostructures, such as nanopillars, slanted posts, zigzag columns, and spirals [104–106]. These nanostructures have a wide range of applications in fields like photonic crystals, magnetic storage media, and optoelectronic devices [107–112]. What sets ATT apart is its compatibility with various materials, allowing for a diverse range of combinations. By adjusting the angle of incidence, the tilt of the columns and the porosity of the film can be controlled through shadowing effects. Rotating the substrate during deposition dramatically transforms the film's morphology, and altering the substrate's angular frequency and adjust its position in relation to the evaporation source that can further manipulate the columns' characteristics. Complex structures can be achieved by employing predefined cycles with different rotation frequencies. Additionally, ATT offers the possibility of developing nanostructures consisting of multiple components by changing material present in the source during deposition, enhancing the functionality of conventional single-component layers.

Figure 5a provides the experimental setup which involves two stepper motors for manipulating the substrate. One motor adjusts the incident angle denoted as "α," while the other controls the azimuthal rotation of the substrate relative to its surface normal. The evaporator beam used in this system has a notable incident angle "α," and the substrate can be rotated at an angle "ϕ" that determines its azimuthal position relative to a chosen reference point. The incoming vapor is represented by the vector F, which consists of two components: a transverse component $F^{\perp} = F \cos\alpha$ and an axial component $F\| = F \sin\alpha$ (Fig. 5b). While the substrate receives vapor from both components, the growth of vertically aligned nanocolumns relies solely on the $F\|$ component, effectively cancelling out the $F^{\perp}$ component (Fig. 5c). In traditional deposition processes, thin films exhibit homogeneity, density, and uniformity. In contrast, ATT films grow as anisotropic and porous structures and random deposition

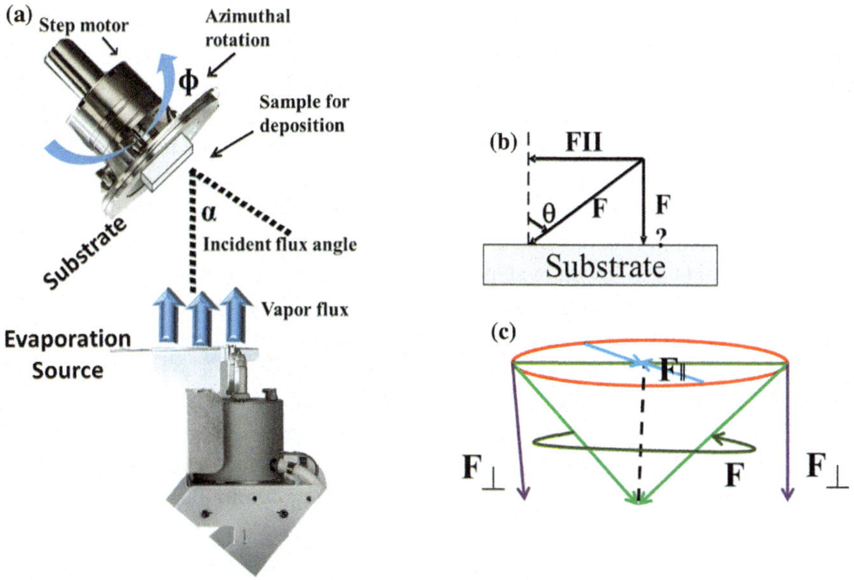

Fig. 5 **a** ATT experimental setup, **b** material and its components, **c** cancellation of the parallel component

of seeds is influenced by the atomic shadowing effect. The diffusivity of atoms that land on the substrate is contingent upon the evaporation rate. When the evaporation rate is decreased, adatoms could diffuse over greater distances prior to encountering additional atoms. This extended diffusion results in an augmented capture radius and an expanded nucleation area for the adatoms.

2.2 Different Patterns by ATT

The morphology of nanostructures in the ATT technique relies heavily on several factors, including the deposition rate, deposition angle, substrate temperature [112], and rotation rate. This unique technique enables the growth of microstructures with diverse shapes, each possessing distinct optical, electrical, and biological properties. By adjusting the ratio of the deposition rate to the rotational rate, the column morphology can be modified to exhibit matchstick, helical, or vertical column structures [99, 113]. The fabrication of column structures using the ATT technique is influenced by the angle of vapor incidence (α), substrate rotation (ϕ), type and direction of rotation, as well as the deposition rate. Both "α" and "ϕ" can be categorized into three types: constant, discrete, and continuous. In the constant category, oblique deposition occurs with the substrate maintaining a fixed angle. In the discrete system, the substrate undergoes periodic changes at a specific angle but remains stationary otherwise. In the continuous system, the substrate moves continuously at the specified angle during deposition. Through the combination of "α" and "β" motions, six common structures can be fabricated using the ATT technique. Here are brief descriptions and illustrative topographies of these basic structures depicted in Fig. 6:

Slanted structure: This structure is obtained when the substrate remains fixed ((α, ϕ), $\alpha = \phi = 0$) during deposition. For isolated posts, α is typically greater than or equal to 70°, as shown in Fig. 6a. Chevron: A series of slanted posts form this structure, as presented in Fig. 6b. One arm of the chevron is deposited with the substrate held at (α, ϕ) [114], followed by depositing an additional arm with the substrate held at (α, $\phi + \pi$). Square spirals belong to the same category as chevrons but involve a $\pi/2$ rotation at discrete intervals. Vertical structure: achieving a glancing angle ($\geq$75°) and continuously rotating the substrate azimuthally at a suitable constant speed during deposition produces this structure (Fig. 6c). The rotation rate plays a crucial role in controlling the morphology of vertical posts [115].

Helical thin films or spring structures: Slower rotation rates are employed to create this type of morphology. High–low stacks: Produced by continuous rotation in ϕ and

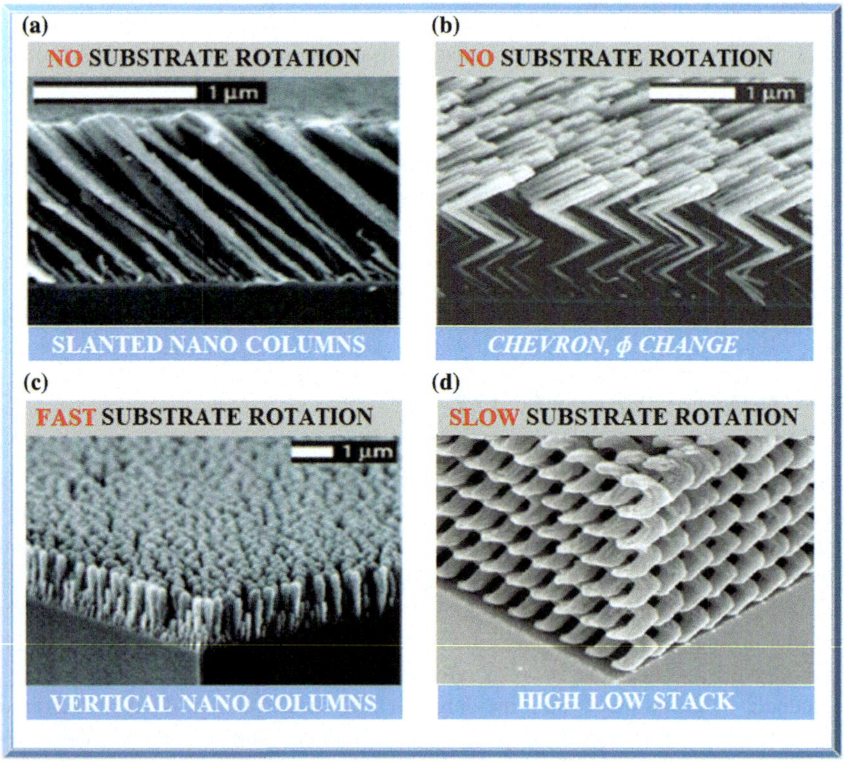

Fig. 6 Pictorial basic topography of ATT image **a** slanted posts, **b** chevrons, **c** vertical post, **d** high–low stack

discrete changes in α, these structures are represented in Fig. 6d. Please refer to Fig. 6 for a visual representation of these structures created using the ATT technique [116].

2.3 Advantages of the ATT Technique

The Angle Toppling Technique (ATT) is a method that utilizes engineering nanoscale thin film microstructures in three dimensions that can be achieved through oblique angle deposition and substrate motion. By precisely controlling the orientation of the substrate with respect to the vapor flux, the columnar microstructure can be customized within a specific range. The extent of this control is influenced by the material properties and deposition conditions. The advantages of the ATT technique are as follows:

1. Microstructural and Nanostructural Control: The ATT technique offers precise control over the microstructure and nanostructure of the fabricated films.
2. Increased Surface Area and Material Flexibility: This technique enhances the surface area of the films and provides flexibility in the choice of materials.
3. Control of Film Porosity: By simply adjusting the incident angle, the porosity of the film can be controlled, allowing for customization based on specific requirements.
4. Growth of Three-Dimensional Nanostructures: The ATT technique enables the straightforward growth of three-dimensional nanostructures, expanding the design possibilities.
5. Modifiable Shape and Alignment of Columns: The shape and in-plane alignment of columns can be easily modified using this technique, providing versatility in structure engineering.
6. Wide Range of Material Options: Since the growth process involves thermal evaporation, there are almost no limitations on the choice of materials, making it applicable to various material systems.
7. Non-toxic Fabrication for Biomedical Research: The ATT technique is particularly advantageous for biomedical research as it ensures non-toxic fabrication, aligning with safety requirements.

These advantages collectively make the ATT technique highly promising for the fabrication of nanostructures.

3 Result and Discussions

The findings presented above demonstrate a new pathway for the development of memristor devices, highlighting the morphological and electrical characteristics of these devices. The results are as follows, indicating promising prospects for hardware neuromorphic computing:

(a) **X-ray Diffraction**: The X-ray diffraction analysis reveals that the grown titanium dioxide (TiO_2) thin film (TF) is amorphous, while the titanium dioxide (TiO_2) nanoparticles (NPs) exhibit a crystalline nature.
(b) **Photoluminescence (PL) and Raman Analysis**: The TiO_2 NPs exhibit a higher photoluminescence (PL) intensity compared to the TiO_2 TF. This enhancement is attributed to the annealing process performed after depositing the TF layer. The annealing process also leads to an increased presence of defect traps between the TiO_2 NPs and TiO_2 TF. Additionally, the Raman scattering intensity analysis indicates higher intensity in the TiO_2 NPs compared to the TiO_2 TF.
(c) **Structural Analysis and Element Mapping**: The growth of both the thin film and nanoparticles is characterized through Field Emission Gun Scanning Electron Microscopy (FEGSEM) and Atomic Force Microscopy (AFM). Furthermore, element mapping and Energy-Dispersive X-Ray Analysis (EDX) confirm the presence of specific elements in the sample.

These findings provide valuable insights into the properties and behavior of the developed memristor devices, paving the way for further advancements in hardware neuromorphic computing (Fig. 7).

(a) **Electrical Characteristics**: The electrical characteristics of the developed devices exhibit distinct I–V (current–voltage) behavior, showcasing hysteresis patterns depending on the electrodes used. The devices demonstrate resistive switching under different voltage conditions, revealing a negative differential resistance (NDR) behavior. Notably, the device with Ag (silver) contact exhibits more pronounced NDR and enhanced charge retention, which holds significant commercial potential.
(b) **Endurance**: The manufactured memristor devices demonstrate stable endurance over extended operation cycles. The devices exhibit the generation of oxygen vacancies and the formation of a strong conductive path beneath the device structure. This approach is crucial for emulating the artificial linkage or connection between presynaptic and postsynaptic neurons, mimicking synaptic functionality.
(c) **Applications**: The crossbar arrays formed by the grown memristors show promise in developing artificial synaptic emulators. These arrays, acting as artificial neurons, can replicate the stochastic behavior of the brain. Experimental training using three conventional letters over 18 iterations resulted in increased membrane potential, eventually achieving a predefined threshold. Such applications highlight the potential of these memristor devices in neuromorphic computing and cognitive systems (Fig. 8).

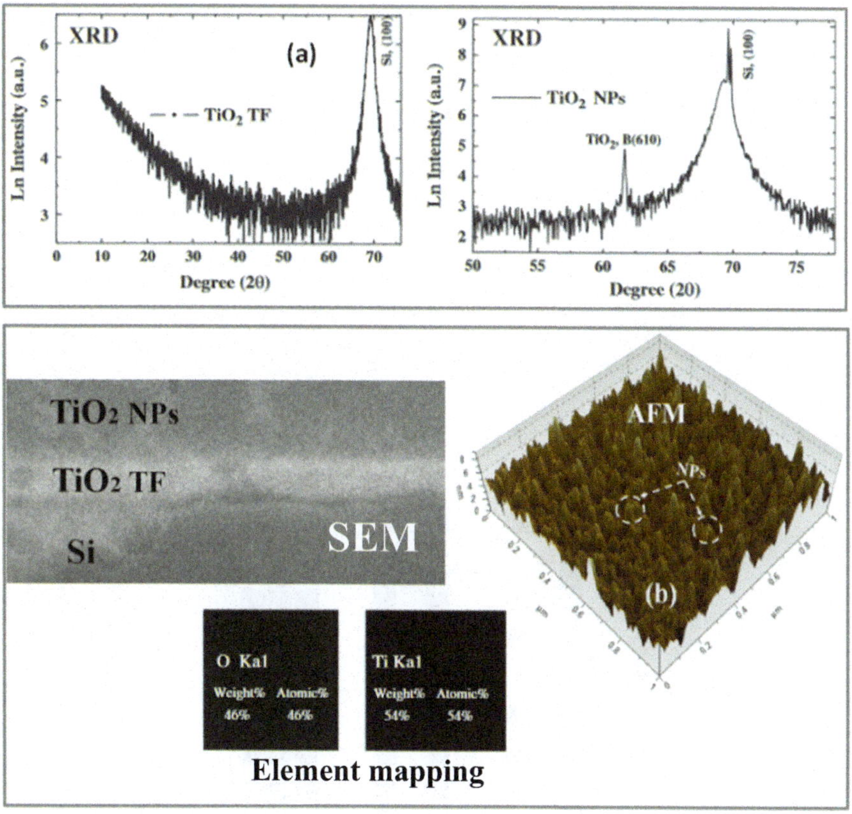

Fig. 7 **a** XRD analysis of both the devices and **b** morphological characteristics of the grown sample with element mapping [40]

4 Conclusions

The findings of this study demonstrate the developmental trajectory of manufacturing memristor devices using the Catalytic Free Angle Toppling Technique (ATT) process. It was observed that memristors were created using various nanomaterials. These manufactured memristors play a crucial role in the advancement of hardware neuromorphic systems. The study also sheds light on the emulation of both postsynaptic and presynaptic components.

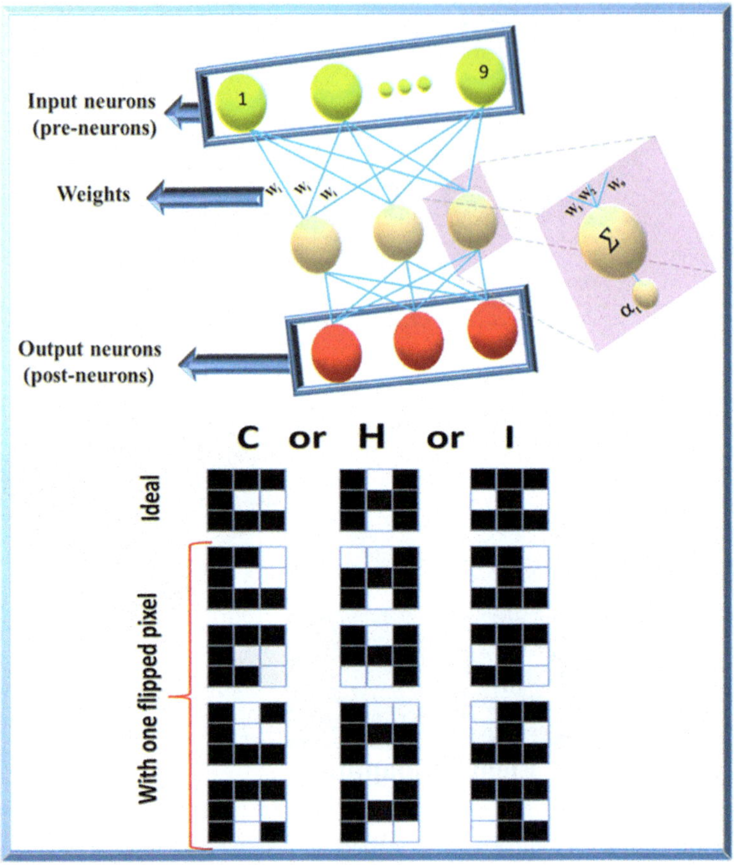

Fig. 8 Three layers of neural network and the set of used patterns [117]

References

1. Chua LO (1971) Memristor—the missing circuit element. IEEE Trans Circuit Theory CT-18:507–519
2. Strukov DB, Snider GS, Stewart DR, Williams RS (2008) The missing memristor found. Nat Lett 453
3. Strukov DB, Snider GS, Stewart DR, Williams RS (2008) The missing memristor found. Nature 453:80–83
4. Strukov DB, Borghetti JL, Williams RS (2009) Coupled ionic and electronic transport model of thin-film semiconductor memristive behavior. Wiley Inter-Science, no 1–6
5. Yang JJ, Picket MD, Li XM, Ohlberg DAA, Stewart DR, Williams RS (2008) Memristive switching mechanism for metal/oxide/metal nanodevices. Nat Nanotechnol 3:429–433
6. Chua LO (1980) Device modeling via basic nonlinear circuit elements. IEEE Trans Circuits Syst CAS-27:1014–1044
7. Chua L (1984) Nonlinear circuits. IEEE Trans Circuit Theory CAS-31(1):69–87
8. Kumar U (1995) A detailed analytical study of non-linear semiconductor device modeling. Act Passiv Electron Compon 18:211–245

9. Chua L (2003) Nonlinear circuit foundations for nanodevices, part I: the four-element torus. Proc IEEE 91(11):1830–1859
10. Ventra MD, Pershin YV, Chua LO (2009) Circuit elements with memory: memristors, memcapacitors, and meminductors. Proc IEEE 1717–1724
11. Itoh M, Chua L (2008) Memristor oscillators. J Bifurc Chaos 18(11)
12. Wang FY (2008) Memristor for introductory physics. Preprint arXiv:0808.0286
13. Ho Y, Huang G, Li P (2009) Nonvolatile memristor memory: device characteristics and design implications. In: IEEE/ACM international conference on computer aid design (ICCAD), Nov 2009
14. Al-Shidaifat A, Kumar S, Song H (2017) Voltage regulation of retina neuron model with dynamic feedback for biological acquisition image. Biochip J 11:300–307
15. Zhang Q, Yu H, Barbiero M, Wang B, Gu M (2019) Artificial neural networks enabled by nanophotonics. Light Sci Appl 8:42
16. McCulloch WS, Pitts W (1943) A logical calculus of the ideas immanent in nervous activity. Bull Math Biophys 5:115–133
17. Hodgkin AL, Huxley AF (1952) Currents carried by sodium and potassium ions through the membrane of the giant axon of Loligo. J Appl Physiol 116:449–472
18. Sun W, Gao B, Chi M, Xia Q, Yang JJ, Qian H, Wu H (2019) Understanding memristive switching via in situ characterization and device modeling. Nat Commun 10:3453
19. Jo S, Sun W, Kim B, Kim S, Park J, Shin H (2019) Memristor neural network training with clock synchronous neuromorphic system. Micromachines 10:384
20. Upadhyay NK, Jiang H, Wang Z, Asapu S, Xia Q, Joshua YJ (2019) Emerging memory devices for neuromorphic computing. Adv Mater Technol 4:1800589
21. Hasegawa T, Ohno T, Terabe K, Tsuruoka T, Nakayama T, Gimzewski JK, Aono M (2010) Learning abilities achieved by a single solid-state atomic switch. Adv Mater 22:1831–1834
22. Choi H, Jung H, Lee J, Yoon J, Park J, Seong D-J, Lee W, Hasan M, Jung G-Y, Hwang H (2009) An electrically modifiable synapse array of resistive switching memory. Nanotechnology 20:345201
23. Yang JJ, Kobayashi NP, Strachan JP, Zhang MX, Ohlberg DAA, Pickett MD, Li Z, Medeiros-Ribeiro G, Williams RS (2011) Dopant control by atomic layer deposition in oxide films for memristive switches. Chem Mater 23:123–125
24. Umar A (2009) Growth of comb-like ZnO nanostructures for dye-sensitized solar cells applications. Nanoscale Res Lett 4:1004–1008
25. Hu SG, Liu Y, Chen TP, Liu Z, Yu Q, Deng LJ, Yin Y, Hosaka S (2013) Emulating the Ebbinghaus forgetting curve of the human brain with a NiO-based memristor. Appl Phys Lett 103:133701
26. Fan Z, Fan X, Li A, Dong L (2013) In situ forming, characterization, and transduction of nanowire memristors. Nanoscale 5:12310–12315
27. Waser R, Dittmann R, Staikov G, Szot K (2009) Redox-based resistive switching memories—nanoionic mechanisms, prospects, and challenges. Adv Mater 21:2632–2663
28. Seok JY, Kim GH, Kim JH, Kim UK, Chung YJ, Song SJ, Yoon JH, Yoon KJ, Lee MH, Kim KM et al (2012) Resistive switching in TiO_2 thin films using the semiconducting In-Ga-Zn-O electrode. IEEE Electron Device Lett 33:582–584
29. Kwon D-H, Kim KM, Jang JH, Jeon JM, Lee MH, Kim GH, Li X-S, Park G-S, Lee B, Han S et al (2010) Atomic structure of conducting nanofilaments in TiO_2 resistive switching memory. Nat Nanotechnol 5:148
30. McFarland EW, Tang J (2003) A photovoltaic device structure based on internal electron emission. Nature 421:616–618
31. Kim H-S, Lee J-W, Yantara N, Boix PP, Kulkarni SA, Mhaisalkar S, Grätzel M, Park N-G (2013) High-efficiency solid-state sensitized solar cell-based on submicrometer rutile TiO_2 nanorod and $CH_3NH_3PbI_3$ perovskite sensitizer. Nano Lett 13:2412–2417
32. Kim I-D, Rothschild A, Lee BH, Kim DY, Jo SM, Tuller HL (2006) Ultrasensitive chemiresistors based on electrospun TiO_2 nanofibers. Nano Lett 6:2009–2013
33. (2018) The memristor revisited. Nat Electron 1:261

34. Kolka Z, Biolek D, Biolkova V (2015) Improved model of TiO_2 memristor. Radioengineering 24:378–383
35. Sawa A (2008) Resistive switching in transition metal oxides. Mater Today 11:28–36
36. Yang JJ, Pickett MD, Li X, Ohlberg DAA, Stewart DR, Williams RS (2008) Memristive switching mechanism for metal/oxide/metal nanodevices. Nat Nanotechnol 3:429
37. Strachan JP, Pickett MD, Yang JJ, Aloni S, David Kilcoyne AL, Medeiros-Ribeiro G, Stanley Williams R (2010) Direct identification of the conducting channels in a functioning memristive device. Adv Mater 22:3573–3577
38. Borghetti J, Snider GS, Kuekes PJ, Yang JJ, Stewart DR, Williams RS (2010) 'Memristive' switches enable 'stateful' logic operations via material implication. Nature 464:873–876
39. Zhu LQ, Wan CJ, Guo LQ, Shi Y, Wan Q (2014) Artificial synapse network on inorganic proton conductor for neuromorphic systems. Nat Commun 5:3158
40. Chakrabartty S, Kumar S, Song H, Jeon M (2018) Ag/TiO_2 NPs/TiO_2 TF/Si based non-volatile memristor device for neuromorphic computing applications. J Nanosci Nanotechnol 18:7912–7916
41. Garcia V, Fusil S, Bouzehouane K, Enouz-Vedrenne S, Mathur N, Barthelemy A, Bibes M (2009) Nature 460:81
42. Shang S, Shi L, Sun JR, Shen BG, Zhuge F, Li RW, Zhao YG (2010) Appl Phys Lett 96:072103
43. Li Y, Sinitskii A, Tour JM (2008) Nat Mater 7:966
44. Inoue H, Yasuda S, Akinaga H, Takagi H (2008) Phys Rev B 77:035105
45. Shen W, Dittmann R, Breuer U, Waser R (2008) Appl Phys Lett 93:222102
46. Kim DK, Suh DS, Park J (2010) IEEE Electron Device Lett 31:600
47. Song S, Cho B, Kim TW, Ji Y, Jo M, Wang G, Choe M, Kahng YH, Hwang H, Lee T (2010) Adv Mater 22:5048
48. Jung R, Lee MJ, Seo S, Kim DC, Park GS, Kim K, Ahn S, Park Y, Yoo IK, Kim JS, Park BH (2008) Appl Phys Lett 93:042115
49. Gao B, Kang J, Liu L, Liu X, Yu B (2011) Appl Phys Lett 98:232108
50. Jeong DS, Thomas R, Katiyar RS, Scott JF, Kohlstedt H, Petraru A, Hwang CS (2012) Rep Prog Phys 75:076502
51. Puglisi RA, Lombardo S, Corso D, Crupi I, Nicotra G, Perniola L, De Salvo B, Gerardi C (2006) J Appl Phys 100:86704
52. Ng TH, Ho V, Teo LW, Tay MS, Koh BH, Chim WK, Choi WK, Du AY, Tung CH (2004) Thin Solid Films 46:462
53. Shi Y, Saito K, Ishikuro H, Hiramoto T (1998) J Appl Phys 84:2358
54. Friedrich JM, Bavykin DV, Walsh FC (2006) Adv Mater 18:2807
55. Kim SK, Kim KM, Jeong DS, Jeon W, Yoon KJ, Hwang CS (2013) J Mater Res 28:313
56. Choi BJ, Jeong DS, Kim SK, Rohde C, Choi S, Oh JH, Kim HJ, Hwang CS, Szot K, Waser R, Reichenberg B, Tiedke S (2005) J Appl Phys 98:033715
57. Jeong DS, Schroeder H, Breuer U, Waser R (2008) J Appl Phys 104:123716
58. Yang JJ, Pickett MD, Li X, Ohlberg DAA, Stewart DR, Williams RS (2008) Nat Nanotechnol 3:429
59. Strukov DB, Snider GS, Stewart DR, Williams RS (2008) Nature 453:80
60. Kim KM, Choi BJ, Shin YC, Choi S, Hwang CS (2007) Appl Phys Lett 91:012907
61. Kim S, Jeong HY, Kim SK, Choi S-Y, Lee KJ (2011) Nano Lett 11:5438
62. Song SJ, Seok JY, Yoon JH, Kim KM, Kim GH, Lee MH, Hwang CS (2013) Sci Rep 3:3443
63. Lee C, Kim I, Shin H, Kim S, Cho J (2010) Nanotechnology 21:185704
64. Choudhury B, Dey M, Choudhury A (2013) Int Nano Lett 3:25
65. Sweeney SW, Roseman G, Deming CP, Wang N, Nguyen TA, Millhauser GL, Chen S (2016) Int J Hydrogen Energy 41:18005
66. Hu C, McDaniel MD, Posadas A, Demkov AA, Ekerdt JG, Yu ET (2014) Nano Lett 14:4360
67. Róka, Csapó Á, Reskó B, Baranyi P (2007) Edge detection model based on involuntary eye movements of the eye-retina system. Acta Polytech Hung 4(1):31–46
68. Al-Shidaifat, Kumar S, Song H (2017) Voltage regulation of retina neuron model with dynamic feedback for biological acquisition image. BioChip J 11(4):300–307

69. Zidan MA, Strachan JP, Lu WD (2018) The future of electronics based on memristive systems. Nat Electron 1(1):220
70. Zhou F, Chang Y-F, Fowler B, Byun K, Lee JC (2015) Stabilization of multiple resistance levels by current-sweep in SiO_x-based resistive switching memory. Appl Phys Lett 106(6):063508
71. Wong H-SP, Lee H-Y, Yu S, Chen Y-S, Wu Y, Chen P-S, Lee B, Chen FT, Tsai M-J (2012) Metal-oxide RRAM. Proc IEEE 100(6):1951–1970
72. Waser R, Dittmann R, Staikov G, Szot K (2009) Redox-based resistive switching memories–nano ionic mechanisms, prospects, and challenges. Adv Mater 21(25–26):2632–2663
73. Chua L (2011) Resistance switching memories are memristors. Appl Phys A, Solids Surf 102(4): 765–783
74. Dongale TD, Khot KV, Mohite SV, Desai ND, Shinde SS, Patil VL, Vanalkar SA, Moholkar AV, Rajpure KY, Bhosale PN, Patil PS, Gaikwad PK, Kamat RK (2017) Effect of write voltage and frequency on the reliability aspects of memristor-based RRAM. Int Nano Lett 7(3):209–216
75. Hickmott TW (1962) Low-frequency negative resistance in thin anodic oxide films. J Appl Phys 33(9):2669–2682
76. Hong X, Loy DJ, Dananjaya PA, Tan F, Ng C, Lew W (2018) Oxide-based RRAM materials for neuromorphic computing. J Mater Sci 53(12):8720–8746
77. Kumar N, George BPA, Abrahamse H, Parashar V, Ray SS, Ngila JC (2017) A novel approach to low-temperature the synthesis of cubic HfO_2 nanostructures and their cytotoxicity. Sci Rep 7(1):9351
78. Yuan F-Y, Deng N, Shih C-C, Tseng Y-T, Chang T-C, Chang K-C, Wang M-H, Chen W-C, Zheng H-X, Wu H, Qian H, Sze SM (2017) Conduction mechanism and improved endurance in HfO_2-based RRAM with nitridation treatment. Nanoscale Res Lett 12(1):574
79. Rodriguez-Fernandez, Cagli C, Perniola L, Miranda E, Suné J (2018) Characterization of HfO_2-based devices with an indication of second order memristor effects. Microelectron Eng 195:101–106
80. Cabout T, Buckley J, Cagli C, Jousseaume V, Nodin J-F, de Salvo B, Bocquet M, Müller C (2012) Resistance switching variability in HfO_2-based memory structures with different electrodes. In: Proceedings of the EMRS spring meeting, pp 1–27
81. Jiang H, Han L, Lin P, Wang Z, Jang MH, Wu Q, Barnell M, Yang JJ, Xin HL, Xia Q (2016) Sub-10 nm Ta channel responsible for superior performance of a HfO_2 memristor. Sci Rep 6:28525
82. Eshraghian JK, Kang S-M, Baek S, Orchard G, Iu HH-C, Lei W (2019) Analog weights in ReRAM DNN accelerators. In: Proceedings of IEEE International Conference on Artificial Intelligence Circuits and Systems, Mar 2019, pp 267–271
83. Tan Z, Samanta S, Yoo WJ, Lee S (2005) Self-assembly of Ni nanocrystals on HfO_2 and N-assisted Ni confinement for nonvolatile memory application. Appl Phys Lett 86(1):013107
84. Wang TT-J, Chu C-L, Hsieh I-J, Tseng W-S (2010) Formation of iridium nanocrystals with high thermal stability for the applications of a nonvolatile memory device with excellent trapping ability. Appl Phys Lett 97(14):143507
85. Jeff RC Jr, Yun M, Ramalingam B, Lee B, Misra V, Triplett G, Gangopadhyay S (2011) Charge storage characteristics of ultra-small Pt nanoparticle embedded GaAs based non-volatile memory. Appl Phys Lett 99(7):072104
86. Mikhelashvili V, Meyler B, Yoffis S, Salzman J, Garbrecht M, Cohen-Hyams T, Kaplan WD, Eisenstein G (2009) A nonvolatile memory capacitor based on Au nanocrystals with HfO_2 tunneling and blocking layers. Appl Phys Lett 95(2):023104
87. Tromm TCU, Zhang J, Schubert J, Luysberg M, Zander W, Han Q, Meuffels P, Meertens D, Glass S, Bernardy P, Mantl S (2017) Ferroelectricity in Lu doped HfO_2 layers. Appl Phys Lett 111(14):142904
88. Spiga S, Driussi F, Congedo G, Wiemer C, Lamperti A, Cianci E (2018) Sub-1 nm equivalent oxide thickness Al-HfO_2 trapping layer with excellent thermal stability and retention for nonvolatile memory. ACS Appl Nano Mater 1(9):4633–4641

89. Lin C-C, Kuo Y, Zhang S (2014) Nonvolatile memory devices with AlOx embedded Zr- doped HfO_2 high-k gate dielectric stack. J Vac Sci Technol B, Microelectron 32(3):03D116
90. Chattopadhyay R (2001) Surface wear: analysis, treatment, and prevention. ASM International, Cleveland, OH, USA
91. Gope J, Vandana, Batra N, Panigrahi J, Singh R, Maurya KK, Srivastava R, Singh PK (2015) Silicon surface passivation using thin HfO_2 films by atomic layer deposition. Appl Surf Sci 357:635–642
92. Aoki Y, Kunitake T, Nakao A (2005) Sol-gel fabrication of dielectric HfO_2 nano-films; formation of uniform, void-free layers, and their superior electrical properties. Chem Mater 17(2):450–458
93. Vinod A, Rathore MS, Rao NS (2018) Effects of annealing on quality and stoichiometry of HfO_2 thin films grown by RF magnetron sputtering. Vacuum 155:339–344
94. Hu X, Duan S, Wang L, Liao X (2011) Memristive crossbar array with applications in image processing. Sci Sin Inf 41(4):500–512
95. West DM, Allen JR (2018) How artificial intelligence is transforming the world. Report, 24 Apr 2018
96. Jacob AP, Xie R, Sung MG, Liebmann L, Lee RT, Taylor B (2017) Scaling challenges for advanced CMOS devices. In: Pettersson H, Willander M (eds) Scaling and integration of high-speed electronics and optomechanical systems. Selected topics in electronics and systems, vol 26. World Scientific Publishing Co. Pte. Ltd., Singapore, pp 1–76
97. Chua L (1971) Memristor-the missing circuit element. IEEE Trans Circuit Theory 18(5):507–519
98. Chakrabartty S, Song H (2023) Memristor devices and preparation method. 1023950310000
99. Robbie K, Brett MJ (1997) J Vac Sci Technol A 15:1460–1465
100. Norgaard JM, Lorentzen NB, Petersen R, Madsen MV (2007) Project on growth of nanostructure by angle toppling technique. Institute of Physics and Nanotechnology, Aalborg University, pp 1–77
101. Liu F, Umlor MT, Shen L, Weston J, Eads W, Barnard JA, Mankey GJ (1999) J Appl Phys 85(8):5486–5488
102. Zhao Y-P, Yeb D-X, Wangb G-C, Lu T-M (2003) Process SPIE 5219:59–73
103. Malac M, Egerton RF, Brett MJ, Dick B (1999) J Vac Sci Technol B 17:2671–2674
104. Malac M, Egerton RF (2001) J Vac Sci Technol A 19:158–166
105. Dick B, Brett MJ (2003) J Vac Sci Technol B 21:23–28
106. Kennedy SR, Brett MJ, Miguez H, Toader O, John S (2003) Photonics Nanostructures 1:37–42
107. Dick B, Brett MJ, Smy TJ, Freeman MR, Malac M, Egerton RF (2000) J Vac Sci Technol A 18:1838–1844
108. Alouach H, Fujiwara H, Mankey GJ (2005) J Vac Sci Technol A 23:1046–1050
109. Singh JP, Tang F, Karabacak T, Lu TM, Wang GC (2004) J Vac Sci Technol B 22:1048–1051
110. Lakhtakia A (2002) Mater Sci Eng C 19:427–434
111. Hawkeye MM, Brett MJ (2007) J Vac Sci Technol A 25:1317–1335
112. Chinnamuthu P, Mondal A, Singh NK, Dhar JC, Chattopadhyay KK (2012) J Appl Phys 112:054315
113. Messier R, Gehrke T, Frankel C, Venugopal VC, Otaño W, Lakhtakia A (1997) J Vac Sci Technol A 15:2148–2152
114. Schmidt D, Müller C, Hofmann T, Inganäs O, Arwin H, Schubert E, Schubert M (2011) Thin Solid Films 519:2645–2649
115. Dick B, Brett MJ, Smy T (2003) J Vac Sci Technol B 21:2569–2575
116. Konig H, Helwig G (1950) Optik (Stuttgart) 6:111
117. Al-Shidaifat A et al (2020) A novel characterization and performance measurement of memristor devices for synaptic emulators in advanced neuro-computing. Micromachines 11(1):89. https://doi.org/10.3390/mi11010089

Chapter 11
Machine Learning and Deep Learning Techniques

Reena Thakur, Prashant Panse, and Parul Bhanarkar

Abstract The classification of images is a multifaceted task influenced by numerous factors. This chapter looks at a variety of difficulties, challenges, machine learning algorithms, deep learning algorithms, and applications of image classification. The purpose of this chapter is to summarize the most important advanced classification methodologies as well as the techniques for improving classification accuracy. In addition to classification performance, other crucial issues are discussed. According to this review, developing an image-processing strategy that can classify remotely sensed data successfully is critical. Remote sensing data must be used effectively, and appropriate algorithms must be selected to improve classification accuracy. As multisource data classification has become increasingly complex, non-parametric classifiers such as artificial neural networks and decision trees have become more and more relevant. Other approaches include random forest classifiers, support vector machines, and object-based classification tree analysis. Identifying and reducing uncertainty in the image-processing chain is needed to improve classification accuracy. Finally, future research directions for improving image classification are discussed. Machine learning techniques and deep learning techniques have great potential, but they also face several challenges and issues. Addressing these challenges, such as data availability, interpretability, overfitting, scalability, fairness, and staying up to date, is crucial for the successful application and further development of these techniques in various domains.

Keywords Image classification · Accuracy · Machine learning · Satellite images · Deep learning

---

R. Thakur (✉)
Department of Computer Science and Engineering, Jhulelal Institute of Technology, Nagpur, India
e-mail: en19cs601002@medicaps.ac.in

P. Panse
Department of Information Technology, Medi-Caps University, Indore, India
e-mail: prashant.panse@medicaps.ac.in

P. Bhanarkar
Department of Computer Science and Engineering, SGBAU Amravati University, Amravati, India

U. N. Dulhare and E. H. Houssein (eds.), *Machine Learning and Metaheuristics: Methods and Analysis*, Algorithms for Intelligent Systems,
https://doi.org/10.1007/978-981-99-6645-5_11

1 Introduction

In recent years, remarkable progress has been made in domains like deep learning, neural networks, and machine learning, including its subsets. Artificial intelligence's growing importance, volatility, and applicability have driven this progress. To facilitate training, the use of software and tools like classifiers has become essential. These classifiers process vast amounts of data, analyzing it and extracting valuable characteristics. During the classification process, the purpose is to allocate specific pixels to predefined classes in a digital image. This is commonly achieved by utilizing multispectral data, where the spectral characteristics of each pixel are used for numerical classification. The fundamental objective of image classification is to recognize and characterize the various features depicted in an image, linking them to specific objects on the ground by representing them as distinct gray levels or colors. Image classification is crucial in computer vision, as distinguishing between various objects can be complex. It involves labeling images into predefined categories, with a single image potentially belonging to multiple classes. Manually checking and classifying a large number of images can be time-consuming. Therefore, automating the process through computer vision can be highly beneficial. Classifiers are essential in various applications, including stock photography, automated image organization, visual search, and video websites to enhance product discoverability, image and face recognition, and managing large visual databases on social networks. Their role is crucial in achieving high accuracy and efficiency in these domains.

2 Performing Image Classification Using a Structured Approach

1. ***Image Pre-processing***: By suppressing unwanted distortions and enhancing essential image features, this process improves the image data (features) for computer vision models so that the machines can benefit from it. Images are pre-processed by reading, resizing, and enhancing their data (Gaussian blurring, gray scaling, histograms, equalization, translation, and rotation).
2. ***Detection of an object***: Once detected, the image undergoes segmentation to isolate the object, allowing for the identification of its specific position of interest.
3. ***Training and Extraction of features***: The most prominent patterns within the image are identified through statistical and deep learning methods during this phase. These patterns may include distinctive features indicative of a particular class, ultimately enhancing the model's capacity to differentiate between them. Throughout the training process, the model assimilates these features from the dataset.
4. ***Classification of the object***: By comparing the detected objects with reference targets, the detected objects are classified into predefined classes.

3 Supervised Classification

An image can be supervised by selecting a representative set of pixels from the image that represents a particular class accurately. This selected set, known as training sites, is then utilized as reference data by the image-processing software to classify other image pixels. The user's skill is influential in selecting the training sites, also known as input classes or testing sets. Furthermore, the user determines a threshold for similarity to group pixels with similar characteristics. Typically, the spectral properties of the training areas determine these thresholds. Specifying how many classes the image should be divided into is also possible.

An image is then classified by comparing the reflectance signature of each pixel with the signatures of each information class after they have been statistically characterized. Supervised classification uses regression techniques and classification algorithms to develop predictive models. Popular methods used in this process include k-nearest neighbor, decision trees, naive Bayes, neural networks, support vector machines, linear regression, and logistic regression.

4 Unsupervised Classification

Without user-provided sample classes, the software independently analyzes an image and generates results by grouping pixels with similar features. Using algorithms, the computer identifies corresponding pixels and performs classification. While the user has the freedom to choose the algorithm and specify the desired number of output classes, it is important to take additional factors into consideration for achieving accurate classification.

Matching the computer-generated clusters of pixels with real-world features on the ground requires the user's familiarity with the classified region. Unsupervised learning relies on algorithms such as anomaly detection, cluster analysis, and neural networks. These algorithms play a vital role in analyzing image data autonomously, eliminating the need for explicit user guidance.

4.1 Convolutional Neural Network

In contrast, convolutional neural networks (CNNs) are a specialized class of multilayer neural networks developed for directly identifying visual patterns in pixel-based images without significant pre-processing. As depicted in Fig. 1, CNNs possess a unique architecture within artificial neural networks. They leverage properties inspired by the visual cortex and have demonstrated up-to-the-minute performance in various computer idea tasks.

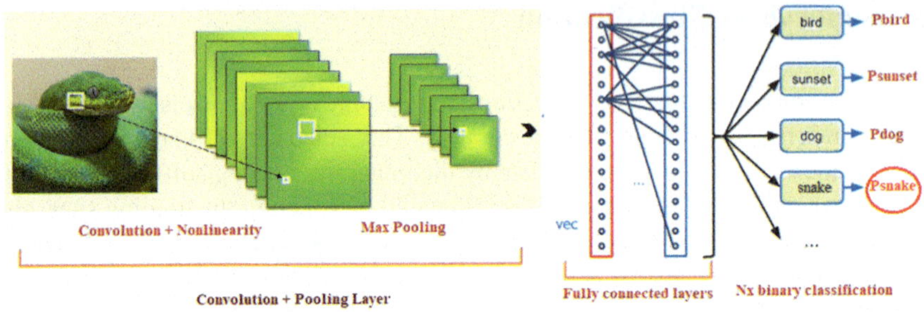

Fig. 1 An overview of convolutional neural networks

CNNs are built upon two fundamental building blocks: convolutional layers and pooling layers. These components and other elements specific to CNNs offer flexibility in constructing architectures tailored to specific computer vision problems. While the basic concepts behind convolutional and pooling layers are relatively easy to grasp, effectively harnessing their potential in practical applications can be challenging.

CNNs have gained significant popularity due to their unique architecture, particularly since explicit feature extraction is not required. The core concept behind how the system learns to extract features lies in the convolutional operation between images and filters, which generates invariant features passed on to subsequent layers. These features are combined with various filters in subsequent layers to produce increasingly invariant and abstract representations. This iterative process continues until a final set of features invariant to occlusions is obtained.

Several well-known CNN architectures have emerged and found widespread use, including LeNet, AlexNet, ZFNet, GoogLeNet, VGGNet, and ResNet. These designs are effective templates for implementing CNNs in various computer vision tasks.

Among the diverse range of deep neural networks (DNNs), convolutional neural networks (CNNs) have emerged as remarkably powerful in computer vision applications. They have demonstrated exceptional efficacy, particularly in the area of image classification. CNNs are specialized multi-layer neural network that draws inspiration from the functioning of the human visual and brain systems. The development of CNN models, such as VGGNet, GoogleNet, ResNet, DenseNet, and others, gained significant momentum in 2012 when a deep convolutional neural network called AlexNet achieved exceptional act in the ImageNet Large-Scale Visual Recognition Challenge (ILSVRC).

Buildings and Layers: A neuron is the fundamental building block of a CNN architecture. Human neurons are the foundation of the concept of neurons. These statistical procedures add an activation function to the output after computing the weighted average of the inputs. A layer is a group of neurons, and each layer serves a specific purpose. A depiction of a neural network showcasing the weights (in red) and input values (in green) (represented in blue). A CNN system could include anywhere from 3 to 150 layers, or possibly more: The number of layers is what is referred to as "deep" in deep neural networks. The output of one layer serves as the input of another layer. Resnet50 (50 layers) and ResNet101 are two examples of deep multi-layer neural networks (101 layers). Convolutional neural network concept (CNN): Convolution layer, ReLU layer, fully connected layer, and pooling layer are the four primary types of CNN layers. In the convolution layer, an activation is generated by applying a filter directly to input through convolution. This layer comprises a set of trainable filters that effectively utilize the depth of the input data, even with a limited receptive field.

Rectified Linear Unit (ReLU) Layer: Rectified Linear Units (ReLU) serve as an activation function in convolutional neural networks (CNNs), improving accuracy and efficiency while reducing overfitting. By making models easier to train, ReLU layers contribute to more precise predictions.

Pooling Layer: The pooling layer collects and processes the outputs of all neurons in the layer forgoing it. Its main objective is to decrease the number of variables under consideration and provide a streamlined output, thus reducing dimensionality.

Fully Connected Layer: Positioned at the top of the CNN models, the fully connected layer receives input from the layers below it. It flattens this input and generates the final output of the model.

The authors [1] have pointed out that CNNs can be used for classification tasks on major image categories, including hyperspectral images. They highlighted that 2D CNN is more powerful as it has more classification performance as compared to other algorithms.

It has also been noted by the authors [2] that the CNN and the ELM model can outperform the classification tasks for image classification. With CNN, it also has been noted that a large amount of training data takes a substantial amount of time to process. Several fusion methods can solve this issue, and the performance can be obtained for classification. Data scientists have often used CNN, as pointed out by the authors [3], CNN enables learning to be data driven and highly representative. The transfer learning method for fine-tuning the CNN model is much more effective in training the model. The authors highlighted three main techniques to employ the CNN training from scrape or using the pre-trained model with supervised fine-tuning of the model.

4.2 Artificial Neural Network

Artificial neural networks (ANN) are statistical learning algorithms that find extensive application across various tasks, ranging from basic classification to advanced fields such as speech recognition and computer vision. These networks draw inspiration from the characteristics of biological neural networks. In ANNs, nodes act as connected processing components that emulate the behavior of biological neurons. As depicted in Fig. 2, the network iteratively adjusts the weights, and numerical values representing the connections between nodes, to estimate the desired function.

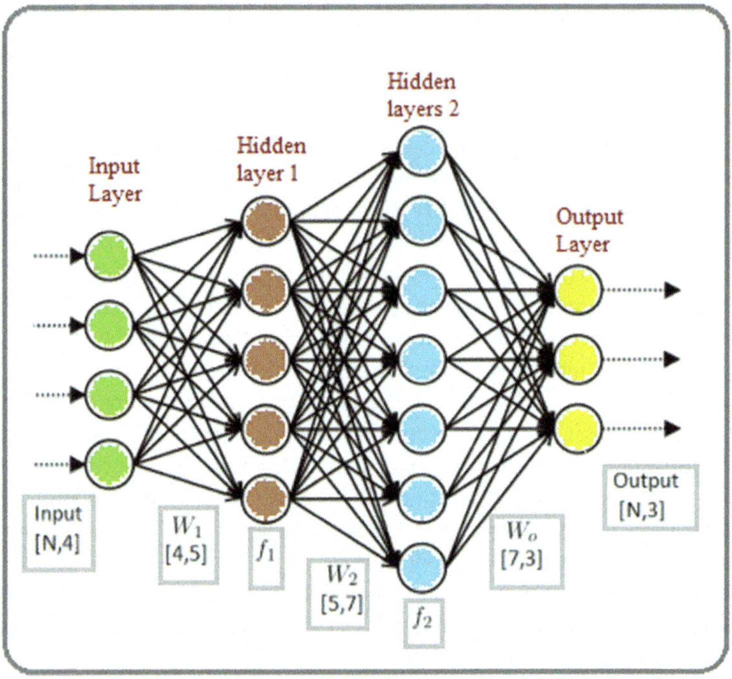

Fig. 2 Artificial neural network model

Hidden layers detect distinct features within the data as it passes through the network, gradually identifying intricate patterns. For example, in the context of face recognition, the initial hidden layer may detect lines, the subsequent hidden layer combines these lines to form a nose, and the subsequent hidden layer matches the nose with an eye, and so forth, ultimately reconstructing the entire face.

Several distinct types of artificial neural networks are tailored to specific problem domains and learning needs. These include convolutional neural networks, probabilistic neural networks, feedforward neural networks, deep stacking networks, time-delay neural networks, recurrent neural networks, and radial basis function networks. Each network type has unique characteristics that make it well suited for specific tasks and learning scenarios.

Few authors have utilized the features of artificial neural networks to perform more complex classification tasks like identifying spinal deformity [4]. A multi-layer feedforward and backpropagation ANN helps implement this problem. These applications of artificial neural networks have been effective in assisting in medical decision-making where precision is of utmost importance.

The authors [5] have concentrated on multiclass image classification techniques involving artificial neural networks. Artificial neural networks have achieved the highest classification efficiency, mainly for image classification tasks. The comparison between the K-NN and ANN methods has been applied, and a comparative study has been represented, which directly specifies the better performance of the ANN model over the image dataset.

The authors [6] have worked on classifying natural objects with the help of artificial neural networks. The feature extraction is done with the help of Hermite functions, which help recognize the images from the dataset. The Hermite function used here has helped identify the texture measurements of the input image. This is applicable in independent scene analysis.

4.3 Support Vector Machine

Known for their effectiveness and adaptability in regression and classification, support vector machines (SVMs) are supervised machine learning algorithms. Unlike other machine learning methods, SVMs have a distinctive implementation approach that sets them apart. They excel in handling both continuous and categorical variables, making them highly favored in various domains.

A support vector machine model can be visualized as a multidimensional hyperplane representing multiple classes. SVMs aim to minimize errors by creating a hyperplane iteratively. The dataset is divided into classes to identify the optimal hyperplane, and the maximum margin hyperplane is determined. In a high-dimensional space, this hyperplane or group of hyperplanes is formed.

It is illustrated in Fig. 3 that the ideal separation between classes can be achieved by identifying the hyperplane that maximizes the distance between any two training data points. The selection of kernel function is crucial for the SVM algorithm's efficacy. Commonly employed kernels include the linear kernel, polynomial kernel, and Gaussian kernel. Selecting the appropriate kernel function substantially affects the performance of SVM.

The authors have utilized the features of Support Vector Machine (SVM) in pattern recognition algorithms [7]. It has been observed that SVM algorithms work best and are good at robustness and stability. The authors have observed that the multispectral images have similar classes, and the methods available for image classification need to generate good accuracy levels in the classification. The method of fuzzy SVM has been applied to the classification of multispectral images. The fuzzy SVM can be further improved by checking on more bands in multispectral images.

The authors in [8] have specified that supervised learning methods are mainly used to classify hyperspectral images. Computational complexity and accuracy play an essential role in the case of hyperspectral image classification. Higher processing speed has effectively helped in achieving better classification accuracy. The authors developed two categories of composite kernel SVM implementation for hyperspectral image classification on FPGA. These implementations offer both fast processing speed and improved classification accuracy. The FPGA-based implementation only necessitates additional buffers, enabling achieving the desired processing speed while conserving memory resources.

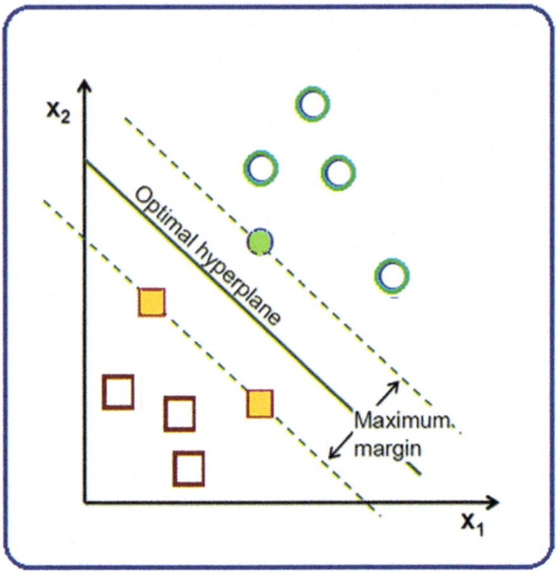

Fig. 3 Support vector machine

The authors in this work have worked on fish detection and classification through SVM classification [9]. They have considered the classification complexities like segmentation error in an image, noise presence, distortion in an image to be classified, etc. The SVM method is used to recognize and categorize fish species with the help of various kernel functions. The linear, polynomial, sigmoid, and radial substructure functions have been used for the classification. Fish identification, in this case, is based on the traits of the fishes where similar attributes of the fish species have been used. The classification accuracy is presentable and good at this task.

4.4 K-Nearest Neighbor

K-Nearest Neighbors (K-NN) is an important non-parametric technique in classification and regression. Based on the feature space, it considers the closest k training instances. Its simplicity comes from the fact that it does not assume anything about the underlying distribution of the data, and it takes a lazy approach to learning.

Unknown data points are categorized based solely on the distance between the feature vectors of the k-closest examples. To enable K-NN classification, a distance metric or similarity function is constructed, commonly using the Manhattan distance or the Euclidean distance. The final class membership is determined through majority voting among the k-nearest neighbors.

As shown in Fig. 4, when k is set to 1, an object is assigned to the class of its nearest neighbor. To streamline the dataset used in K-NN classification, Condensed Nearest Neighbor (CNN) or Hart algorithms are used.

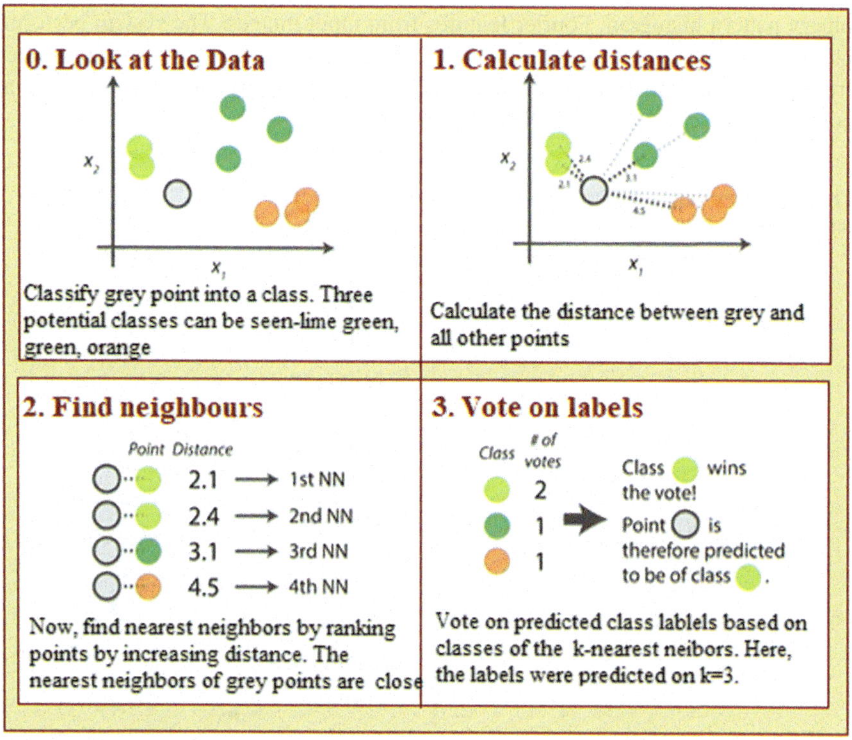

Fig. 4 k-nearest neighbor process

The authors have focused on the classification of hyperspectral images in their work on the FPGA acceleration of a composite kernel SVM.

In order to classify images, the authors have used principal component analysis [10]. Rainfall recognition is a task that depends upon the quality of the image-capturing device. The rainfall recognition problem also depends on the radar image quality. The principal component analysis used in this case has been done for dimension reduction and feature extraction. The K-NN method has been used to perform the classification task and can achieve better classification accuracy. Capturing the image can improve classification performance further.

In their work, the authors [11] filtered the spam emails based on image and text-based emails in the inbox. They have applied the K-NN algorithm for the same. TESSERACT's OCR library has been applied to text mining and text recognition. The system's advantages include high filtering accuracy, but disadvantages like the identification or filtering could be better, and the system implementation could be expensive. The system can also identify the threats in the emails received.

The authors in [12] this work have identified that content-based image retrieval systems are essential for medical image analysis as they are responsible for retrieving images of the same type. K-NN classifiers are developed that extract local binary pattern histogram Fourier features from input images. The system performs better by applying the K-NN classifier over the histogram intersection generated from processing the Fourier features. The generated output has better accuracy as compared to the authors. The system is shown in Fig. 5.

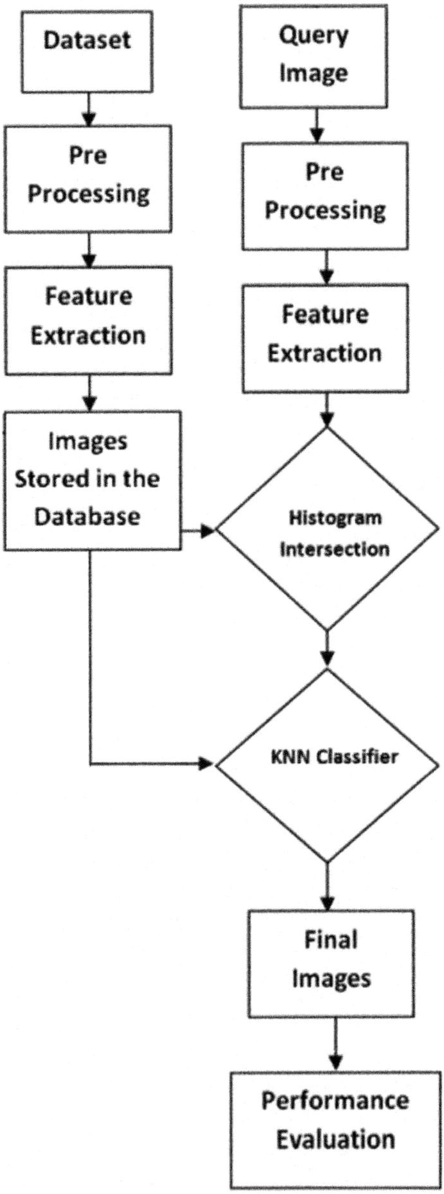

Fig. 5 Content-based image retrieval

In recent years, CBIR systems have become increasingly important. A two-stage mechanism is proposed in this study for combining two classifiers to improve performance. A two-stage classifier combining Histogram intersections and K-NNs outperforms single-stage classifiers. According to the results, the techniques are more effective than conventional methods.

4.5 Naive Bayes Algorithm

Naive Bayes classifiers [18], derived from Bayes' theorem, comprise a collection of classification algorithms. Instead of being a single method, they form a family of algorithms sharing a common principle: the assumption of independence between pairs of classified features. Naive Bayes is known for its simplicity in constructing classifiers.

The class labels in naive Bayes classifiers are selected from a finite list of class labels for each problem case. Assuming that the class variable is independent of other features, these classifiers assume that a certain feature has an independent value. This approach applies to both binary and multiclass classification tasks.

Naive Bayes is a reckless and greatly scalable approach, relying on performing various counts. It has gained popularity in applications such as text categorization and spam email classification, among others. Using a modest dataset, it is simple to train. It is limited because it accepts that all features are independent, making it unable to understand how they relate. Naive Bayes can understand the relative relevance of individual features but need help understanding the connections between features. Gaussian, multinomial, and Bernoulli naive Bayes are different varieties of naive Bayes algorithms.

For single image classification of smear images, the authors have investigated the integration of features from naive Bayes classifiers and weighted principal component analysis (PCA) [13]. The objective of this classification is to detect cells accurately. Two classes and seven classes were used to evaluate the classification accuracy. The weighted (PCA) method significantly improved classifying the input dataset. In contrast to the large-scale dataset test, where the accuracy of the naive Bayes algorithm decreased, it exhibited high accuracy when handling small-scale datasets rapidly and accurately. It is important to compare the naive Bayes algorithm with the weighted PCA model to achieve higher accuracy levels, resulting in accuracy improvements of 67.45% and 87.28%, respectively.

In their work, the authors [14] concentrated on detecting breast tissues of grade 4 cancer cells. Several morphological operations, pixel subtraction, and statistical feature extraction methods have been used for classification. The authors have tested the method over MRI images and achieved a better 100% detection rate for standard images. The REMBRANDT database has been used in the research. The method can detect cancer or tumor in the higher stage or more with the help of the maximum entropy method. On tumor images, the algorithm produces a detection rate of 81.25%, and on non-tumor images, it produces a detection rate of 100% with an overall accuracy of 94%.

The authors in [15] have concentrated on pixel-level plant segmentation to classify the plant and the background by extracting statistical information from the images. The extraction is done with the help of single image and optimization techniques which help reduce the segmentation cost. The method implemented is better for a limited number of classes used for classification. A good dataset is needed to achieve the required accuracy. The method is called sub-optimal classification because it accurately approximates the probability distribution functions (PDFs) of pixels and calculates each pixel's posterior probability.

According to the simulation results, the uncertainty region in pixel categorization shrank when more details and features were recovered from an image. Users can train the naive Bayes classifier to recognize plants in images from their experimental system by incorporating this method into the PlantCV platform, eliminating or drastically reducing the requirement to parameterize image-processing operations manually.

4.6 Random Forest Algorithm

Random forest is a supervised learning technique for classification and regression tasks. It follows the approach of creating an ensemble of decision trees, forming a forest that comprises multiple trees. The random forest algorithm builds decision trees using data samples, generates predictions from each tree, and combines the outcomes through voting to determine the ultimate result. By leveraging this ensemble method and averaging the predictions, random forest reduces the risk of overfitting and outperforms a single decision tree. By averaging the predictions and leveraging the ensemble method, random forest mitigates overfitting issues often associated with individual decision trees.

The random forest model consists of multiple decision trees working together, forming a robust and resilient forest. In order to create a diverse and uncorrelated set of trees, which collectively provide more accurate predictions, the random forest algorithm incorporates bagging (bootstrap aggregating). It introduces randomness in feature selection during the construction of each tree. This ensemble approach is illustrated in Fig. 6. Uses for image classification before a few years ago, security applications were the principal use cases for image classification.

Nonetheless, picture classification use cases are becoming more common in various industries, including health care, smart cities, industrial manufacturing, insurance, and space exploration. The exponential increase in visual data availability and the rapid development of superior computing technology contribute to the rise in applications. This data can be valuable extracted via image classification. Visual data has equity when used strategically because the value derived from applications throughout the company exceeds the expense of storing and maintaining it.

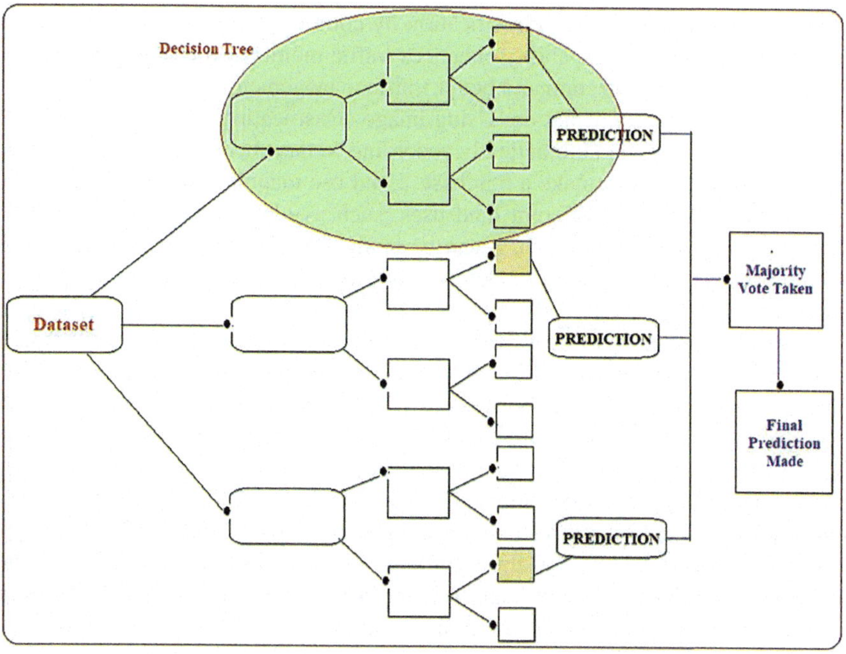

Fig. 6 Random Forest algorithm

Image classification finds applications in various domains, and notable examples of its use cases are as follows:

Application #1: Control and inspection of quality by automated means.
Application #2: A driverless car that recognizes objects.
Application #3: A pathology slide can be used to detect cancer cells.
Application #4: Security applications of face recognition.
Application #5: Detection of traffic congestion and traffic monitoring.
Application #6: Segmentation of retail customers.
Application #7: Land use mapping.

Automated product inspection and quality control benefit from image classification, as it enables the automated assessment of products on an assembly line to identify any deviations or defects that do not meet the specified quality criteria. Artificial vision for pharmaceuticals: Image processing for visually inspecting imprinted tablet packaging. Autonomous vehicles' ability to recognize items on the road is essential to their ability to maneuver safely. For this, image classification might be employed. AI vision classification of skin cancer:

A Medical Use Case Example of Image Classification for Cancer Detection Face recognition in security: Picture classification can be used, for instance, to make face identification at airports or other public locations. This involves automatically identifying persons from surveillance footage. Traffic bottlenecks can be detected automatically through image classification by counting the number of vehicles on a particular route. This capability enhances traffic monitoring systems by enabling accurate congestion detection and better traffic management.

Retail customer segmentation: Using image classification, it is possible to categorize retail consumers automatically according to their behavior, such as whether or not they are likely to make a purchase. Land use mapping: Image classification can automatically map different land uses, such as identifying areas utilized for farming or forested areas. Also, it can be applied to track environmental change, such as spotting urbanization or deforestation or estimating yields in agricultural applications.

With the emergence of deep learning and the availability of powerful AI hardware and GPUs, significant advancements have been made in image classification. Deep learning has revolutionized areas such as image recognition, image classification, and face recognition, enabling real-time object detection and surpassing human presentation [19]. In recent years, there has been a remarkable improvement in the performance of inference algorithms. For example, in 2017, the Mask R-CNN technique achieved the fastest real-time object detection on the MS COCO benchmark, with an inference time of 330 ms per frame. Compared to popular deep learning algorithms like YOLOv3 and YOLOv4, the newly published YOLOR algorithm achieves inference times as low as 12 ms on the same benchmark. In July 2022, the introduction of YOLOv7 sets a new state of the art, surpassing all previous models, including YOLOR, in speed and accuracy.

One of the advantages of deep learning over conventional image-processing methods used about two decades ago is that it requires less specialized knowledge in machine vision engineering. Traditional computer vision approaches rely on manual feature engineering, whereas deep learning algorithms can automatically learn relevant features from the data without explicit feature engineering. This shift has made deep learning more accessible and efficient for image classification tasks. Deep learning necessitates picture annotation and manually labeling data to differentiate between excellent and bad samples. Supervised learning is learning from or deriving insights from data humans have classified. Acquiring labeled data for training AI models is a labor-intensive process that often relies on human effort. Tasks such as annotating millions of high-resolution images across various categories using datasets like ImageNet, LabelMe, Google OID, or MS COCO require substantial human involvement.

In this study [16], the authors propose a biologically inspired approach to enhance the recognition of perceptually salient contours. The key innovation lies in utilizing a multi-feature-based center-surround framework, which incorporates surround inhibition weights from various parameters such as orientation, brightness, and luminance contrast. The effectiveness of this approach is compared to single-cue-based models and other existing methods, primarily those motivated by biological principles. The results demonstrate that integrating multiple cues significantly improves contour detection performance compared to models relying on a single cue. Particularly, brightness and luminance contrast exhibit substantial contributions to contour extraction in grayscale natural photos, surpassing the influence of orientation.

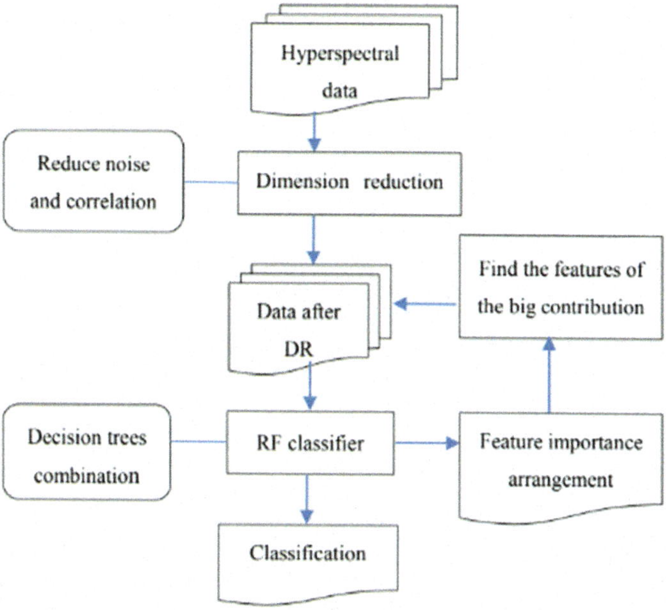

Fig. 7 Classification process

In their research [16], the authors have considered the benefits of hyperspectral images with a broad spectral range and high spectral resolution, as illustrated in Fig. 7. Airborne hyperspectral image classification methods are used as a result. Since the hyperspectral image consists of multiple bands and there is redundancy among the bands, the redundant bands and noisy bands are removed using the principle component analysis. The random forest decision tree technique is then used to classify the image. The contribution of each band to picture categorization is calculated using the significance coefficient. Lastly, we build and develop the classification software to classify the hyperspectral image of Zhangye City acquired by an aerial AISA EAGLET sensor using the C + + programming language and the OpenCV library. The experimental results indicate the validity and dependability of the random forest approach.

5 Discussion

The various machine learning and deep learning techniques have been discussed. These techniques have been compared in Table 1 in detail.

The advent of machine learning and deep learning techniques has brought about a transformative impact in numerous domains by facilitating the creation of intelligent systems that can acquire knowledge from data. Despite their remarkable progress,

Table 1 Comparative Analysis. As shown in Fig. 1, various techniques with their advantages and disadvantages have been described

Sr. no	Author	Technique	Description	Pros	Cons
1	Mekha and Teeyasuksaet [1]	CNN	2D CNN is applied for image classification	Performance is good	Accuracy can be further improved
2	Kannojia and Jaiswal [2]	CNN, ELM	Feature representation of CNN, classification learning using ELM	Improved classification accuracy	Accuracy can be further improved with a hybrid method
3	Shin [3]	Deep CNN	Transfer learning method for image recognition	High performance in computer-aided detection	Performance can be further improved
4	Lin [4]	ANN	Curvature analysis for spinal deformity classification	Very effective decision-making	Decision-making efficiency can be further improved
5	Patidar et al. [5]	K-NN, ANN	Performance analysis based on wavelet features	ANN outperforms in efficiency	K-NN can be improved further on wavelets
6	Singh et al. [6]	ANN	Hermite functions for feature extraction	Robust classification, reasonable image recognition rate	Efficiency can be further improved
7	Wang and Ma [7]	SVM	Fuzzy SVM has been applied for multispectral images	Robust, stable, and better accuracy than SVM	Hybrid methods can be used for further classification corrections
8	Tajiri and Maruyama [8]	SVM	FPGA-based composite kernel method	Computation is simplified, with reasonable accuracy	Processing speed can be better
9	Dewan et al. [9]	SVM	SVM kernel function linear, polynomial, sigmoid, and radial	Fish identification is booming, and better accuracy is achieved	The cost of implementation is high
10	Wei et al. [10]	K-NN	Principal component analysis	Rainfall detection is successful	It can be further improved for image classification
11	Harisinghaney et al. [11]	K-NN	OCR Library TESSERACT	Better accuracy	Time-consuming, work on certain fronts only and is expensive to implement

(continued)

Table 1 (continued)

Sr. no	Author	Technique	Description	Pros	Cons
12	Bharathi et al. [12]	K-NN	Extraction of features by local binary pattern, K-NN classifier	Less computational time and complexity	Multi-layer classification methods are expensive
13	Dewi et al. [13]	Naïve Bayes algorithm	weighted-principal component analysis along with naïve Bayes classification	Can deal with larger datasets, outputs better accuracy	Time complexity, coding complexity
14	Zaw et al. [14]	Naïve Bayes algorithm	Morphological operations, pixel subtraction, statistical feature extraction	Better detection accuracy	The maximum entropy method can be improved
15	Abbasi and Fahlgren [15]	Naïve Bayes algorithm	Single image and optimization along with naïve Bayes classifier, pixel distribution function	The less complex method of classification	Increased complexity
16	Jayaraj and Sathiamoorthy [16]	Random Forest	Multi-feature-based center-surround framework	Accuracy is better	Higher efficiency can be achieved
17	Wang et al. [17]	Random Forest	Principal component analysis	Performs better for small datasets	Improvement is needed; accuracy can be further improved

these techniques also encounter challenges and issues requiring careful consideration. This article presents a comprehensive overview of the challenges and issues in machine learning and deep learning, offering insights into the areas that require attention and further exploration.

One of the major challenges is the availability and quality of labeled training data. Supervised learning algorithms heavily rely on labeled data, and obtaining a large and diverse dataset can be time-consuming and expensive. Furthermore, ensuring the accuracy and reliability of the labeled data poses additional challenges.

Another challenge is related to the interpretability and explainability of machine learning and deep learning models. Understanding the reasoning behind their decisions becomes increasingly difficult as these models become more complex and sophisticated. This lack of interpretability can hinder their adoption, particularly in domains where explainability is crucial.

Additionally, overfitting and generalization issues can arise when training deep learning models. Deep neural networks are prone to overfitting and memorizing the training data instead of generalizing patterns. Finding the right balance between model complexity and generalization capability is crucial to avoid overfitting.

Moreover, scalability and computational requirements pose significant challenges in applying machine learning and deep learning techniques to large-scale datasets and real-time applications. Training complex models on massive datasets requires substantial computational resources, and deploying these models on resource-constrained devices can be challenging.

Ethical considerations and fairness are also emerging issues in machine learning and deep learning. Biases in the training data can lead to biased models, perpetuating discrimination and inequality. Addressing these biases and ensuring fairness in decision-making processes are important concerns.

Lastly, advancements in machine learning and deep learning techniques present a challenge in keeping up with the latest developments and best practices. Continuous learning and staying updated with new algorithms, architectures, and methodologies are essential to leveraging the full potential of these techniques.

In conclusion, machine learning and deep learning techniques have great potential but face several challenges and issues. Addressing these challenges, such as data availability, interpretability, overfitting, scalability, fairness, and staying up to date, is crucial for the successful application and further development of these techniques in various domains [17].

References

1. Mekha P, Teeyasuksaet N (2021) Image classification of rice leaf diseases using random forest algorithm. In: 2021 joint international conference on digital arts, media, and technology with ECTI northern section conference on electrical, electronics, computer and telecommunication engineering. Cha-am, Thailand, pp 165–169. https://doi.org/10.1109/ECTIDAMTNCON51128.2021.9425696
2. Kannojia SP, Jaiswal G (2018) Ensemble of hybrid CNN-ELM model for image classification. In: 2018 5th international conference on signal processing and integrated networks (SPIN). Noida, India, pp 538–541. https://doi.org/10.1109/SPIN.2018.8474196
3. Shin H-C et al (2016) Deep convolutional neural networks for computer-aided detection: CNN architectures, dataset characteristics and transfer learning. IEEE Trans Med Imaging 35(5):1285–1298. https://doi.org/10.1109/TMI.2016.2528162
4. Lin H (2008) Identification of spinal deformity classification with total curvature analysis and artificial neural network. IEEE Trans Biomed Eng 55(1):376–382. https://doi.org/10.1109/TBME.2007.894831
5. Patidar D, Jain N, Parikh A (2014) Performance analysis of artificial neural network and K Nearest neighbors image classification techniques with wavelet features. In: Proceedings of IEEE international conference on computer communication and systems ICCCS14. Chennai, India, pp 191–194. https://doi.org/10.1109/ICCCS.2014.7068192
6. Singh S, Markou M, Haddon J (2000) Natural object classification using artificial neural networks. In: Proceedings of the IEEE-INNS-ENNS international joint conference on neural networks. IJCNN 2000. Neural computing: new challenges and perspectives for the New Millennium. Como, Italy, vol 3, pp 139–144. https://doi.org/10.1109/IJCNN.2000.861294
7. Wang HB, Ma JH (2008) A classification method of multispectral images which is based on fuzzy SVM. In: 2008 international conference on computer science and software engineering. Wuhan, China, pp 815–818. https://doi.org/10.1109/CSSE.2008.784

8. Tajiri K, Maruyama T (2023) FPGA acceleration of a composite kernel SVM for hyperspectral image classification. IEEE Access 11:214–226. https://doi.org/10.1109/ACCESS.2022.3230066
9. Dewan J, Gele A, Fulari O, Kabade B, Joshi A (2022) Fish detection and classification. In: 2022 6th international conference on computing, communication, control and automation (ICCUBEA). Pune, India, pp 1–5.https://doi.org/10.1109/ICCUBEA54992.2022.10010836
10. Wei Y, Liu Y, Song H, Lu Z, A method of rainfall detection from x-band marine radar image based on the principal component feature extracted. IEEE Geosci Remote Sens Lett. https://doi.org/10.1109/LGRS.2023.3235714
11. Harisinghaney A, Dixit A, Gupta S, Arora A (2014) Text and image-based spam email classification using KNN, Naïve Bayes and Reverse DBSCAN algorithm. In: 2014 international conference on reliability optimization and information technology (ICROIT). Faridabad, India, pp 153–155. https://doi.org/10.1109/ICROIT.2014.6798302
12. Bharathi P, Reddy KR, Srilakshmi G (2014) Medical image retrieval based on LBP histogram Fourier features and KNN classifier. In: 2014 international conference on advances in engineering & technology research (ICAETR - 2014). Unnao, India, pp 1–4. https://doi.org/10.1109/ICAETR.2014.7012840
13. Dewi YN, Riana D, Mantoro T (2017) Improving Naïve Bayes performance in single image pap smear using weighted principal component analysis (WPCA). In: 2017 international conference on computing, engineering, and design (ICCED). Kuala Lumpur, Malaysia, pp 1–5. https://doi.org/10.1109/CED.2017.8308130
14. Zaw HT, Maneerat N, Win KY (2019) Brain tumor detection based on Naïve Bayes classification. In: 2019 5th international conference on engineering, applied sciences and technology (ICEAST). Luang Prabang, Laos, pp 1–4. https://doi.org/10.1109/ICEAST.2019.8802562
15. Abbasi A, Fahlgren N (2016) Naïve Bayes pixel-level plant segmentation. In: 2016 IEEE western New York image and signal processing workshop (WNYISPW). Rochester, NY, USA, pp 1–4.https://doi.org/10.1109/WNYIPW.2016.7904790
16. Jayaraj D, Sathiamoorthy S (2019) Random forest-based classification model for lung cancer prediction on computer tomography images. In: 2019 international conference on smart systems and inventive technology (ICSSIT). Tirunelveli, India, pp 100–104. https://doi.org/10.1109/ICSSIT46314.2019.8987772
17. Wang S, Dou A, Yuan X, Zhang X (2016) The airborne hyperspectral image classification based on the random forest algorithm. In: 2016 IEEE international geoscience and remote sensing symposium (IGARSS). Beijing, China, pp 2280–2283.https://doi.org/10.1109/IGARSS.2016.7729589
18. Houssein EH, Sayed A (2023) Dynamic candidate solution boosted beluga whale optimization algorithm for biomedical classification. Mathematics 11(3):707
19. Arif F, Dulhare UN, A machine learning based approach for opinion mining on social network data. In: Computer communication, networking and internet security proceedings of IC3T 2016. Lecture Notes in Networks and Systems (LNNS, vol 5)
20. Dulhare UN, Ali H (2023) Underwater human detection using faster R-CNN with data augmentation. Materials Today: Proceedings 80:1940–1945. ISSN 2214-7853

Chapter 12
Classification Models in Education Domain Using PSO, ABC, and A2BC Metaheuristic Algorithm-Based Feature Selection and Optimization

Uma N. Dulhare, D. Naga Jyothi, Bhavana Balimidi, and Rama Rithika Kesaraju

Abstract During the process of creating a machine learning model for real-life data, we often encounter numerous features in the dataset, but not all of them are essential. Incorporating irrelevant features during model training can lead to a decrease in overall accuracy, an increase in complexity, a decrease in generalization capability, and bias in the model. Therefore, selection of features is a crucial initialization step when building a machine learning model. The primary objective of feature selection is to identify the optimal set of features by reducing the dimensionality of the data to improve the model's accuracy. Recently, swarm intelligence-based optimization techniques have gained popularity due to their effectiveness in solving dimensionality optimization problems. Techniques like PSO, ABC, and A2BC (automated ABC) have been explored, and feature selection methods based on these algorithms have been implemented and evaluated using various classifiers like SVM, Random Forest, XGBoost, and Voting (SVM + RF + DT) classifiers. The education science domain is used to evaluate these algorithms, which tackled a classification problem which is binary in nature related to freshman student's success. The aim is to compare the accuracies and the performances of the three machine learning algorithms, namely, PSO, ABC, and automated ABC (A2BC), and conclude the best algorithm.

Keywords Swarm intelligence · PSO · ABC · Advanced automated ABC · Agglomerative clustering · XGBoost · Voting classifier · Support vector machine

---

U. N. Dulhare (✉) · D. N. Jyothi
Muffakam Jah College of Engineering & Technology, Hyderabad, Telangana, India
e-mail: prof.umadulhare@gmail.com

D. N. Jyothi
e-mail: dnagajyothi_cseaiml@cbit.ac.in

B. Balimidi · R. R. Kesaraju
Chaitanya Bharathi Institute of Technology, Hyderabad, Telangana, India

U. N. Dulhare and E. H. Houssein (eds.), *Machine Learning and Metaheuristics: Methods and Analysis*, Algorithms for Intelligent Systems,
https://doi.org/10.1007/978-981-99-6645-5_12

Abbreviations

PSO Particle Swarm Optimization
ABC Artificial Bee Colony
A2BC Automated Artificial Bee Colony
SVM Support Vector Machine
FS Feature Selection

1 Introduction

This study is a comparison of swarm intelligence-based optimization algorithms [1–15], like Particle Swarm Optimization (PSO), Artificial Bee Colony (ABC), and Automated Artificial Bee Colony (A2BC). The comparison is done based on the accuracy they provide when used for feature selection to classify student success. The accuracy is calculated using support vector machine (SVC), Random Forest, XGBoost, and Voting classifiers. The Automated Artificial Bee Colony algorithm is an improved version of ABC that includes extensions to the ABC algorithm (Fig. 1).

Feature selection is the important process for choosing the relevant features and fed into machine learning model [26]. This section of the chapter presents a comprehensive survey of existing algorithms used to perform feature selection.

- A framework for selecting the features using distinct soft computing algorithms is proposed by Singla [16]. This paper explaining the importance of feature selection in machine learning. In the realm of machine learning, feature selection pertains

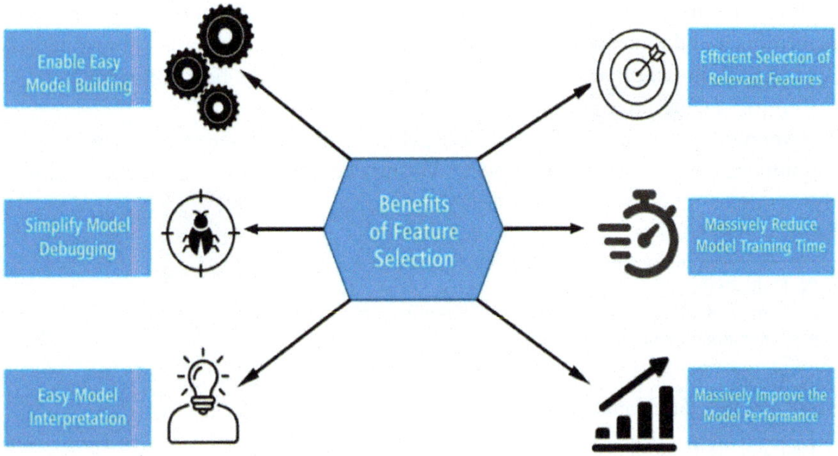

Fig. 1 Benefits of feature selection

to the act of selecting a subset of significant features from a larger pool of input features to enhance the potential of machine learning algorithms. According to the authors, this process can aid in mitigating overfitting, increasing accuracy, and reducing computational expenses. They proposed Grouped-ABC to optimize feature selection and compared the results with other algorithms like PSO and ABC. To optimize feature selection, they suggested Grouped-ABC, and they evaluated its performance against that of PSO and ABC. Their findings show that the G-ABC with neural network method performs well in terms of feature selection (91 features were chosen from 100 records), execution time (three seconds), and accuracy (98.2) [16].

- A New Co-Evolution BPSO (binary particle swarm optimization) with multiple inertia weight strategy for feature selection is proposed by Jingwei Too et al. [17]. In the research paper, a novel method for feature selection is introduced, which involves utilizing co-evolution binary particle swarm optimization and a multiple inertia weight strategy. This approach aims to improve the accuracy and effectiveness of feature selection in the data analysis process. The proposed approach aims to improve the accuracy of feature selection by using co-evolutionary techniques and multiple inertia weights. In conclusion, the paper presents a novel approach to feature selection using BPSO. The results demonstrate its effectiveness and superiority compared to other feature selection methods. The proposed approach has potential applications in various domains, such as medical diagnosis and image recognition, where accurate feature selection is crucial for achieving high performance in machine learning models.
- A2BCF: An automated ABC-based feature selection algorithm for classification models in an education application is proposed by Leila Zahedi et al. [18]. The scholarly article presents a new feature selection method called A2BCF, which utilizes the ABC algorithm to enhance the classifier's performance. The suggested approach is investigated using a binary classification issue and assessed on a learning domain to forecast student achievement in 20 universities throughout the US. By adding OBL learning techniques, the suggested method enhances the fundamental ABC algorithm's exploration capabilities. Additionally, the population variety increases and the likelihood of finding potential locations decreases. The outcomes demonstrate the strength of A2BCF when the acquired results are contrasted with those of a latest study. Experimental findings demonstrate that the process enhances precision and could be used to resolve educational issues with a fairly large dimensionality. The behavior of A2BCF is investigated under various scenarios. In conclusion, the paper presents an automated feature selection algorithm called A2BCF based on the ABC algorithm for classification models in an education application. The results demonstrate its effectiveness and superiority compared to other feature selection methods. The proposed algorithm can be applied in various educational domains, such as predicting student performance or identifying at-risk students, where accurate feature selection is crucial for achieving high performance in classification models.

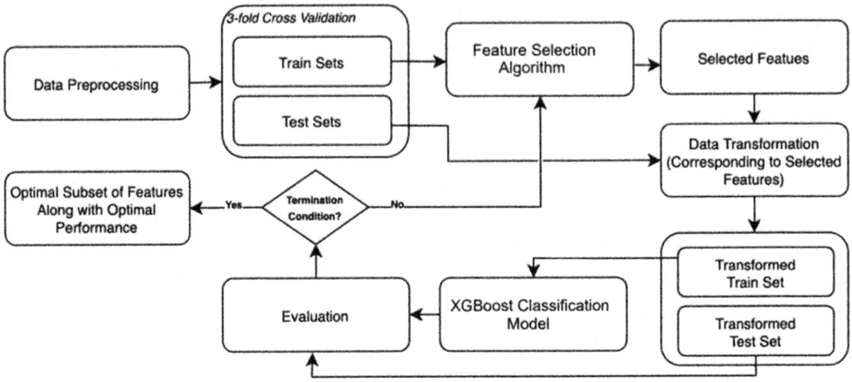

Fig. 2 The complete structure of feature selection

2 12.2 Implementation

2.1 Algorithmic Approach of the Machine Learning Model

This section focuses on outlining the approach along with describing the purpose of each step followed with the help of a flow diagram or flowchart. The approach is classifying student success using XGBoost classifier by performing feature selection using above-mentioned algorithms (Fig. 2).

The phases of data preprocessing, data transformation, optimization of features, training, and testing with the modified data which make up the study's overall procedure are thoroughly detailed below.

2.2 Data Preprocessing

To make the data comprehensible for the ML classifier [27], the raw data is altered into a format that can be precisely handled. This step includes tasks including feature engineering, feature scaling, and data cleaning. These tasks are summarized as follows:

- **Data cleaning**: Data cleaning entails finding and eliminating from the dataset any inaccurate records. In this investigation, we identified and eliminated from the dataset any characteristics that had more than 60.
- **Feature engineering**: This term refers to the process of generating fresh features from unprocessed data to boost the learning algorithm's capacity for prediction. One of the elements that was designed is the admission term for students. The work also utilizes another feature engineering technique, which involves one-hot

encoding of the variables. This method converts categorical features into their corresponding binary representations.

- **Feature scaling**: The method used to standardize the variety of characteristics in data is known as feature scaling, sometimes referred to as data normalization. An important step in data preprocessing is to avoid the algorithm from becoming biased toward the features with greater values in scale because the range of values for various features can vary greatly.

2.3 Feature Selection Optimization Using PSO, ABC, and A2BC

In Automated ML, feature selection is a key process that aids the model in achieving two main objectives: improved performance and shorter computation times.

2.3.1 Particle Swarm Optimization (Binary)

A method created specially to handle binary optimization tasks is called binary particle swarm optimization (BPSO). The binary-variable-compatible variant of particle swarm optimization (PSO). Like the PSO, the method updates the particle's position and velocity based on the personal best (pbest) and global best (gbest) solutions. The following formula is used to revise every particle's velocity (solution):

$$v_i^d(t+1) = wv_i^d(t) + c_1 r_1 \left(\text{pbest}_i^d(t) - x_i^d(t)\right) + c_2 r_2 \left(\text{gbest}^d(t) - x_i^d(t)\right) \quad (1)$$

$$S\left(v_i^d(t+1)\right) = \frac{1}{1 + \exp\left(-v_i^d(t+1)\right)} \quad (2)$$

$$x_i^d(t+1) = \begin{cases} 1, & \text{if rand} < S\left(v_i^d(t+1)\right) \\ 0, & \text{otherwise} \end{cases} \quad (3)$$

The formula used to calculate a particle's velocity (v), which determines its position, takes into account a number of factors, including the particle's inertia weight (v), acceleration factors (c1 and c2), two autonomous random numbers (r1 and r2) amid 0 and 1, the particle's (element's) personal best solution (pbest), the population's overall best solution (gbest), the particle's position in the population (i), and the size of the search space (d) A maximum velocity (vmax) and a minimum velocity (vmin) are further constraints on the velocity. In this investigation, the vmax and vmin values were chosen at 6 and 6, respectively. Then, as shown in Eq. (3), the particle's location is updated.

In BPSO, the personal best (pbest) and global best (gbest) solutions play a crucial role in directing the particle toward the global optimum, as the goal is to minimize the function. The pbest and gbest are revised iteratively, as shown below, where rand is a uniformly distributed random number between 0 and 1:

For pbest: If the current fitness value of a particle is better than its previously stored value, then the current fitness value becomes the new pbest value.

For gbest: If the current fitness value of a particle is better than the previously stored value of gbest, then the current particle becomes the new gbest.

$$\text{pbest}_i(t+1) = \begin{cases} x_i(t+1), & \text{if } F(x_i(t+1)) < F\left(\text{pbest}_i(t)\right) \\ \text{pbest}_i(t), & \text{otherwise} \end{cases} \tag{4}$$

$$\text{gbest}(t+1) = \begin{cases} \text{pbest}_i(t+1), & \text{if } F\left(\text{pbest}_i(t+1)\right) < F(\text{gbest}(t)) \\ \text{gbest}(t), & \text{otherwise} \end{cases} \tag{5}$$

A population-based optimization algorithm's fitness evaluation of a solution is represented by the equation. Here, x stands for the solution, pbest stands for the individual's best solution, gbest stands for the population's best solution, F(.) stands for the fitness function that is used to assess the solution's fitness, and t stands for the current iteration number (Fig. 3).

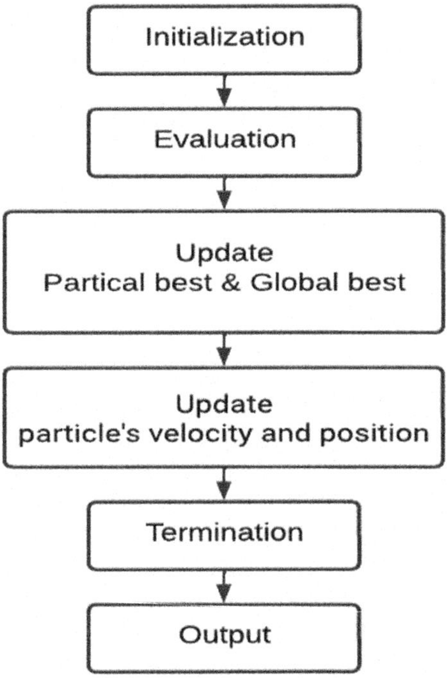

Fig. 3 The flow of PSO for feature selection

2.3.2 Artificial Bee Colony Algorithm

A number of real-world issues have been successfully solved using the Artificial Bee Colony (ABC) algorithm that simulates honeybee foraging behavior to optimize the features. Three different types of bees—worker, observer, and scout bees—make up ABC algorithm, which functions as a swarm to complete tasks. The worker bees look for food nearby the food source and communicate their findings to the observer bees. From among the food sources found by the hired bees, the observer bees pick the best ones. They are more likely to choose a food source with more capability than one with less capability. The scout bees are the ones which leave their current food source in search of new ones.

Initialization:

$$V_{i,j} = X_{i,j} + \text{rand}(-1, 1)\left(X_{i,j} - X_{k,j}\right) \tag{6}$$

The formula consists of a random number with a uniform distribution in the range [16] and an index k for one of the attributes. The superior food source is then chosen using a greedy selection strategy after comparing Vi,j and Xi,j. The employed bees then inform the spectator bees of the knowledge they have gathered. According to Eq. (2), observer bees select a roulette wheel to determine the probability values.

$$P_i = \frac{f_i}{\sum_{j=1}^{PS} (f_j)} \tag{7}$$

$$fit_i = \begin{cases} \frac{1}{1+f_i}, & f_i \geq 0 \\ 1 + abs(f_i), & f_i < 0 \end{cases} \tag{8}$$

Equation (3) demonstrates how the objective function (fi) is used to calculate fitness (fiti). We always have the first function in the equation since the classifier performs well, which is always in the positive. Therefore, in this situation, we are minimizing fitness or we are maximizing the objective function. If observing bees choose a food source, new food sources are generated and selected using greedy selection and Eq. (1) once more. The scout bee in ABC algorithm is picked from bees that are already working. This decision is based on the limit control parameter. As a result, if a particular food source is not improved for a predetermined number of iterations (limits), the hired bee will depart (or exhaust) that food source and become a scout. Equation (4) is then used to replace the left food source with at random generated food source, which can lead to the discovery of rich, undiscovered food sources.

$$X_{i,j} = x_{\min} + \text{rand}(0, 1)(x_{\max} - x_{\min}) \tag{9}$$

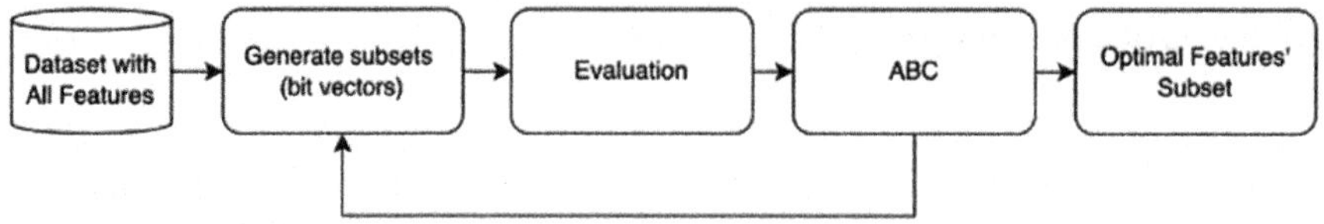

Fig. 4 The general schematic approach to ABC-based feature selection [18]

In the end, the algorithm is reset to the beginning after memorizing the position of the finest food supply. Up until the stopping requirement is satisfied, this process is continued. The stopping criteria could be a cutoff for a range of measures, including time, the maximum number of estimations or repetitions, or desired performance (Fig. 4).

In ABC, a colony of synthetic hunter bees (agents) looks for abundant synthetic food supplies (ideal results to a particular problem). The quality of each food source is equal to the classifier's accuracy across the validation set with the associated subset of features in feature selection problems involving ABC. The following are main actions to implement feature selection based on ABC algorithm:

1. Create the initial population.
2. Calculate fitness.
3. Exploitation by employed bees.
4. Exploitation by observer bees.
5. Memorizing the best food source.
6. Exploration by scout bees.

2.3.3 Automated Artificial Bee Colony Algorithm

The A2BC (automated artificial bee colony) algorithm has enhanced the ABC approach. Because A2BC increases classification precision while utilizing the least number of computational resources, it may be the ideal preprocessing technique to speed up the feature selection process. By including OBL (opposition-based learning) learning algorithms in the scout bee phase, this technique enhances the fundamental ABC algorithm's exploration potential. Additionally, the population variance is increased and the likelihood of finding potential locations is decreased due to the clustering technique utilized in the first step. In the initialization step, a clustering algorithm is included to partition the data into natural groups based on their similarity, and the ABC method is then used it to find the best characteristics. Since the data is grouped beforehand and the search is enhanced to keep relevant features in an acceptable period of time, the new expansion increases the accuracy of the ABC approach. The following is an explanation of the A2BC algorithm's steps:

Initialization: An agglomerative clustering method using the cosine similarity function is used to initially place repeated or similar individuals in the same cluster. We produce a population of arbitrary food sources (bit vectors), as illustrated in the algorithm. As a result, we only include one representative (random member) from each cluster in our new population for the objective function to evaluate. After selecting members from each cluster, subpar food sources are eliminated from the new population (based on a predetermined threshold). This is accomplished by filtering out the subpar representations to prevent additional scrutiny of subpar food sources. To focus on the food sources that can improve in a smaller number of iterations, the low-quality food sources are filtered out in this study. To utilize the search space, the method only updates one bit of the vector, as explained in the ABC steps for feature selection. As a result, if a bit vector has poor quality, it takes longer to exploit the food source and reach a high-quality food source. To increase the rate of convergence during the exploitation phase, such food sources should be stopped as soon as possible. To increase the classifier's dependability, classification accuracy is calculated using a three-folded cross-validation method. The dataset is partitioned into k-folds of similar size for k-fold cross validation, which involves running the ML algorithm three times. Cross validation is superior to a single train-test split as it trains across multiple train-test splits and provides a better knowledge of how well a model performs on missing data. The fitness function assesses the quality of food sources using **accuracy**.

$$f_{it_i} = \frac{1}{1 + f_i} \tag{10}$$

$$f_i = \text{Accuracy} = \frac{TP + TN}{TP + TN + FP + FN} \tag{11}$$

True positives, true negatives, false positives, and false negatives, respectively, are denoted by the letters TP, TN, FP, and FN.

Employed Bee Phase: Each bit vector is used by a worker bee to regulate the food sources by flipping just one of the bit values in the vector. If the neighboring (new) vector has a greater fitness, the preceding vector gets swapped with it and its associated accuracy is preserved. In other words, the employed bee evaluates the model's accuracy by including (bit 1) or omitting (bit 0) only one of the features and communicating the relevant information to observers.

Onlooker Bee Phase: The observers are informed of the quality of the food sources, and they pick the ones that are most likely to be investigated (according to Eq. (7)). Then, a few food sources are put through the same procedure as step 3 to be used for profit. The best food source is remembered once everyone has finished playing their roles.

$$P_i = \frac{\text{Fit}_i - \min(\text{Fit})}{\max(Fit) - \min(\text{Fit})} \tag{12}$$

Scout Bee Phase: A2BCF has an added step for this phase. In the initial ABC, a scout bee (tasked with finding any deserted food sources) creates a novel food source at random that is N squared in size. Scout bees alter every bit in the bit vector, as opposed to onlooker/employed bees who alter just one bit. Scout bees discover low-quality food sources on average, as we already explained. To create an opposing food source location based on Eq. (8), we created a step in which the abandoned food supply is subjected to Opposition-Based Learning (OBL). The program then advances with improved food sources derived from random and OBL techniques. The ML classifier is then presented with the created bit vector, and its accuracy is evaluated and stored.

$$\tilde{x}_{i,j} = 1 - x_{i,j} \tag{13}$$

Termination Process: The employed, observer, and scout bee stages will continue to run until the algorithm has received the set maximum number of evaluations or runs (Fig. 5).

2.3.4 Data Transformation

Data is transformed via this technique into the format needed by our destination system. As shown in Fig. 2, the initial dataset is preprocessed and then divided into training and test sets (using k-fold cross validation). The A2BCF optimization procedure then begins. New train and test sets are substituted for the originals based on the subset of features that the algorithm has picked. Once the modified training and test sets are ready, we apply the ML technique on the train set. Following that, the testing set is used to assess and record the results. Up until the stopping condition is satisfied, this procedure keeps going. The best feature set and an average accuracy that is more than three times greater are reported finally.

2.3.5 Train the Transformed Data

This stage involves submitting the XGBoost classifier and training the model. The final features and the XGBoost classifier's accuracy are returned. A created kind of gradient boosting known as extreme gradient boosting (XGBoost) makes use of an ensemble of gradient boosting frameworks. The effectiveness and speed of the parallelization-supporting model are one of XGBoost's main areas of focus. Ridge and Lasso's regularization is another strategy used by XGBoost to avoid overfitting. In each iteration of the training process, XGBoost corrects or fixes the more recent models.

3 Classifiers

The machine learning classifiers are summarized as follows[23–25].

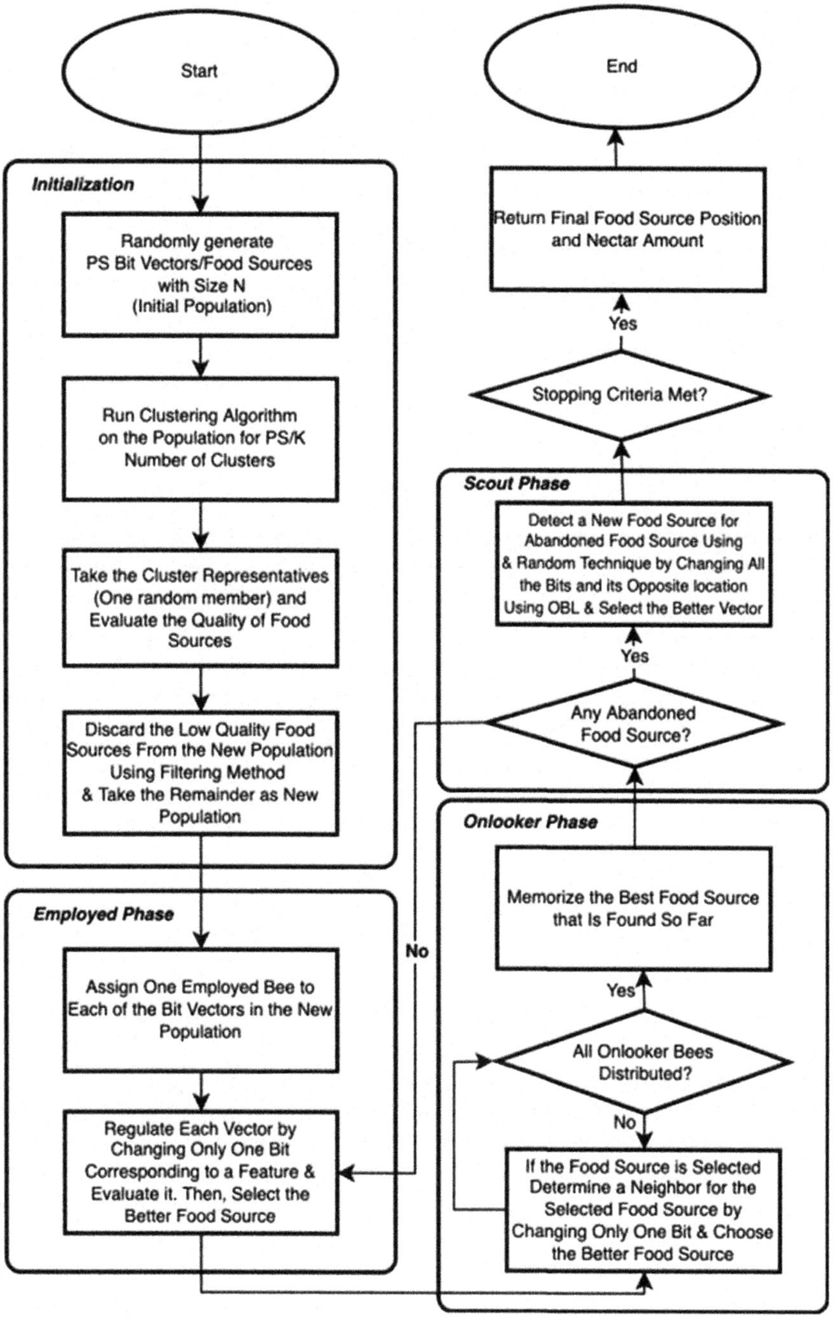

Fig. 5 The flowchart of A2BC algorithm

3.1 Support Vector Machine (SVM)

The Support Vector Machine (SVM) is a widely used machine learning technique for carrying out classification and regression analysis. SVM operates as a supervised learning algorithm and can be used for both multi-class and binary classification. The algorithm functions by identifying a hyperplane that effectively separates the data points into their respective classes, aiming to widen the gap as much as possible between the two classes. For the classification of new data points, SVM projects them onto the hyperplane and determines their class based on which side of the hyperplane they fall on. To handle datasets with numerous features and non-linearly separable datasets, SVM employs kernel functions.

3.2 Random Forest

Random Forest is an ensemble learning algorithm widely utilized for tasks like feature selection, regression, classification, and outlier detection. It works by training many decision trees on various subsets of data, then merging all the trees' predictions to produce the final prediction. This approach enhances the accuracy of the model and mitigates the problem of overfitting. The subsets used for training are produced through bootstrapping, where random samples are drawn from the initial dataset with replacement. In addition, at each decision tree split, a random subset of features is selected, resulting in a more varied and robust model. Because of the fact that it can manage high-dimensional datasets and complex interactions between features, Random Forest is a versatile algorithm and is commonly used in various applications.

3.3 XGBoost

XGBoost, also known as Extreme Gradient Boosting, is a versatile machine learning algorithm suitable for classification, regression, and ranking applications. The algorithm utilizes an ensemble learning approach that combines several decision trees, with each subsequent tree learning from the errors of the previous one. To prevent overfitting and enhance generalization, XGBoost includes a regularization term in the model. The loss function, which calculates the difference between the anticipated and actual values, is optimized by applying a gradient boosting framework. XGBoost is capable of handling high-dimensional datasets and feature interactions, and it also has built-in mechanisms for managing missing values and imbalanced datasets. It is a well-established algorithm in the industry and has demonstrated its effectiveness in various applications such as fraud detection and recommender systems.

3.4 Voting Classifier (DT + RF + SVM)

The Voting Classifier is an ensemble learning technique in machine learning that involves integrating the predictions of multiple classifiers to boost the accuracy of the final prediction. To achieve this, several individual classifiers are trained on the same dataset, and their predictions are combined. The Voting Classifier can be applied to classification tasks by computing the majority vote of the individual classifiers, or to regression tasks by calculating the average of their predictions. This method helps to enhance the overall accuracy and robustness of the model. For this application, the selected classifiers are Decision Tree, Support Vector Machine (SVM), and Random Forest.

3.5 Dataset Description

In education domain ,various tools, methods, mechanisms used to monitor the students for enhancing their education [28, 29]. The dataset used for students results pass or fail which is extracted from Kaggle. The dimensions of the dataset are 395 × 30 meaning there are 395 records of students and 30 features in total. The features include school, sex, age, address, mother's and father's education, family time, study time, grades, pass or fail, etc. All the data is numerical and not categorical. In the fields where the values are just 0's and 1's, 0 stands for no and 1 is for yes. And in sex, 0 is for females and 1 is for males. And for the remaining fields, the values range from 0 to 4 and 15 to 21 (for age).

4 Results

In the education science domain, all three algorithms were evaluated using various classifiers, including SVM, Random Forest, XGBoost, and Voting (SVM + RF + DT). The results showed that A2BC outperformed both PSO and ABC in terms of accuracy, sensitivity, and specificity, indicating that it is the best algorithm for feature selection in this particular domain. In summary, while all three swarm intelligence-based optimization techniques are effective in feature selection, A2BC demonstrated superior performance compared to PSO and ABC in the education science domain.

PSO:

Model with selected features

Accuracy: 94.11764705882352.
Feature Size: 17

ABC:

Model with selected features

Accuracy: 97.47899159663865
Feature Size: 16

A2BC:

N = 20

Model with selected features
Accuracy: 100.0
Feature Size: 3
Comparison between PSO, ABC, and A2BC

5 Conclusion

The present study focuses on the analysis of feature selection techniques applied to high-dimensional binary classification tasks in the field of education. Additionally, the study reviews the limitations of swarm intelligence-based feature selection approaches for high-dimensional classification found in the existing literature. The study introduces the concept of feature selection and compares the performance of three meta-heuristic algorithms, namely, PSO, ABC, and A2BC, used in the feature selection process. The results indicate that the selection of features is a critical stage in machine learning applications and can effectively analyze high-dimensional data by removing irrelevant and redundant information. The task of selecting the appropriate feature selection (FS) algorithm can be a daunting challenge in machine learning. To tackle this issue, numerous FS approaches have been proposed in recent years. This chapter focuses on the application of PSO, ABC, and A2BC meta-heuristic algorithms for high-dimensional education datasets. The study draws on previous literature and provides a comprehensive overview of the most used FS methods for high-dimensional classification.

To sum up, the advanced ABC-based feature selection algorithm offers a hopeful strategy for enhancing classification models in the field of education. It is proficient in selecting the most pertinent features from a considerable pool of potential predictors, resulting in the reduction of data dimensionality and the enhancement of the classification model's accuracy. Based on the comparison of accuracy and performance among Particle Swarm Optimization (PSO), Artificial Bee Colony (ABC), and Automated Artificial Bee Colony (AABC) machine learning models, the automated ABC outperforms the other two models with higher accuracy and better performance.

References

1. Houssein EH, Helmy BE, Rezk H, Nassef AM (2021) An enhanced Archimedes optimization algorithm based on local escaping operator and orthogonal learning for PEM fuel cell parameter identification. Eng Appl Artif Intell 103:104309
2. Hassanien AE, Kilany M, Houssein EH, AlQaheri H (2018) Intelligent human emotion recognition based on elephant herding optimization tuned support vector regression. Biomed Signal Process Control 45:182–191
3. Ismaeel AAK, Elshaarawy IA, Houssein EH, Ismail FH, Hassanien AE (2019) Enhanced elephant herding optimization for global optimization. IEEE Access 7:34738–34752
4. Houssein EH, Mahdy MA, Fathy A, Rezk H (2021) A modified marine predator algorithm based on opposition based learning for tracking the global MPP of shaded PV system. Expert Syst Appl 183:115253
5. Hamad A, Houssein EH, Hassanien AE, Fahmy AA (2018) Hybrid grasshopper optimization algorithm and support vector machines for automatic seizure detection in EEG signals. In: The international conference on advanced machine learning technologies and applications (AMLTA2018). Springer International Publishing, pp 82–91
6. Houssein EH, Mahdy MA, Shebl D, Manzoor A, Sarkar R, Mohamed WM (2022) An efficient slime mould algorithm for solving multi-objective optimization problems. Expert Syst Appl 187:115870
7. Houssein EH, Abdelminaam DS, Hassan HN, Al-Sayed MM, Nabil E (2021) A hybrid barnacles mating optimizer algorithm with support vector machines for gene selection of microarray cancer classification. IEEE Access 9:64895–64905
8. Hamad A, Houssein EH, Hassanien AE, Fahmy AA (2016) Feature extraction of epilepsy EEG using discrete wavelet transform. In: 2016 12th international computer engineering conference (ICENCO). IEEE, pp 190–195
9. Shaban H, Houssein EH, Pérez-Cisneros M, Oliva D, Hassan AY, Ismaeel AA, AbdElminaam DS, Deb S, Said M (2021) Identification of parameters in photovoltaic models through a runge kutta optimizer. Mathematics 9(18):2313
10. Abdelminaam DS, Said M, Houssein EH (2021) Turbulent flow of water-based optimization using new objective function for parameter extraction of six photovoltaic models. IEEE Access 9:35382–35398
11. Houssein EH, Hassaballah M, Ibrahim IE, AbdElminaam DS, Wazery YM (2022) An automatic arrhythmia classification model based on improved marine predators algorithm and convolutions neural networks. Expert Syst Appl 187:115936
12. Houssein EH, Neggaz N, Hosney ME, Mohamed WM, Hassaballah M (2021) Enhanced Harris hawks optimization with genetic operators for selection chemical descriptors and compounds activities. Neural Comput Appl 33:13601–13618
13. Ahmed MM, Houssein EH, Hassanien AE, Taha A, Hassanien E (2018) Maximizing lifetime of wireless sensor networks based on whale optimization algorithm. In: Proceedings of the international conference on advanced intelligent systems and informatics 2017. Springer International Publishing, pp 724–733
14. Houssein EH, Sayed A (2023) Dynamic candidate solution boosted beluga whale optimization algorithm for biomedical classification. Mathematics 11(3):707
15. Dulhare UN (2018) Prediction system for heart disease using Naive Bayes and particle swarm optimization. Biomed Res 29(12):2646–2649. ISSN 0970-938X
16. Singla J (2022) A framework for selecting features using various soft computing algorithms. In: 6th international conference on trends in electronics and informatics (ICOEI). IEEE. 978-1-6654-8328-5/22/31.00 ©2022. https://doi.org/10.1109/ICOEI53556.2022.9776721
17. Abdullah AR, Mohd Saad N (2019) A new co-evolution binary particle swarm optimization with multiple inertia weight strategy for feature selection. Informatics 6:21. https://doi.org/10.3390/informatics6020021

18. Ghareh Mohammadi F, Amini MH (2022) A2BCF: an automated ABC-based feature selection algorithm for classification models in an education application. Citation: Zahedi L, Ghareh Mohammadi F, Amini MH, A2BCF: An automated ABC-based feature selection algorithm for classification models in an education application. Appl Sci 12:3553. https://doi.org/10.3390/app12073553
19. Kumar V, Minz S (2014) Feature selection: a literature review. SmartCR 4:211–229 [CrossRef]
20. Y. Feature selection and classifier ensembles: a study on hyperspectral remote sensing data. Ph.D. Thesis, University of Antwerp, Antwerp, Belgium, 2003
21. Koller D, Sahami M (1996) Toward optimal feature selection; technical report; Stanford InfoLab: Bari, Italy, pp 284–292
22. Ahmed EF, Yang WJ, Abdullah MY (2009) Novel method of the combination of forecasts based on rough sets. J Comput Sci 5:440 [CrossRef]
23. Arif F, Dulhare UN (2017) A Machine Learning Based Approach for Opinion Mining on Social Network Data. In: Satapathy S, Bhateja V, Raju K, Janakiramaiah, B (eds) Computer Communication, Networking and Internet Security. Lecture Notes in Networks and Systems, vol 5. Springer, Singapore. https://doi.org/10.1007/978-981-10-3226-4_13
24. Niha SAR, Dulhare UN (2014) Extraction of high utility rare itemsets from transactional databases, International Conference on Computing and Communication Technologies, Hyderabad, India, pp 1–6. https://doi.org/10.1109/ICCCT2.2014.7066754
25. Dulhare UN, Ayesha M (2016) Extraction of action rules for chronic kidney disease using Naïve bayes classifier, IEEE International Conference on Computational Intelligence and Computing Research (ICCIC), Chennai, India, pp 1–5. https://doi.org/10.1109/ICCIC.2016.7919649
26. Dulhare UN, Ghori I (2018) An efficient hybrid clustering to predict the risk of customer churn, 2nd International Conference on Inventive Systems and Control (ICISC), Coimbatore, India, pp 673–677. https://doi.org/10.1109/ICISC.2018.8398883
27. Dulhare UN, Ahmad K, Ahmad KAB (eds) (2020) Machine learning and big data: concepts, algorithms, tools and applications, John Wiley & Sons
28. Kudikyala UK, Dulhare UN, (2015) Using Scrum and Wikis to manage student major projects, IEEE 3rd International Conference on MOOCs, Innovation and Technology in Education (MITE), Amritsar, India, pp– 15-20. https://doi.org/10.1109/MITE.2015.7375279
29. Kumar KU, Dulhare UN (2014) Enhancing Engineering Education in an Indian Education System: WebBlogs, Int J Comp Sci and Information Tech (IJCSIT), 5 (4): 5828–5830

Chapter 13
Metaheuristic Algorithm's Role in Medical Care and Diagnostics

Kiran Sharma

Abstract Metaheuristics is a strong optimisation tool that is currently being utilised more widely across medical disciplines to solve health-related optimisation issues. It can provide practical solutions to currently trending challenges in healthcare data to identify problems more effectively and efficiently than earlier approaches. Medical information from hospital databases and public health datasets are used to analyse anomalies via Internet of medical things and obtained large data is optimised using a metaheuristic search technique. A classified list of over a hundred metaheuristic algorithms was present to tackle any feature selection issues, and some recent variations of metaheuristic algorithms were also available to generate optimal feature subset using various methods. Nature-inspired metaheuristic algorithms are one of the sub-category in metaheuristic algorithms that was mostly demonstrated to be very flexible and effective in tackling complex optimisation issues in medical research. For improved performance, it is a standard practise to hybridise metaheuristics with another suitable algorithm. The fundamental purpose is to demonstrate the investigator's contributions by presenting their methods for predicting illnesses and solving health-related concerns utilising the metaheuristic approaches. Subsequently, their research can be compared and assessed using reliable, precise, specific, sensitive and other metrics to assist researchers in selecting the best field and techniques to anticipate ailments in the near future.

Keywords Metaheuristics · Algorithms · Machine learning · Optimisation · Dataset

K. Sharma (✉)
School of Dentistry and Medical Sciences, Charles Sturt University, Orange, NSW, Australia
e-mail: kiransharmaa1383@gmail.com

U. N. Dulhare and E. H. Houssein (eds.), *Machine Learning and Metaheuristics: Methods and Analysis*, Algorithms for Intelligent Systems,
https://doi.org/10.1007/978-981-99-6645-5_13

1 Introduction

The 'training' of machines is the method of machine learning, and the output created is known as the 'model'. Machine learning is a form of data analytics technology that automates the development of analytical models. It is a system that learns from data with little human interaction. Data is sent into the algorithm, which then generates new knowledge based on what it has already learnt [1, 46]. The categorical model is intended to evaluate a category as 'it's one of two things'. During the dataset classification, the model is trained. The clustering model is developed when a large amount of data is provided but no result has been established and the user just wants to identify unique trends in the data [2]. The regression model was designed to detect values. The programme will use data to identify a correlation among two factors, and the outcome will be predicted appropriately. Deep learning (DL) methods have been widely used to handle complicated issues in a number of domains, including healthcare applications like medical imaging processing or medical information mining, due to the quick development of algorithms, technology and the massive expansion in the volume of data[47]. Neuro-evolution and metaheuristic optimisation techniques have lately been employed to handle increasingly complicated issues and to optimise DL models [3]. With the proliferation of smart sensory media, objects and cloud technologies, innovative health care is gaining traction among government agencies, academia, industry and healthcare organisations. The Internet of Things (IoT) has recently made the idea of a smart world a reality by providing a plethora of services and a massive amount of data. Because it provides a customisable stack of computer, programmes and storage offerings at a cheaper cost, cloud computing is ideally suited for enabling innovations in the scenarios. Cloud-based services might provide clinics, carers and physicians with a superior quality of experience at all times and from any location. However, the cloud and IoT convergence may open up new avenues for these two strategies. The Internet of Things cloud convergence plays an important role in intelligent medical care by recognising diverse healthcare contents and therefore promoting excellent and inexpensive health care [4].

A metaheuristic optimisation technique is a more advanced method created to discover, generate, tune or opt for a heuristic that could offer an adequately good solution to an optimisation problem or a machine learning obstacle, specifically when information is inadequate or deficient or computation capacity is limited. Metaheuristics sample a subdivision of possibilities that could else be too huge to be fully computed or inspected [5]. Metaheuristics can make minimal assumptions regarding the optimisation issue being addressed and hence may be applicable to a wide range of problems. The behaviours of swarming, birds and animals have inspired several metaheuristic optimisation techniques. Hybrid metaheuristic algorithms have also been used as a sophisticated solution to more complicated situations. Because of the capacity of metaheuristic techniques to identify optimum solutions, hybridisation approaches have been developed to mix classic machine learning methods or sophisticated deep learning methods with metaheuristic algorithms. As a result, these techniques have a lot of potential for medical applications such as picture segmentation, ageing monitoring, and text evaluation and classification [6].

Medical data is being acquired on a regular basis from many systems, including hospitals, health organisations, wearable devices (for example, to follow the behaviours of the geriatric), smartphone sensors, home automation systems and air quality records, to name a few. As a result, more robust systems utilising optimised DL techniques are required to deal with the enormous dimensions of data. Traditional approaches confront significant obstacles when dealing with various types of medical data, such as photos, text and others [7]. Because of the high dimensionality and vast amount of this data, which demand more time, hybrid deep learning and neuro-evolution metaheuristic optimisation algorithms may deliver more efficient solutions. With the use of machine learning, healthcare practitioners may make better judgements regarding treatment alternatives for patients, resulting in overall changes in healthcare services. Machine learning algorithms are also beneficial in supplying medical practitioners with real-time data statistics and sophisticated analysis regarding the patient's condition, laboratory test results, blood pressure, familial background, data from clinical trials and so on [8].

2 Classification of Metaheuristics

Metaheuristic algorithms search for the ideal (feasible) answer to an optimisation issue among all conceivable elucidations. To that function, they analyse probable resolutions and run a successions of processes on them with the objective to uncover alternative, superior answers. Metaheuristics work on the presentation or coding of a solutions, an information which can be kept in computer memory and readily handled by the different operators. On the basis of how solutions are changed, three basic groups of metaheuristics may be recognised. Local search metaheuristics relatively made modest adjustments to a solo solution. Constructive metaheuristics build elucidations from their constituent elements. Population-based metaheuristics integrate elucidations into different solutions repeatedly [9]. However, mentioned categories are not mutually reinforcing, and various metaheuristic algorithms mix principles from many categories. Such strategies are known as hybrid metaheuristics. Metaheuristics are classified in a variety of ways, including the following:

- Nature-inspired metaheuristics versus non-nature-inspired metaheuristics. It is based on a metaheuristic's roots.
- Memory-based metaheuristics versus memory-less metaheuristics. They are determined by how they use the search memory, i.e. did they employ memory or not. Memory is regarded as one of the most important components of a successful metaheuristic.

- Population-based metaheuristics versus single-point metaheuristics. A metaheuristic determines how many solutions are used at any one moment. At each iteration, population metaheuristics (Fig. 1) alter a collection of solutions from that the sample of the following iteration is generated. Evolutionary algorithms and scatter searches are two examples, as are construction-oriented approaches like ant colony optimisation and the greedy randomised adaptable search strategy. Trajectory metaheuristics are metaheuristics that work with just one solution at a time, and the search procedure specifies a trajectory in the area of search [10].

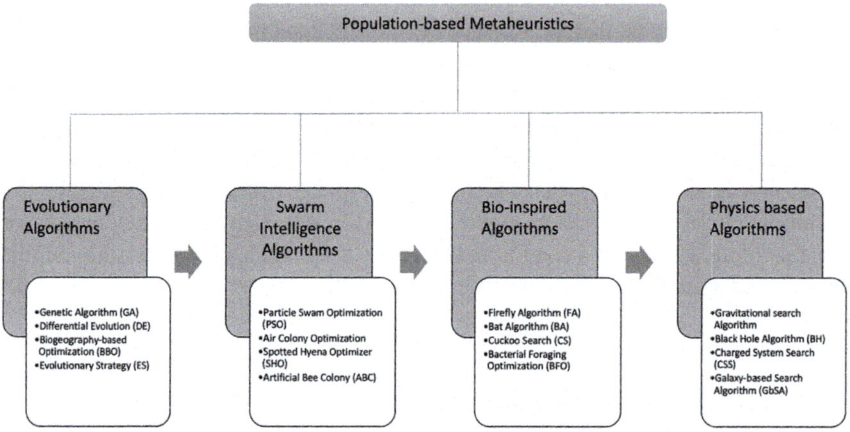

Fig. 1 Schematic representation of metaheuristics classification based on population

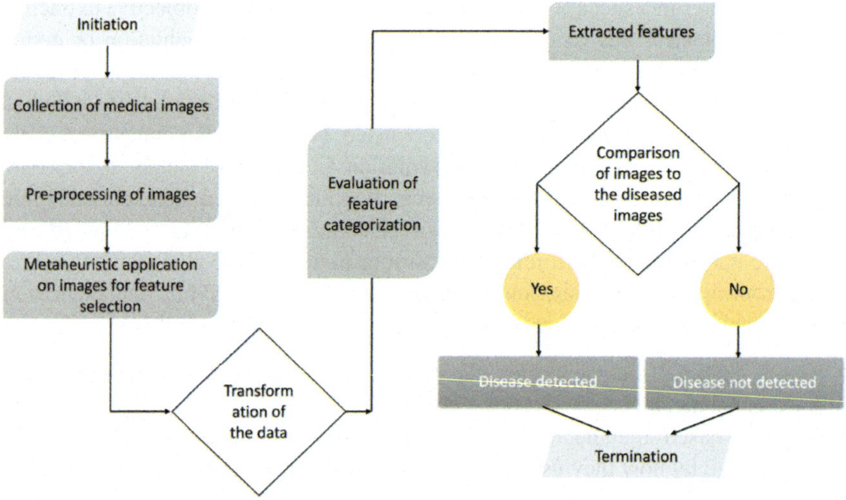

Fig. 2 Schematic representation of medical image segmentation using metaheuristic algorithms [18]

Encouraging results were recently accomplished by application of metaheuristic algorithms in tackling various sorts of optimum design issues for various uses in the empirical research. For example, particle swarm optimisation (PSO)/an altered type of PSO utilised to identify multiple kinds of optimal designs for medical applications, such as minimax optimal designs or standardised maximin criterion, in which the former requires completing three- or four-level nested optimisation loops over extremely distinct search spaces. PSO basis algorithms have also been demonstrated to be capable of handling discrete optimisation challenges [11]. In improved versions of PSO, for example, discovered (a) best saturated designs with an extensive sum of features, (b) optimal measurements repetitive over the course of time in a model known as Michaelis–Menten, and (c) a category of second stage adaptable designs which includes the honoured Simon two-Stage designs; the expanded designs resolved complex optimisation challenges containing 10-integer variables that faced several nonlinear limitations from the mode I and II errors. Other forms of metaheuristic algorithms have recently been used in statistical methods to solve optimisation issues, such as utilising Imperialist Competitive Algorithm in identifying Bayesian design options for a pharmacometric simulation or a differential evolution (DE) method to search best designs in chemometrics. Utilising a DE and a form of PSO termed quantum PSO, ideal designs for nonlinear models along with numerous related components in a regression scenario are also discovered [12].

Nature-inspired metaheuristic algorithms have grown in reputation over the short period of time [31–45], and now hold a dominating position in both academic and industry applications. They currently represent a crucial component of artificial intelligence. Their modesty, ease of implementation and capability to deliver a practical solution to an optimisation issue quickly are all appealing. Another advantage of these methods is that they frequently make no assumptions about the function to be optimised, with scholars from a diverse field of specialisations finding greater success rates in identifying an ideal or nearly ideal solutions to various groups of complex optimisation issues in medical and engineering, in which there are often thousands or hundreds of variables to optimise [13]. Hybridisation is a frequent and often successful strategy for improving algorithmic efficiency by integrating two or more algorithms to solve a difficult optimisation issue. Because of model nonlinearity, numerous mixed effects, various optimisation targets or user-imposed limitations, optimisation issues in pharmacometrics can be highly difficult. Practical factors, for example, may demand that assessments from a topic not be obtained too close to one another, or that cost structures from various places differ. Metaheuristics has the ability to provide appropriate answers to such issues when standard approaches fail. PSO was hybridised with arbitrary forest to foresee disease development in individuals with idiopathic pulmonary fibrosis, for example, and hybrid methods for developing optimum designs for distinguishing multiple nonlinear models underneath different error distributional presumptions were developed [14].

3 Applications of Algorithms in Health Care

Artificial intelligence is a branch of computing concerned with producing machines that function like humans. It jeopardises expert systems, video games, natural language processing and cybernetics. Soft computing (SC), on the other hand, differs from hard (traditional) computation in that it is tolerant of partial truth, ambiguity, imprecision and estimation, and so reflects the way humans think. Artificial neural networks, Internet of medical things, deep learning and metaheuristic algorithms are among the most prevalent SC approaches. Soft computing technologies have appealed the interest of various academics in the last few years for solving challenges in health-related data applications [15]. Segmentation of pictures is the process of dividing a picture into groups based on its similarities. This method is used to detect the nature of a disease using anomalies volumetric analyses in medical pictures. Recently, metaheuristic algorithms have been developed to aid segmentation procedures. Image assessments, recognition of patterns and image sciences are among the most noteworthy zones of medical science and engineering in a range of fields like medicine, defence and computer science with their practical applications. Image-guided management is a critical approach for precise diagnosis in healthcare settings. Medical examination image computation is becoming progressively more important in healthcare diagnostics. Image analysis systems use new image handling algorithms to isolate measurable features from these images to help in identification of a disease and providing best medical therapy. Automated and robust medical image processing systems have been an important research topic in order to produce precise clinical processes [16].

Model-based image evaluation and image-based modelling approaches are becoming important tools for reliable and accessible analysis of items in medical images. These approaches need prior knowledge of the architecture of medical imaging, such as bones, cancers, arteries and different organs. Following that, image-based modelling algorithms were used to automatically determine the relevant characteristics. Preparation of digital indicative imaging information is done to facilitate illness progression tracking, diagnosis and pre-operative planning for complicated visualisation and quantitative metrics. However, good image investigation requires optimised and advanced processing systems, which is a challenging factor [17]. At present, medical image evaluation technology is being practiced by only a few reputable clinics in order to produce effective solutions grounded on computer approaches.

The primary processing of images methodologies are medical picture segmentation and categorization. The segmentation of magnetic resonance (MR) images with high soft tissue contrast plays an important role in assessing brain tumour treatment. Manual segmentation performed by doctors is still the old standard for abnormal brain imaging, although it is prone to human biasness and is a time-consuming operation. As a result, an effort for trustworthy automation of medical picture segmentation is highly sought. This necessitates the employment of clustering algorithms to identify the clarity of a medical picture into a certain amount of clusters, wherein the total quantity of clusters is called a priori in the majority of anatomical forms. In addition, medical image categorisation is often utilised to distinguish between aberrant and normal pictures [19]. A computerised categorization procedure is widely wanted to assist physicians with the tedious work of evaluation.

3.1 Genetic Algorithm

Most often used algorithm in medical sector is the genetic algorithm (GA) because the outcome of a genetic algorithm for optimisation issue may be expressed as integers or binary values, and several researches have employed GA for medical scheduling challenges, like minimising waiting times for patients time. Another emphasis is on using the evolutionary algorithm to the timing issue faced by nurses in order to create an improved programme to enhance the movement of patients in the emergency section of the hospital, thereby dramatically reducing patient queuing time as contrast to the schedule plan manually. A later research evaluated multiple goals at exactly the same time (like total patient wait time and doctor schedule time) in order to maximise the impact of medical resources while minimising wasted expenses [20]. A potential way to apply metaheuristics for healthcare challenges is to utilise GA to identify best parameters to regulate the classifiers or learn the categorisation algorithms. Another technique to improve the effectiveness of an algorithm for classification utilising GA is to choose appropriate characteristics from the raw information (i.e. lower the total amount of components) to increase the categorisation algorithm's efficiency. For instance, GA was utilised to pick the six most essential variables for predicting cardiovascular disease from a set of 13. Since the method for choosing features can significantly decrease the complexity of the data, this reduces over 50 percent of the computing time for similar data. A comparable investigation employed a parallel genetic algorithm to minimise not just facts complexity but also GA time to respond. Furthermore, the cardiovascular diagnostic technique was created as an embedded procedure in the identical research so that it may be utilised as a diagnostic device for cardiovascular diseases in everyday situations [21].

The parameters with which each chromosome encodes, such as the random starting weight generator, the amount of hidden nodes and chosen attributes, are particularly fascinating aspects of the study. As a consequence, within the convergence phase, GA may concurrently determine improved starting weights, the amount of hidden nodes and the characteristics to be picked for multi-layer perception. GA has also been used effectively in the management of medical resources and information. A watermarking strategy that employs GA for exclusive rights protection and authorisation of medical photographs could be identified, where GA and PSO are employed to minimise watermark embedding calculation time. Of course, regardless of the approach chosen, whether GA or PSO, the outcomes are all superior compared to those produced by discrete cosine transform (DCT) and least significant bit (LSB) in regards to attack resilience. Another investigation found that GA may be utilised for identifying medical fraud. In a study, GA was utilised to calculate improved weights of data characteristics for a system of classifications. Unlike the research presented in, the investigation presented in first modelled the medical management of resources issue as a maximum coverage location challenge and then utilised GA to identify an effective solution for RFID reader placements in order to maximise the potential coverage of RFID readers on the service's area [22]. GA may identify microcalcifications in mammography that indicate breast cancer. The boundary of the breast and the nipple location were discovered by the GA in an approach after augmentation and normalisation of the mammograms. The mammography pictures were aligned and deducted from one another utilising the boundary and nipple location of the right and left breasts as a point of reference to determine the asymmetry image indicative of breast cancer. The Az-value, referred to as the area underneath the receiver operating characteristic (ROC) curve, is often used to evaluate a system's diagnosis efficiency. Their recommended algorithm's Az-value was approximately 0.9.

3.2 Ant Colony Optimization (ACO)

The ACO seeks an ideal path based on the behaviour of ants finding a path among their colonies and food sources. It is commonly acknowledged that in nature, ants roam aimlessly at first and then returned to their home while leaving pheromone trails. Pheromone trails are a key role in guiding other ants. If additional ants discover such a track, they will most likely following it, returning and strengthening the pheromone trails as they ultimately locate equivalent food sources. Nevertheless, it is important to remember the fact that pheromone trails will fade with time. Then, the longer it requires for an ant to go along the route and come back more of the pheromone trails would fade. As a result, if the route is shorter in length the ant will make more choices and the density of pheromones will continue to increase. The lengthier trail, on the other hand, may be travelled less resulting in lower pheromone concentrations. Assuming that there was no evaporation, the pathways chosen by the very first ant could be overly appealing to the succeeding ones. However, in this instance, the investigation of the solution's space may be limited. As a result, if one ant looks for a quick path across their colony to the source of food, other ants will more likely to follow in its footsteps. The ACO is designed and used to solve challenges that utilise these behaviours.

ACO employs a novel approach to simulating the natural behaviour of ants in order to discover a satisfactory solution to the medical optimisation problem. Although there aren't many researches that employ ACO for medical care, it does offer limitless potential that may be discovered in the investigations. Kuo and Shih employed ACO to determine the principles of association for medical insurance datasets and to cluster related data. These findings demonstrate that ACO may be utilised as analytics for various data mining jobs in health care. Furthermore, Kuo and Shih discovered several intriguing and relevant association rules that may be used to improve the efficiency of preventive immunisation. The research suggests that ACO can enhance the outcomes of a classification algorithm for medical classification challenges [23]. ACO was also used with a fuzzy rule to categorise hepatitis components. A further study paired ACO with linear discriminant analysis (LDA) to better comprehend endoscopic data on stomach cancer. More specifically, in this hybrid technique known as ACO-LDA, LDA does clustering and ACO performs categorisation. A recent research found that ACO may also be used for predicting cardiac disease. Uma and Kirubakaran coupled ACO and GA and used these two algorithms at every stage of the development of convergence to choose the best characteristics for the classification algorithm used in this work [18].

3.3 Particle Swarm Optimization (PSO)

Eberhart and Kennedy suggested the PSO, which was inspired by the movement of swarms. The fundamental system of PSO is given in a supplementary file, and as with other similar algorithms, it has undergone several revisions since its first implementation. According to studies, PSO has the same level of efficiency as the GA but PSO demands considerably fewer function assessments and, as a result, less processing time. Another explanation for PSO's prevalence is that the tuning parameters' default settings that have been proved to function effectively for a wide range of issues, particularly estimation of parameters in non-pharmacometrics contexts. Comprehensive studies of PSO and its numerous versions, as well as their different uses, including model selection and parameter estimates, are available. PSO was also used to analyse data from single nucleotide polymorphisms (SNPs) in renin-angiotensin system gene in order to better comprehend the relationship between SNPs and hypertension. Another intriguing research area is the application of PSO as a breast cancer screening method. The study employed statistical approaches to determine important characteristics before using PSO to divide the entire population into two groups: those with breast cancer and those without. It is worth noting that the study represented the PSO algorithm using a discrete encoding. As a consequence, the medical system discovers certain valuable decision criteria (i.e. breast cancer pattern) that will benefit clinicians in identifying breast cancer [24]. PSO is occasionally utilised to calculate the weight attributes for the technique of categorisation or to choose the beneficial characteristics from the available information to increase the reliability rate of a system for classification. Chowdhury et al. employed the PSO to identify the best pathophysiological component weights for a diagnostic system, which they subsequently implemented on the field programmable gate array (FPGA). It is clear from the research's results that an adaptive approach was utilised to actively alter the perceptual selection (i.e. the influence of further components on the location of a component) of every particle of PSO that may be utilised to increase the reliability level of the categorisation findings. Another investigation found that PSO may be used to choose relevant characteristics for other classification algorithms. PSO can be integrated with a neural network in the present research to choose parameters for support vector machine (SVM) analysis of lymph nodes in ultrasound scans [25].

In adjunction to being utilised as a categorisation method, several researches have shown that PSO may be used for a variety of health-related applications.

PSO has been utilised to direct the weight of a neural network for a touch floor technology that tracks a user's location. The study presented employed PSO to determine the ideal position of MRT stations for screening individuals to avoid the spread of infectious illness. For optimising the quality of the samples that were gathered, the key features of concerned examination are in modelling the challenge as maximising the number of individuals which can be inspected per important node and regulating the quantity of individuals on every development guide plan area. Similar to how GA and ACO were used to establish an improved hospital scheduling plan, the research project employed PSO to identify a decent sequencing for all of the patients' surgeries by addressing it as a scheduling challenge. According to the findings, PSO can develop an improved scheduling strategy than standard scheduling methods like the first-come, first-served system (FCFS). PSO was used in conjunction with an evolutionary algorithm (EA) which can run alternatively at each iteration of the process of convergence in this hybrid algorithm, implying that PSO may be employed in conjunction with such other metaheuristics so as to perform analysis of a medical facility [26].

3.4 Firefly Algorithm

FA ensued initially suggested by Xin-She Yang, who was instigated by the flashing behaviour and patterns of fireflies. It is a widely recognised and effective nature-inspired swarm intelligence and has grown to be well appreciated among scientists due to its competence and has turned out to be a fascinating approach for professionals as well as scientists for resolving issues across multiple fields of investigation like classifications, grouping, neural networks, biomedical technology, medical treatment and other areas of study since its inception. Furthermore, FA has an established record of tackling biomedical engineering (BE) and healthcare (HC) difficulties. So many challenges are being solved with the help of FA and their derivatives. Considering these details into account, this work provides detailed examination of the variations, significance, uses and upgrades of FA medical care. Al-Thanoon et al. introduced a unique approach for identifying characteristic features by tuning in PSVM (penalised support vector machine) using the blending of FA and statistical PSO in 2019. Experiments using various chemical datasets to evaluate the efficiency of the researchers-suggested approach revealed that the system is capable of managing appropriate augmentation and modification, and it has been demonstrated to be successful when competed with arithmetic PSO, FA and cross validation (CV). Various finding's also discovered that the experiments had a greater degree of precision in terms of categorisation than the comparisons [27]. Wang et al. suggested incorporating a combined mutation process in an enhanced hybrid FA to effectively taking both local and universal search abilities while resolving optimisation challenges. Because chaotic search (CS) demonstrates greater ergodicity, an evolution technique for arbitrary linkage of all fireflies based on CS had been presented, which increases the algorithm's capacity for negotiating the whole search space. Zhang et al. proposed a new method for extracting end-members (EE) from hyper-spectral pictures using discrete FA (EE-DFA). To find optimal results, the researchers used the end-members from the hyper-spectral picture and viewed it as an issue of combinational optimisation which can be handled by DFA. For the entire set of tests, they investigated a variety of datasets. Yuan et al. created a Modified Firefly Algorithm (MFA) in 2016 for the grouping of Chinese Medicine prescriptions that is a mix of FA and SAM (simulated annealing technique). It has been noticed that when fitness levels fluctuate, so does the capability of the global search technique [28]. Subsequently the projected approach was examined in relation to the basic K-Means clustering algorithm, which was discovered to enhance time for executing the technique with better clustering analysis was discovered when compared to other evaluated methods.

3.5 Gravitational Search Algorithm

Raahedi et al. provided the gravitational search algorithm (GSA) in 2009. It is an easy to construct swarm intelligence global optimisation method that produces pretty excellent performance in worldwide search and convergence. It is widely utilised in many domains, including planning paths, classification of images, neural networks, data estimation, scheduling and the estimation of parameters. BGSA, the binary counterpart of GSA, has already been utilised for feature selection and has demonstrated promising outcomes in a variety of areas. GSA, which is conceptually similar to PSO, is even more effective than the latter, it uses less storage spaces. Convergence occurs faster, and features an effective exploration step. Consequently, scientists have effectively used the binary variant of GSA, i.e. BGSA, for feature selection, yielding promising results in areas such as power distribution networks, natural scene categorization, vowel identification and losses that are not technical in nature. Asha Gowda Karegowda et al. employed GA for feature selection on healthcare datasets. They used a variety of classifiers and compared their findings [29]. Reliability has improved, and a smaller selection of characteristics has been picked in each case. Amir Rajabi Behjat et al. have also utilised it for feature selection in intrusion detection. GSA, on the other hand, has received less attention in the medical field. GSA has been employed on a variety of datasets, including medical ones, by S. Sarafrazi et al. and J. Xiang. Sarafrazi et al. proposed a hybrid GSA-SVM approach for feature selection, whereas J. Xiang merged the 1-nearest neighbour technique with BGSA to get the best feature subset [30]. For highly dimensional gene selection situations, a pyramid variant of GSA is utilised. The suggested approach is a hybrid strategy that decreases the total amount of genes cyclically and picks the fewest genes for accurate classification. The name pyramid refers to the downward-sloping method for feature selection using PGSA, the depth of which is dictated by the nature of the issue and the number of characteristics that may be required.

4 Summary

Artificial Intelligence is infiltrating many parts of our existence, including medical care. Scientists are observing and establishing a better understanding of the challenges and possibilities offered by AI as a smart technology in the area of health-related identification of illnesses utilising metaheuristic algorithms. Metaheuristics offer a feasible answer to medical care, which requires intelligent techniques to give essential services to worldwide organisations, governments, companies, hospitals, universities, homes and individuals. Metaheuristics often identify an approximate answer faster than neural networks and exhaustive search techniques. When competed to deterministic and rule-based algorithms, metaheuristics generally produce better findings since metaheuristics can give an accurate and reliable solution having considerably few calculation time. They have emerged as a potential healthcare study trend in the past few years, particularly for systems that require real-time results. Since metaheuristics may be utilised to improve the execution of a data mining algorithm, various metaheuristic-based algorithms have been proposed over the past decade. Aside from the data mining challenge in medical care, certain latest findings have employed metaheuristics to tackle various optimisation difficulties in the healthcare industry, such as maximising medical resources to minimise overall patient waiting time. A novel approach for acquiring cine-MR (magnetic resonance) brain pictures has just been developed to synchronise the MRI signal with the ECG (electrocardiographic) signal. As a whole, metaheuristic-based techniques lead the search process, successfully explore the field of medical search to locate the best solutions, provide estimations and are not problem-specific.

References

1. Najafabadi MM, Villanustre F, Khoshgoftaar TM, Seliya N, Wald R, Muharemagic E (2015) Deep learning applications and challenges in big data analytics. Journal of big data 2(1):1–21
2. Patibandla RL, Veeranjaneyulu N (2018) Survey on clustering algorithms for unstructured data. In: Intelligent engineering informatics: proceedings of the 6th international conference on FICTA, pp 421–429
3. Wang X, Zhao Y, Pourpanah F (2020) Recent advances in deep learning. Int J Mach Learn Cybern 11:747–750
4. Maksimović M, Vujović V (2017) Internet of things based e-health systems: ideas, expectations and concerns. Handbook of large-scale distributed computing in smart healthcare 1:241–280
5. Nesmachnow S (2014) An overview of metaheuristics: accurate and efficient methods for optimisation. Int J Metaheuristics 3(4):320–347
6. Ting TO, Yang XS, Cheng S, Huang K (2015) Hybrid metaheuristic algorithms: past, present, and future. Recent Adv Swarm Intell Evol Comput 1:71–83
7. Tsanas A, Little MA, McSharry PE (2012) A methodology for the analysis of medical data. In: Handbook of systems and complexity in health, pp 113–125
8. Abd Elaziz M, Dahou A, Abualigah L, Yu L, Alshinwan M, Khasawneh AM, Lu S (2021) Advanced metaheuristic optimization techniques in applications of deep neural networks: a review. Neural Comput Appl 1:1–21

9. Ezugwu AE, Shukla AK, Nath R, Akinyelu AA, Agushaka JO, Chiroma H, Muhuri PK (2021) Metaheuristics: a comprehensive overview and classification along with bibliometric analysis. Artif Intell Rev 54:4237–4316
10. Agushaka JO, Ezugwu AE (2022) Initialisation approaches for population-based metaheuristic algorithms: a comprehensive review. Appl Sci 12(2):896
11. Mokarram V, Banan MR (2018) A new PSO-based algorithm for multi-objective optimization with continuous and discrete design variables. Struct Multidiscip Optim 57:509–533
12. Chen PY, Chen RB, Wong WK (2022) Particle swarm optimization for searching efficient experimental designs: a review. Wiley Interdisc Rev: Comput Stat 14(5):e1578
13. Nanda SJ, Panda G (2014) A survey on nature inspired metaheuristic algorithms for partitional clustering. Swarm Evol Comput 1(16):1–8
14. Shi Y, Wong WK, Goldin JG, Brown MS, Kim GH (2019) Prediction of progression in idiopathic pulmonary fibrosis using CT scans at baseline: a quantum particle swarm optimization-Random forest approach. Artif Intell Med 100:101709
15. Sharma KS (2023) Artificial intelligence assisted fabrication of 3D, 4D and 5D printed formulations or devices for drug delivery. Curr Drug Deliv 20:752–769
16. Ma X, Niu Y, Gu L, Wang Y, Zhao Y, Bailey J, Lu F (2021) Understanding adversarial attacks on deep learning based medical image analysis systems. Pattern Recogn 110:107332
17. Tajdari M, Pawar A, Li H, Tajdari F, Maqsood A, Cleary E, Saha S, Zhang YJ, Sarwark JF, Liu WK (2021) Image-based modelling for adolescent idiopathic scoliosis: mechanistic machine learning analysis and prediction. Comput Methods Appl Mech Eng 374:113590
18. Tsai CW, Chiang MC, Ksentini A, Chen M (2016) Metaheuristic algorithms for healthcare: open issues and challenges. Comput Electr Eng 53:421–434
19. Ramesh KK, Kumar GK, Swapna K, Datta D, Rajest SS (2021) A review of medical image segmentation algorithms. EAI Endorsed Trans Pervasive Health Technol 7(27):e6
20. Ghaheri A, Shoar S, Naderan M, Hoseini SS (2015) The applications of genetic algorithms in medicine. Oman Med J 30(6):406
21. Arabasadi Z, Alizadehsani R, Roshanzamir M, Moosaei H, Yarifard AA (2017) Computer aided decision making for heart disease detection using hybrid neural network-genetic algorithm. Comput Methods Programs Biomed 141:19–26
22. Suriya A, Porter JD (2014) Genetic algorithm based approach for RFID network planning. In: TENCON 2014–2014 IEEE region 10 conference, pp 1–5
23. Guo X, Zhou HY, Guo S, Luan XX, Cui WK, Ma YF, Shi L (2014) Design of broadband omnidirectional antireflection coatings using ant colony algorithm. Opt Express 22(104):A1137–A1144
24. Purzycka-Bohdan D, Nedoszytko B, Sobalska-Kwapis M, Zabłotna M, Żmijewski MA, Wierzbicka J, Gleń J, Strapagiel D, Szczerkowska-Dobosz A, Nowicki RJ (2023) Assessment of the potential role of selected single nucleotide polymorphisms (SNPs) of genes related to the functioning of regulatory T cells in the pathogenesis of psoriasis. Int J Mol Sci 24(7):6061
25. Lokanayaki K, Malathi A (2013) Exploring on various prediction model in data mining techniques for disease diagnosis. Int J Comput Appl 77(5):234
26. Aruna M, Bhanu D, Karthik S (2019) An improved load balanced metaheuristic scheduling in cloud. Clust Comput 22:10873–10881
27. Li J, Wei X, Li B, Zeng Z (2022) A survey on firefly algorithms. Neurocomputing 500:662–678
28. Nayak J, Naik B, Dinesh P, Vakula K, Dash PB (2020) Firefly algorithm in biomedical and health care: advances, issues and challenges. SN Comput Sci 1(6):311
29. Karegowda AG, Jayaram MA, Manjunath AS (2012) Cascading k-means clustering and k-nearest neighbor classifier for categorization of diabetic patients. Int J Eng Adv Technol 1(3):147–151
30. Rashedi E, Rashedi E, Nezamabadi-Pour H (2018) A comprehensive survey on gravitational search algorithm. Swarm Evol Comput 41:141–158
31. Houssein EH, Helmy BE, Rezk H, Nassef AM (2021) An enhanced archimedes optimization algorithm based on Local escaping operator and orthogonal learning for PEM fuel cell parameter identification. Eng Appl Artif Intell 103:104309

32. Hassanien AE, Kilany M, Houssein EH, AlQaheri H (2018) Intelligent human emotion recognition based on elephant herding optimization tuned support vector regression. Biomed Signal Process Control 45:182–191
33. Ismaeel AAK, Elshaarawy IA, Houssein EH, Ismail FH, Hassanien AE (2019) Enhanced elephant herding optimization for global optimization. IEEE Access 7:34738–34752
34. Houssein EH, Mahdy MA, Fathy A, Rezk H (2021) A modified marine predator algorithm based on opposition based learning for tracking the global MPP of shaded PV system. Expert Syst Appl 183:115253
35. Hamad A, Houssein EH, Hassanien AE, Fahmy AA (2018) Hybrid grasshopper optimization algorithm and support vector machines for automatic seizure detection in EEG signals. In: The international conference on advanced machine learning technologies and applications (AMLTA2018). Springer International Publishing, pp 82–91
36. Houssein EH, Mahdy MA, Shebl D, Manzoor A, Sarkar R, Mohamed WM (2022) An efficient slime mould algorithm for solving multi-objective optimization problems. Expert Syst Appl 187:115870
37. Houssein EH, Abdelminaam DS, Hassan HN, Al-Sayed MM, Nabil E (2021) A hybrid barnacles mating optimizer algorithm with support vector machines for gene selection of microarray cancer classification. IEEE Access 9:64895–64905
38. Hamad A, Houssein EH, Hassanien AE, Fahmy AA (2016) Feature extraction of epilepsy EEG using discrete wavelet transform. In: 2016 12th international computer engineering conference (ICENCO). IEEE, pp 190–195
39. Shaban H, Houssein EH, Pérez-Cisneros M, Oliva D, Hassan AY, Ismaeel AA, AbdElminaam DS, Deb S, Said M (2021) Identification of parameters in photovoltaic models through a runge kutta optimizer. Mathematics 9(18):2313
40. Abdelminaam DS, Said M, Houssein EH (2021) Turbulent flow of water-based optimization using new objective function for parameter extraction of six photovoltaic models. IEEE Access 9:35382–35398
41. Houssein EH, Hassaballah M, Ibrahim IE, AbdElminaam DS, Wazery YM (2022) An automatic arrhythmia classification model based on improved marine predators algorithm and convolutions neural networks. Expert Syst Appl 187:115936
42. Houssein EH, Neggaz N, Hosney ME, Mohamed WM, Hassaballah M (2021) Enhanced Harris hawks optimization with genetic operators for selection chemical descriptors and compounds activities. Neural Comput Appl 33:13601–13618
43. Ahmed MM, Houssein EH, Hassanien AE, Taha A, Hassanien E (2018) Maximizing lifetime of wireless sensor networks based on whale optimization algorithm. In: Proceedings of the international conference on advanced intelligent systems and informatics 2017. Springer International Publishing, pp 724–733
44. Houssein EH, Sayed A (2023) Dynamic candidate solution boosted beluga whale optimization algorithm for biomedical classification. Mathematics 11(3):707
45. Dulhare UN (2018) Prediction system for heart disease using Naive Bayes and particle swarm optimization. Biomed Res 29(12):2646–2649. ISSN 0970-938X
46. Arif F, Dulhare UN (2017). A Machine Learning Based Approach for Opinion Mining on Social Network Data. In: Satapathy S, Bhateja V, Raju K, Janakiramaiah B (eds) Computer Communication, Networking and Internet Security. Lecture Notes in Networks and Systems, vol 5. Springer, Singapore. https://doi.org/10.1007/978-981-10-3226-4_13
47. Rasool S, Dulhare UN, Khan MN,Gangodkar D, Rana A, Kalra R (2022) Automated Multiclass Classification Using Deep Convolution Neural Network on Dermoscopy Images, 2nd International Conference on Technological Advancements in Computational Sciences (ICTACS), Tashkent, Uzbekistan, 711–716. https://doi.org/10.1109/ICTACS56270.2022.9988394.

Chapter 14
Biomedical Applications of Chiral Nanoplasmonics

Disha M. Dhabarde, Monika P. Maske, and Jagdish R. Baheti

Abstract This chapter deals with the chiral nanoplasmonics which is an extremely interesting research field that is attractive due to various backgrounds. The discovery of nanoscale chirality in plasmonic nanoparticles has resulted in a significant revitalization of the chirality field. Its potential uses include chiral separation and synthesis, molecular sensing and optical polarising components. Cells and all sizes of biomolecules are affected by these nanoscale fields in both optical and thermal ways. It is feasible to employ this characteristic to sense, analyse, trap, move and manipulate various cells and molecules by managing these interactions. As a result, instruments like analytical tools for biology and nanomedicine have been developed. Wide range of applications based on optical, thermal and mechanical molecule-plasmon interactions. Chiral nanoplasmonics, an emerging subject that sits between plasmonics and molecular nanotechnology, has made incredible strides in the development of research for early sickness diagnosis and enantioselective sensing.

Keywords Chiral plasmonics · Biomedical applications · Nanoplasmonics · Plasmon · Nanofabrication

1 Introduction

Plasmonics-engineered materials are constitutionally created compounds that use multiphoton, optically behaved collective free electron oscillations in nature. The ability to manipulate and regulate light, improve radiations and contain even in nanometric dimensions, energy are main characteristics of plasmon. The characteristics vary on their size, form, composition and environment in nanostructure. Fabrication method allows for the management of these characteristics. A range of plasmonic structures with various topologies have been created throughout the past few decades from nanodiscs, nanorods or nanorings to sophisticated designs. As very sensitive

D. M. Dhabarde (✉) · M. P. Maske · J. R. Baheti
Kamla Nehru College of Pharmacy, Butibori, Nagpur, MS 441108, India
e-mail: dishamandave20@gmail.com

U. N. Dulhare and E. H. Houssein (eds.), *Machine Learning and Metaheuristics: Methods and Analysis*, Algorithms for Intelligent Systems,
https://doi.org/10.1007/978-981-99-6645-5_14

biological sensors or to improve photovoltaic systems, these structures have been widely exploited as plasmonic devices. The study of electron oscillations in metallic nanostructures and nanoparticles is the focus of plasmonics. It has been challenging to integrate the components with macroscopic size to nanoscale for optical components, plasmonic metal nanoparticles are roughly of 5–100 nm in size [1, 2].

Recent advances study the resonance light interaction and electrons have been significantly governed by preambling up novel and interesting possibilities for nanoscale light manipulation [3]. With its exceptional capacity to dramatically amplify evanescent fields close to metal surfaces, surface plasmon resonance (SPR) has created new opportunities to identify the refractive index without labels. This ultimately led to the SPR detection method's commercialisation and wide adoption, which is currently routinely used for assessing biomolecular interactions [4].

A key characteristic of molecules, as well as nanoparticles and macroscopic objects, is chirality. An item is achiral if it can be rotated or translated in three dimensions and superimposed onto its mirror counterpart. Chirality results from the structures' lack of mirror symmetry. In general, induced and structural chiralities are the two main types. More than half of the medications used today are chiral, and the majority of them are marketed as racemic mixtures [4, 5].

Chiral nanoplasmonics is an extremely interesting research field that is attractive due to various backgrounds. Since advanced nanofabrication methods and soft-chemistry synthesis made it possible to create intricate and precisely controlled chiral metallic nanostructures, chiral plasmonics has received increasing attention. Its potential uses include chiral separation and synthesis, molecular sensing and optical polarising components. It is also used to detect DNA, micro-RNA, cancer biomarkers and glycoproteins [6]. Glycoproteins which describes recent developments in biomedical applications of chiral plasmonic reactions. While many uses, such as chiral sensing, circular polarisers, and catalysis, have been suggested, plasmon-enhanced chiral signals offer the greatest promise for use in disease detection and treatment. Prior to 1998, the majority of chiral materials were synthesised and investigated using organic molecules, including systems based on chiral organic molecules that displayed microscale length chiral structures [7].

At the nexus of the areas of optics and condensed matter, the oscillation of electrons in a plasma or around the atomic nucleus of metals is known as a plasmon [8]. When light causes the free electrons that belong to the metal nanoparticle to oscillate, the resonance of surface plasmons occurs. In relation to the lattice ions, the incident light moves the electrons. The electromagnetic fields around the metal nanoparticles are strengthened upon stimulation of the surface plasmon resonance. The photons are then transformed into heat in this increased field, producing thermal energy. Cells and all sizes of biomolecules are affected by these nanoscale fields in both optical and thermal ways. It is feasible to employ this characteristic to sense, analyse, trap, move and manipulate various cells and molecules by managing these interactions. Analytical tools for biology and nanomedicine have been developed as a result. Metal nanoparticles are very sensitive to detecting biological activities occurring at the molecular level. Further studies demonstrated that laser excitation can be utilised to selectively kill tumour cells which can be applied to plasmonic photothermal cancer therapies [9, 10].

The study of molecular interactions with metal nanostructure surfaces through molecular plasmonics offers several options for biological and medical research. Numerous applications are built on the optical, thermal and mechanical molecule-plasmon interactions [10]. Molecular plasmonics, a new field straddling molecular nanotechnology and plasmonics, has made remarkable strides as a result of developments. Because of the ongoing developments in molecular plasmonics, new standalone devices will undoubtedly appear, but the future is likely to see a shift towards optimisation [11]. Such a system, which may incorporate sensing, spectrum analysis, molecular manipulation, drug administration, gene switching and photothermal therapy, will benefit from the multifunctionalities of localised surface plasmon resonance-based phenomena. Experts from a variety of disciplines, including physics, engineering, chemistry, biology and pharmacology, will need to work closely together to accomplish this. Using nanoparticles with both therapeutic and diagnostic properties, theranostic nanomedicine is one of the early advancements in this approach. It is possible to detect ions, antigens, acids and other tiny compounds using plasmonic metal nanostructures [12].

2 Need for Nanoplasmonics

It will be possible for molecular plasmonics to move from laboratory research to therapeutic applications if each of these factors is properly addressed. The development of lab-on-a-chip devices, also known as point-of-care biosensing devices, which have the potential to lead to the development of next-generation healthcare systems, calls for improved sensing capabilities [13, 14].

3 Principle of Operations

Fabrication of chiral plasmonic nanomaterials and plasmonic chirality modification are the primary areas of interest in chiral plasmonics. The two categories of chiral nanomaterials include discrete chiral nanoparticles and nanostructures. Nanoparticles or chiral nanoparticles are guided by chiral molecules. The straightforward chiral molecule-attaching technique frequently yields modest plasmonic optical activity as a result of weak molecule dipole-plasmon dipole interactions. Chiral plasmonic nanoparticles have demonstrated a great deal of potential for use in biosensing, imaging, and the detection and treatment of illnesses. The most popular plasmonic nanoparticles are made of gold and silver due to their ability to scatter and absorb light in the 350–900 nm spectral range [15].

4 Nanoplasmonic Properties

A chiral object lacks mirror planes and inversion symmetry in terms of symmetry elements. It is not exact in every way [16]. Enantiomeric forms typically share similar physical characteristics, making it challenging to distinguish between them. Analysing a molecule's response to polarised light or seeing how it interacts with another chiral molecule reveals that it is chiral [17].

Optical rotational dispersion (ORD) is the rotation of light with a linear polarisation with regard to its original polarisation. Between the orthogonal polarisations, a phase shift, which results in ORD, causes the rotation of linearly polarised light. ORD is caused by this phase shift. For the measurement of optical rotation and circular dichroism (CD), biophysical techniques are used [18].

The extraordinarily high dipole strength connected to surface plasmon resonances in plasmonic nanomaterials, which also results in strong CD signals, is an intriguing feature of these materials. Additionally, individual nanoparticle resonances may couple to one another, resulting in collective modes that include the entire assembly [18, 19].

4.1 Optical Properties of Chiral Plasmonic Nanomaterials:

The chirality-related optical features have been developing quickly in chiral plasmonic nanostructures and superstructures. It is a technology that is surface sensitive. Using monolayer transition metal dichalcogenides and chiral plasmonic antenna, SERS demonstrates a larger dependence on nearby electromagnetic fields than field-enhanced emission. Overall, these chirality-related optical features offer additional tools for imaging and bio-detection [20].

4.2 Magnetoplasmonics

The effect of magnetic fields on the transmission of light is known as magneto-optics (MO). Typically magneto optics which alters the medium's physical characteristics, which light flows through. In nanofabricated devices, magnetic metals like cobalt or nickel are combined with noble metals like gold to produce magnetoplasmonics, a linked phenomenon [21].

5 Nanofabrication of Plasmonic Structures

5.1 Chiral Plasmonic Nanostructure Synthesis

The chirality of nanoparticles has been created using a variety of techniques, such as.

(i) the creation of nanomaterials with chiral shapes [19–22];
(ii) when chiral chemicals and achiral plasmonic nanoparticles interact, the nanoparticles' electronic states exhibit optical activity [23–25];
(iii) arranging achiral plasmonic components into chiral geometry using chiral chemicals or other templates [12, 26–28].

Individual particles produced by methods (i) and (ii) are chiral by nature, whereas nanoparticle assemblies produced by method (iii) have a collective chiral geometry. Strong chiroptical responses have been attained, chiral nanostructures are often created either (i) top-down using nanofabrication methods like lithography or (ii) bottom-up using self-assembly techniques [22].

The major methods used to create individual chiral nanostructures are top-down methods which produce spirals and helices as structural elements. However, asymmetric chiral nanoparticles with built-in optical activity have recently been created chemically [23].

5.2 Synthesis of Inherently Chiral Nanostructures

The potential for the manufacturing of nanostructures with various shapes has been revealed advanced in lithographic techniques. As a result, light with a right circular polarisation between 4000 and 8000 nm had a good transmittance. An efficient circular polarizer, the helical nanoarray, may transform incident linearly polarised light into circularly polarised light [24-0.26].

Plasmonic spirals have been created and have great material quality. The created arrays displayed chiral characteristics at optical frequencies and had a very wide operating bandwidth. While lithographic methods are used to create the majority of chiral-shaped plasmonic nanostructures, recent attempts have been successful in creating intrinsically chiral nanoparticles [26].

Chiral plasmons can be created by coupling the plasmon modes of interacting nanoparticles in addition to being created by building intrinsically chiral nanostructures. With this technique, nanoparticles are assembled in various sizes and shapes, starting with the smallest (the dimer). These helicoidal nanoparticles' high anisotropy factor made it possible to discern macroscopically between circular polarisation-dependent colour variations [27].

5.3 Synthesis of Chiral Nanoassemblies on Basis of Templates

Chiral plasmons can be created by connecting the plasmon modes of interacting nanoparticles in addition to being created by building intrinsically chiral nanostructures. Starting with the simplest (the dimer), this technique enables the synthesis of nanoparticles in a variety of shapes and sizes. Flexible templates can be utilised to adsorb nanoparticles via covalent or non-covalent interactions. Au nanorods were arranged into abysmally helical patterns, but these structures produced strong chiral signals. As a result, the spectrum location and strength of chiral plasmonic signals can be adjusted [28, 29].

5.4 Chiral Nanostructures Synthesis from Achiral Constituents

Chiral plasmonic signals can be derived from nanosystems with poorly defined geometries in addition to the classic chiral structures like the helix or spiral, which have a distinctive twist. This category includes a wide variety of morphologies and assemblages of deformed nanoparticles. Although at first glance they don't appear to be chiral, the mere inability to superimpose them with their mirror copies causes them to become optically active. The relative orientation, dimers and direction of excitation will all have a big impact on the optical response in such dimers. Two Au nanorods constructed on a reversible DNA origami template were joined to generate an optically controlled plasmonic nano-system as a prototype for this class of nanoparticles. DNA origami can be used to construct the most complex and sophisticated two-dimensional DNA nanostructures. The active photoresponsive site on the template is a DNA fragment that has undergone azobenzene modification [17, 30].

6 Recent Advances in Chiral Nanoplasmonics

Recently, there has been a lot of activity in the study of near-infrared (NIR) resonant nanomaterials, such as hollow Au nanoshells, Au nanorods and gold nanoshell particles. The light transmission through tissues is quite effective in the NIR range (800–1200 nm). Because these nanoparticles efficiently absorb NIR light and convert it to heat, they can be employed in photothermal therapy. Even the delivery of a therapeutic heat dose to destroy specific cells without compromising healthy cells due to excessive local heating may be accomplished using silica/Au nanoshell particles. In order to create a versatile platform for concurrent MRI diagnosis, Au nanoshells were also combined with magnetic nanoparticles, cancer cell-specific antibodies and NIR photothermal treatment. Local temperature rise caused by photothermal conversion in Au nanoshells can make target cells more susceptible to harmful chemicals by enhancing membrane permeability and blood vessel dilation. Therefore, a significantly better therapeutic efficacy is anticipated if photothermal treatment and chemotherapy are used together [31].

7 Biomedical Applications

7.1 Biosensing Based on Plasmonic Chiral Responses

Numerous chiral plasmonic nanoparticle-based biosensors have been created due to strong chiroptical activity, showing tremendous promise for early illness detection and enantioselective sensing. Due to their simple surface bioconjugation with molecular probes and exceptional optical characteristics, metal nanoparticles, primarily gold, are intriguing in biomedicine [32, 33].

7.2 Disease Diagnosis and Treatment

The chiral plasmonic nanoparticles bind with specific disease markers which are then exploited as detection signals for accurate disease diagnosis. Additionally, image-guided diagnosis and treatment uses chiral plasmonic nanoparticles in combination with a range of imaging modalities. Chiral plasmonic superstructures can be created which also makes protein-related disorders detectable and predictable [34].

7.3 Detection of Parkinson's Disease

The end result of several protein aggregation mechanisms are amyloid fibrils, which are strongly related to many neurodegenerative disorders. Chiral plasmonic nanostructures are very optically active and have applications in both illness detection and treatment. For amyloid fibril detection in Parkinson's disease, chiroptical properties in Au nanorods are used [34]. However, the helical structure of the amyloid fibrils allowed Au nanorods to form on their surface through non-covalent interactions, creating intense optical activity at the wavelengths of the surface plasmon resonance. Monomeric-synuclein proteins did not appear to interact with Au nanorods. The approach's broad potential as a detection platform for a range of neurological illnesses was illustrated by the incorporation of the precise detection of infectious amyloids produced by prion proteins [35].

7.4 Targeted Therapy

In order to treat specific tumours, photothermal therapy (PTT) makes use of plasmonic nanoparticle accumulation. The local temperature in the tumour increases due to the outside light being absorbed by the nanoparticles. This selective hyperthermia causes irreparable damage to the tumour tissue while sparing healthy tissue. When compared to traditional therapeutic approaches, PTT is a far more effective and distinctive way for treating diseases since it demonstrates spatiotemporal selectivity and offers benefits including high sensitivity, less side effects, noninvasiveness, and quick and efficient therapy. It's crucial to select the best plasmonic nanoparticles for usage within the body that are externally laser stimulating to maximise in vivo therapeutic effectiveness. Gold nanostructures that were synthesised demonstrated significant NIR light. Therefore, an NIR laser with low biological tissue absorption or scattering, such as blood, water, melanin and fat, is usually used for PTT because it can penetrate the targeted tissue deeply to reach the imbedded plasmonic nanoparticles. Photothermal cancer treatment efficacy based on nanoparticle size. Gold nanostructures coated in polyethylene glycol (PEG) were administered to tumour-bearing mice. This immediately raised the temperature to 50 °C, which was sufficient to specifically kill cancer cells. When compared to nanorods (53.2 0.5 °C) and nanocages (48.7 3.5 °C) among the various morphologies of gold nanostructures, nanohexapods (55.7 2.4 °C) had the best photothermal conversion efficiency in vivo. Following photothermal therapy, 18F-flourodeoxyglucose absorption significantly decreased 24 h later. Compared to the contralateral nonirradiated tumours, tumours treated with an NIR laser showed no glycolic activity, proving that PTT entirely eradicated this activity. According to quantitative research, tumour-selective PTT lowered tumour metabolism in vivo by almost 90% [36].

7.5 Targeted Drug Delivery

Smart drug distribution is crucial for improving treatment efficacy and lowering dosages to decrease unfavourable side effects. Plasmonic metallic nanoparticles have been extensively used as highly spatiotemporal-resolved nanocarriers to remotely release medicines by external light modification. Drug delivery can be combined with photothermal therapy. This is as a result of their highly concentrated heat and adjustable optical properties. Photothermal techniques are used to release oligonucleotides with gold nanoshells. A poly-l-lysine (PLL) peptide with positively charged lysine residues was used to electrostatically load single-stranded DNA (ssDNA) or double-stranded small interfering RNA (siRNA) with negatively charged phosphate using a poly-l-lysine (PLL) peptide with positively charged lysine residues, and backbones were electrostatically loaded. Fluorescently tagged oligonucleotides were employed to observe the time-dependent release process while continuous wave NIR laser irradiation was used to regulate delivery [37]. Using 4T1 breast cancer cells, an in vitro experiment was conducted. The combination of photothermal heat and a medicine that had been released by light caused more cell death than the free drug did. This technique greatly inhibited lung metastasis in vivo in the orthotropic and decreased initial tumour growth [38, 39].

8 Limitations

The primary issue with a geometry-focussed method for chiral plasmonics is that the functionality must be adjusted after it has been developed. Fabricating 3D chiral plasmonic objects is difficult, especially from the top-down. For this method to achieve the requirements for scalable, reproducible, quick, and affordable fabrication, very novel approaches are needed [40, 41].

9 Conclusions and Future Prospects

Plasmon-induced strong evanescent fields have attracted a lot of attention and have thus found many uses. The synthesis of intricate chiral plasmonic nanosystems has been made easier by advancements in nanofabrication techniques. A wide range of chiral nanoarchitectures with various optical responses have been successfully synthesised from chiral biomolecules/templates. This highlights the critical role of stereoselective synthesis, enantiomer detection, and separation play, especially in the treatment of disorders. On the other hand, the intense chiroptical signals generated by plasmonic nanoparticles may detect analyte at exceedingly low concentrations due to their strong dipolar and multipolar coupling. The creation of biocompatible nanoparticles and the distinctiveness of plasmonic chiral fingerprints have opened up possibilities for applications in the biomedical arena. Plasmonic nanorod metamaterials provide a desirable and adaptable platform for fundamental research, the chemical and pharmaceutical sectors, as well as the creation of hot electron-based nanochemical platforms and high-performance optical sensors. With further developments in the synthetic processes, the field of chiral nanoplasmonics is predicted to displace its achiral equivalents and will undoubtedly find use in additional industries. As research into the creation of plasmonic metamaterials continues to grow and research into the interaction of various biomolecules with metasurfaces develops rapidly, it is anticipated that in the near future, highly sensitive molecular chirality detection will be made possible by exceeding the current limit.

References

1. Rodger A, Norden B (1997) Circular dichroism and linear dichroism. Oxford, London
2. Xia Y, Zhou Y et al (2011) Chiral inorganic nanoparticles: origin, optical properties and bioapplications. Nanoscale 3:1374–1382
3. Wang X, Tang Z et al (2017) Circular dichroism studies on plasmonic nanostructures. Small 13:160–165
4. Slocik J, Govorov M et al (2011) Plasmonic circular dichroism of peptide-functionalized gold nanoparticles. Nano Lett 11:701–705
5. Hu Z, Meng D et al (2019) Plasmonic circular dichroism of gold nanoparticle-based nanostructures. Adv Opt Mater 7:180–190
6. Zhao W, Zhang W et al (2019) Photocontrollable chiral switching and selection in self-assembled plasmonic nanostructure. Adv Funct Mater 29:190
7. Wang P, Huh J et al (2020) DNA origami guided self-assembly of plasmonic polymers with robust long-range plasmonic resonance. Nano Lett 20:8926–8932
8. Xu L, Wang X et al (2022) Enantiomer-dependent immunological response to chiral nanoparticles. Nature 601:366–373
9. Ma W, Xu L et al (2019) Chirality-based biosensors. Adv Funct Mater 29:180
10. Wang Y, Xu J et al (2013) Emerging chirality in nanoscience. Chem Soc Rev 42:2930–2962
11. Zheng G, He J et al (2021) Discrete metal nanoparticles with plasmonic chirality. Chem Soc Rev 50:3738–3754
12. Vila-Liarte D, Kotov N et al (2022) Template-assisted self-assembly of achiral plasmonic nanoparticles into chiral structures. Chem Sci 13:595–610
13. Gao G, Zhang M et al (2017) The size-effect of gold nanoparticles and nanoclusters in the inhibition of amyloid-β fibrillation. Nanoscale 9:4107–4113
14. Xu L, Wang X et al (2022) Enantiomer-dependent immunological response to chiral nanoparticles. 601:366–373
15. Feng H (2014) Magnetoplasmonic nanorings as novel architectures with tunable magneto-optical activity in wide wavelength ranges. Adv Opt Mater 2:612–617
16. Neubrech F, Hentschel M (2020) Reconfigurable plasmonic chirality: fundamentals and applications. Adv Mater 190
17. Khlebtsov N, Dykman L (2010) Optical properties and biomedical applications of plasmonic nanoparticles. J Quant Spectr Radiat Transfer 111:1–35
18. Gao X, Han B et al (2022) Self-assembled inorganic chiral superstructures. Nat Rev Chem 6:125–145
19. Strobbia P, Languirand E et al (2015) Recent advances in plasmonic nanostructures for sensing: a review. Opt Eng 54:100
20. Zhao W, Zhang W et al (2019) Photocontrollable chiral switching and selection in self-assembled plasmonic nanostructure. Adv Funct Mater 29:1–10
21. Homola J (2006) Surface plasmon resonance based sensors 4
22. Fan Y, Ou-Yang S et al (2022) Biological applications of chiral inorganic nanomaterials. Chirality 34:760–781
23. Ozbay E (2006) Plasmonics: merging photonics and electronics at nanoscale dimensions. Science 311:189–193
24. Pan J, Wang X et al (2022) Chirally assembled plasmonic metamolecules from intrinsically chiral nanoparticles. Nano Res 15:9447–9453
25. Xu L, Wang, X et al (2022) Enantiomer-dependent immunological response to chiral nanoparticles. Nature 601:366–373
26. Shen X, Asenjo-Garcia A et al (2013) Three-dimensional plasmonic chiral tetramers assembled by DNA origami. Nano Lett 13:2128–2133
27. Martens K, Binkowski F et al (2021) Long- and short-ranged chiral interactions in DNA-assembled plasmonic chains. Nat Commun 12:2025
28. Unck T, Nicoli F et al (2018) Sensing picomolar concentrations of rna using switchable plasmonic chirality. Angew Chem Int Ed Engl 57:13495–13498

29. Cortie M, McDonagh A (2011) Synthesis and optical properties of hybrid and alloy plasmonic nanoparticles. Chem Rev 111:3713–3735
30. Park H, Yang J, et al (2008) Multifunctional nanoparticles for photothermally controlled drug delivery and magnetic resonance imaging enhancement. Small 4:192–196
31. Sotiriou G (2013) Biomedical applications of multifunctional plasmonic nanoparticles. Wiley Interdiscip Rev Nanomed Nanobiotechnol 5:19–30
32. Jin Y (2014) Multifunctional compact hybrid Au nanoshells: a new generation of nanoplasmonic probes for biosensing, imaging, and controlled release. Acc Chem Res 47:138–148
33. Warner S (2004) Diagnostics + therapy = theranostics. Scientist 18:38–39
34. Xue J, Liu T et al (2019) Neuroprotective effect of biosynthesised gold nanoparticles synthesised from root extract of Paeonia moutan against Parkinson disease—In vitro & In vivo model. J Photochem Photobiol 200:1–10
35. Hu K, Chen X et al (2018) Neuroprotective effect of gold nanoparticles composites in Parkinson's disease model. Nanomedicine 14:1123–1136
36. Doane T, Burda C et al (2012) The unique role of nanoparticles in nanomedicine: imaging, drug delivery and therapy. Chem Soc Rev 41:2885–2911
37. Lee S, Park H et al (2010) Synergistic cancer therapeutic effects of locally delivered drug and heat using multifunctional nanoparticles. Adv Mater 22:4049–4053
38. Funck T, Nicoli F et al (2018) Sensing picomolar concentrations of rna using switchable plasmonic chirality. Angew Chem Int Ed Engl 57:13495–13498
39. Gilroy C, Hashiyada S et al (2019) Roles of superchirality and interference in chiral plasmonic biodetection. J Phys Chem 15:195–203
40. Houssein EH, Sayed A (2023) Dynamic candidate solution boosted beluga whale optimization algorithm for biomedical classification. Mathematics 11(3):707
41. Dulhare UN (2018) Prediction system for heart disease using Naive Bayes and particle swarm optimization. Biomed Res 29(12):2646–2649. ISSN 0970-938X